Springer Series in Optical Sciences Volume 6
Editor David L. MacAdam

Springer Series in Optical Sciences

Volume 1 **Solid-State Laser Engineering**
By W. Koechner

Volume 2 **Table of Laser Lines in Gases and Vapors** 2nd Edition
By R. Beck, W. Englisch, and K. Gürs

Volume 3 **Tunable Lasers and Applications**
Editors: A. Mooradian, T. Jaeger, and P. Stokseth

Volume 4 **Nonlinear Laser Spectroscopy**
By V. S. Letokhov and V. P. Chebotayev

Volume 5 **Optics and Lasers** An Engineering Physics Approach
By M. Young

Volume 6 **Photoelectron Statistics** With Applications to Spectroscopy and Optical Communication
By B. Saleh

Volume 7 **Laser Spectroscopy III**
Editors: J. L. Hall and J. L. Carlsten

B. Saleh

Photoelectron Statistics

With Applications to Spectroscopy and Optical Communication

With 85 Figures

Springer-Verlag Berlin Heidelberg GmbH 1978

Dr. BAHAA SALEH
University of Wisconsin-Madison, Department of Electrical and Computer Engineering
Madison, WI 53706, USA

Dr. DAVID L. MACADAM
68 Hammond Street, Rochester, NY 14615, USA

ISBN 978-3-662-13483-2 ISBN 978-3-540-37311-7 (eBook)
DOI 10.1007/978-3-540-37311-7

Library of Congress Cataloging in Publication Data, Saleh, Bahaa. 1944 —. Photoelectron statistics, with applications to spectroscopy and optical communication. (Springer series in optical sciences; v. 6). 1. Photoelectrons — Statistical methods. 2. Light beating spectroscopy. 3. Optical communications. 4. Stochastic processes. I. Title. QC716.15.S24 535'.2 77-9936

Originally published by Springer-Verlag Berlin Heidelberg New York in 1978
Softcover reprint of the hardcover 1st edition 1978

2153/3130–543210

Preface

With the recent great expansion in optics and laser applications, several new areas of research have emerged, among which are: the theory of coherence, photon statistics, speckle phenomenon, statistical optics, atmospheric propagation, optical communications, and light-beating and photon-correlation spectroscopy. A factor common to these overlapping subjects is their basic dependence on the treatment of light as a randomly fluctuating excitation. Moreover, they all necessitate a thorough understanding of the phenomenon of light detection and the additional randomness it introduces. My objective in writing this book is to provide a unified and general presentation of a basic theoretical background central to these areas.

This book has a threefold purpose: to present a systematic treatment of the statistical properties of optical fields, to develop methods for determining the statistics of the photoelectron events that are generated when such fields are intercepted by photodetectors, and to examine methods of estimating unknown field parameters from measurements of the photoelectron events. Emphasis is placed on the photoelectron measurements that yield information pertinent to spectroscopy and optical communication.

Although some books that treat the theory of coherence and the statistical properties of light are available, the vast body of information central to problems of photoelectron statistics and its applications is scattered in various professional journals and conference proceedings.

The book is written primarily for graduate students in electrical engineering and physics who are interested in the theory of coherence and photoelectron statistics and their applications. It should also serve as a reference book for researchers in the areas of optical communication and photon-correlation spectroscopy.

A detailed preview of the book may be found in Chap. 1. Basically, the book is divided into three parts. Part I reviews the mathematical tools that are necessary for the description of random functions (such as optical fields and radiance distributions) and random point processes (such as the photoelectron events). In Part II, these general concepts from probability theory are applied to light and photoelectrons. Few assumptions are necessary to achieve a simple deduction of the theory of coherence and photoelectron statistics. Many statistical models of light are discussed in detail (e.g., thermal light, laser light, scattered light, modulated light, diffused light, partially polarized light). Part III deals with the extraction of information on the nature of the optical field (e.g., its intensity, spectrum, or statistical model) from measurements on the detected photoelectrons. It presents a detailed discussion of the theory of photoelectron-correlation spectroscopy and of the effect of photoelectron statistics on the performance of optical communication and radar systems.

It is a great pleasure to acknowledge the influence of those scholars whose works are cited; their contributions are the foundations of this work.

I appreciate and acknowledge the support and encouragement of Prof. L. DeMaeyer and the Max-Planck-Institut für Biophysikalische Chemie during the preparation of this book. I am indebted to Dr. E. Jakeman for his critical review of the manuscript, and to Dr. S. Provencher for reading parts of the manuscript. I thank the Institute of Electrical and Electronic Engineers, the American Physical Society, the Institute of Physics, North-Holland Publishing Company, and John Wiley & Sons, Inc., for granting me permission to reproduce illustrations from their publications. I also thank the authors of these publications. Ms. G. Klump patiently typed the manuscript and Mr. W. Stöhr prepared the illustrations; their work is gratefully appreciated.

I wish to express my debt to Prof. J. Minkowski who introduced me to this field and who remains a great source of inspiration for me.

I would like to express my gratitude to my wife, Kenna, for her help, patience, and understanding. To her, I dedicate this book.

Berkeley, Cal.
March 1977

Bahaa Saleh

Contents

1. Introduction 1

Part I. *Tools From Mathematical Statistics*

2. Statistical Description of Random Variables and Stochastic Processes 6

2.1 Statistical Description of Random Variables 6
2.1.1 Probability Distribution (PD) 6
2.1.2 Moments of a Random Variable 7
2.1.3 Moment-Generating Functions (mgf) of a Random Variable 9
2.1.4 Some Standard Random Variables 11
2.1.5 Transformation of a Random Variable 11

2.2 Statistical Description of a Set of Random Variables 11
2.2.1 Probability Distributions 11
2.2.2 Moments and Correlations 14
2.2.3 Moment-Generating Functions 16
2.2.4 An Example: A Set of Jointly Gaussian Random Variables 17
2.2.5 Transformation of Random Variables 18

2.3 Statistical Description of Complex Random Variables 20
2.3.1 The Complex Random Variable (CRV) 20
2.3.2 Circularly Symmetric Complex Random Variables 21
2.3.3 The Circularly Symmetric Gaussian CRV 22
2.3.4 A Set of Complex Random Variables 23
A Set of Complex Gaussian Random Variables 24

2.4 Statistical Description of Stochastic Processes 25
2.4.1 Definitions 25
2.4.2 The Random Spectrum of a Stochastic Process. The Power Spectrum of a Stationary Stochastic Process 27
2.4.3 The Gaussian Process 29
2.4.4 Description of a Stochastic Process by the Coefficients of a Karhunen-Loève Expansion 30
The Karhunen-Loève Basis for a Stationary Process with Lorentzian Spectrum 32
The Karhunen-Loève Basis for a Stationary Process with Rectangular Spectrum 33
2.4.5 Description of a Stochastic Process by a Differential Equation 34

2.5 Complex Stochastic Processes 38
2.5.1 Definitions 38
2.5.2 Karhunen-Loève Expansion of a Complex Stochastic Process 40

2.6 Complex Representation of Bandpass Stochastic Processes 41
2.6.1 Complex Representation of a Real Bandpass Signal 41
Complex Envelope (Amplitude) of a Bandpass Signal 42

2.6.2 Complex Representation of a Real Bandpass Stochastic Process 44
2.6.3 Complex Representation of a Stationary Real Bandpass Stochastic Process 46
2.6.4 Processes with Stationary Quadrature Components. Quasi-stationary Processes 49
2.6.5 Complex Representation of Gaussian Bandpass Stochastic Processes 49
2.6.6 Karhunen-Loève Expansion of the Complex Envelope of a Bandpass Stochastic Process 50

2.7 A Short Review of Some Principles of Estimation and Detection Theory 52
2.7.1 Test of Hypotheses 52
The Maximum-Likelihood (ML) Strategy 52
Bayes Strategy 53
2.7.2 Parameter Estimation 54
Maximum-Likelihood (ML) Estimation 54
Bayes Estimation 54
MMSE Nonlinear Fitting 55

3. Point Processes 57

3.1 One-Dimensional Point Processes 57
3.2 Statistics of Times of Occurrence 59
3.2.1 Joint Probability Density of Multicoincidence 59
3.2.2 Joint Probability Density of Forward Recurrence Times . 60
3.2.3 Joint Probability Density of Intervals Between Events . 61
3.2.4 Joint Probability Density of the Number of Events and Their Instants of Occurrence in a Closed Interval 62
3.2.5 Generating Functional 62

3.3 Counting Statistics 63
Triggered Counting 64

3.4 The Poisson Process 65
3.4.1 Definition 65
3.4.2 Statistics of Times 66
3.4.3 Counting Statistics 68
Probabilities 68
Moment-Generating Functions 69
Moments 70

3.5 Doubly Stochastic Poisson Point Processes (DSP.PP) 72
3.5.1 Definitions 72
3.5.2 Counting Statistics 72
Moment-Generating Functions 72
Moments 73
Probability Distribution 76
3.5.3 Statistics of Times 77

3.6 Appendix: The Poisson Transform 79
3.6.1 Definition and Properties 79
3.6.2 Inversion of the Poisson Transform 83
Method I 83
Method II 83
Method III 84

Part II. *Theory*

4. The Optical Field: A Stochastic Vector Field or, Classical Theory of Optical Coherence 86

4.1 Classical Statistical Description of Light 89
4.1.1 The Classical Description of an Electromagnetic Field . 89
4.1.2 The Statistical Description of Light 93
The Joint Probability Densities and the Correlation Functions (Coherence Function) 93
Description in the Temporal-Frequency Domain 95
Normalization of the Coherence Functions. The Degree of Coherence of Light 96
Factorization of the Temporal and Spatial Dependence of the Coherence Function: Cross-Spectral Purity 100
Coherence Time and Coherence Area 101
The Light Intensity: A Real Stochastic Process 102
4.1.3 Statistical Description of Light Propagation in a Linear Optical System 104
Diffraction of Spatially Incoherent Light 106
Diffraction of Partially Coherent Light from Two Pinholes. Young's Interference Experiment 108
Effect of Propagation on the Cross-Spectral Purity 110
Michelson Interferometer 113

4.2 Statistical Properties of some Special Models of Optical Fields 114
4.2.1 Polarized Thermal Light 114
Intensity Fluctuations 116
4.2.2 Partially Polarized Thermal Light 120
Intensity Fluctuations 122
Fluctuations of the Total Intensity 123
4.2.3 Polarized Superposition of Coherent and Thermal Light . 125
Intensity Fluctuations 126
4.2.4 Mixture of Coherent Light and Partially Polarized Thermal Light 131
4.2.5 Quasi-Stationary Gaussian Light 132
Statistical Properties of the Field 132
Statistical Properties of the Intensity 133
Moment-Generating Functions 133
Probability Densities 134
4.2.6 Transient Thermal Light 135
4.2.7 Van der Pol's Nonlinear-Oscillator Classical Model of Laser Light 137
Intensity Fluctuations 141
Field Correlation 143
Intensity Correlation Function 144
4.2.8 The Sum of a Small Number of Independent Modes of Light 145
Joint Statistics 150
Number Fluctuations 151
4.2.9 Phase-Fluctuating (or Diffused) Light 152
Phase-Fluctuating Field Mixed with a Coherent Field ... 154
Phase-Fluctuating Light after Propagation Through an Optical System. The Phenomenon of Speckle 156

5. Photoelectron Events: A Doubly Stochastic Poisson Process or Theory of Photoelectron Statistics 160

5.1 The Photoelectric Detection of Light 161
5.1.1 Semiclassical Derivation of the Photodetection Equation 161
5.1.2 Photoelectrons versus Photons 165

5.2 Single-Fold Photoelectron Statistics of some Special Optical Fields 166
5.2.1 Coherent Light 167

5.2.2 Linearly Polarized Thermal Light 167
Short Sampling Time, Small Detector $T << \tau_c$, $A << A_c$ .. 167
Arbitrary Sampling Time, Small Detector $A << A_c$ 171
Arbitrary Sampling Time and Arbitrary Detector Area. Cross-Spectrally Pure Light 189
Approximate Statistics for Light with an Arbitrary Spectrum. Arbitrary T/τ_c and A/A_c 193
Numerical Techniques 197
5.2.3 Partially Polarized Thermal Light 198
The Limit $T << \tau_c$ 200
Limit $T >> \tau_c$ 201
Partially Polarized Thermal Light with Lorentzian Spectrum 202
5.2.4 Mixture of Polarized Coherent and Polarized Thermal Light 203
$T << \tau_c$, $A << A_c$, $\Delta\omega T << 1$ 204
Arbitrary T/τ_c and $\Delta\omega T$, $A << A_c$ 206
Limit $T << \tau_c$, $A << A_c$, Arbitrary $\Delta\omega T$, Constant Coherent Component 211
Limit $T >> \tau_c$, $A << A_c$, Arbitrary $\Delta\omega T$, Constant Coherent Component 213
The Thermal Part has a Lorentzian Spectrum 215
Effect of the Detector Area 218
The Approximate Formula 220
5.2.5 Mixture of Coherent Light and Partially Polarized Thermal Light 221
Limit of $T << \tau_c$; $A << A_c$; $T_{\Delta\omega} << 1$ 222
N-Modes 222
5.2.6 Quasi-Stationary Gaussian Light 223
Real Amplitude 223
General Complex Amplitude 227
5.2.7 Transient Thermal Light 227
5.2.8 Laser Light Described by a Van der Pol Oscillator Model 229
Statistics of Times 230
5.2.9 Modulated Light Beams 231
Coherent Light Modulated by Thermal Noise 234
Coherent Light Modulated by Real Gaussian Noise 234
Thermal Light Modulated by Thermal Noise (Gaussian-Gaussian Scattering) 234
Thermal Light Whose Intensity is Modulated by Real Gaussian Noise 236
Intensity Modulation with a Periodic Signal Unsynchronized with Sampling 238
Log-Normally Modulated Light 242

5.3 Multifold Photoelectron Statistics 249
5.3.1 Polarized Thermal Light 250
The Limit $T << \tau_c$ and $A << A_c$ 250
Single Small Detector ($A << A_c$), Arbitrary Sampling Time 254
Single Detector of Arbitrary Area. Arbitrary Sampling Time 259
Several Photodetectors Simultaneously Counting. Cross-Spectrally Pure Light 261
An Approximate Formula 263
Statistics of Times. Probability Density of Multicoincidence PDC 263
5.3.2 Mixture of Coherent and Polarized Thermal Light 268
The Limit $T << \tau_c$, $A << A_c$, and $T\Delta\omega << 1$ 268
Arbitrary T/τ_c, A/A_c, and $T\Delta\omega$ 269

5.4 Nonideal Effects in Photodetectors 271
5.4.1 Dark Current 272
5.4.2 Dead-Time Effect 272
The Moments 275
Joint Probability Distributions 278
5.4.3 Photoelectric-Current Relaxation and Photomultiplier-Gain Fluctuations 279

Part III. *Applications*

6. Applications to Optical Communication 284
6.1 Classification of Optical Communication Systems 286
6.2 Estimation of Optical Signals. Direct Detection 288
6.2.1 Estimation of the Intensity of a Stationary Optical Field 289
Coherent Light 289
Thermal Light 289
Coherent Light in Thermal Background 291
Coherent Light in Thermal Background of Unknown Intensity 293
Thermal Light in Thermal Background of Unknown Intensity 294
Separation of the Intensities of a Mixture of Two Optical Fields with Different Correlation-Time Scales .. 296
6.2.2 Estimation of an Optical Signal with a Time-Varying Intensity 297
Coherent Signal. Intensity Modulation (IM) 297
Coherent Signal in Thermal Background 299
Estimation of Parameters of an Intensity Distribution . 300
6.3 Estimation of Optical Signals. Heterodyne Detection.......... 302
6.3.1 Estimation of a Coherent Time-Varying Optical Signal AM, IM, FM, and PM Systems 302
Effect of the Detector Area. The Problem of Alignment . 304
6.3.2 Homodyne Estimation of the Intensity of a Stationary Thermal Optical Signal in a Thermal Background 308
6.4 Detection of Optical Signals. Digital Communication Systems .. 312
6.4.1 Detection of a Coherent Signal in a Wide-Band Thermal Background 313
Pulse-Gated Binary Modulation (PGBM). Detection of a Radar Signal 314
Pulse-Delay Binary Modulation (PDBM). Binary PPM 316
Binary Polarization Modulation (BPLM) 318
Pulse-Position Modulation (PPM) 319
6.4.2 Detection of a Coherent Signal Corrupted by Thermal Background of Arbitrary Bandwidth. Arbitrary Area of Detector 321
Pulse-Gated Binary Modulation (PGBM). Detection of a Radar Signal 322
Pulse-Position Modulation (PPM), Pulse-Delay Binary Modulation (PDBM), Pulse-Polarization Modulation (BPLM) 324
6.4.3 Detection of a Fluctuating Signal Corrupted by Thermal Background 326
6.4.4 Detection in the Presence of Atmospheric Scintillation 327
Pulse-Gated Binary Modulation (PGBM). Detection of a Coherent Radar Signal 328
Pulse-Position Modulation (PPM) 332

Effect of Diversity on the Performance of a Radar-Signal Detector. Array Detectors 333
6.4.5 Heterodyne Detection 336
PGBM. Detection of a Radar Signal 336
Phase-Shift Keying (PSK) 336
Detection of a Fluctuating Radar Signal 337

6.5 Inference Based on Instants of Occurrence of Photoelectrons .. 339

6.5.1 Estimation Problems 339
Parameters of a Coherent Signal 339
Parameters of a Thermal Signal 340
Estimation of Time-Varying Random Parameters of an Optical Signal 342
6.5.2 Detection Problems 345
Coherent Signals 345
Thermal Signals 346

6.6 Concluding Remarks 347

7. Applications to Spectroscopy 348

7.1 Estimation of the Spectrum from Measurement of the Autocorrelation of Photoelectron-Counting Fluctuations 350

7.1.1 Autocorrelation Function 350
Thermal Light 351
Mixture of Thermal and Coherent Light 352
Mixture of Coherent Light and Light with Unspecified Statistics 353
Sum of N Independent Components of Light 354
Coherent Light Modulated by Gaussian Noise 354
7.1.2 Measurement of the Autocorrelation Function 354
Software Correlator 355
Hardware Correlator 356
Sequential-Processing Correlator 357
7.1.3 Normalized Autocorrelation Function 358
7.1.4 Clipped Correlation 359
Thermal Light 363
Mixture of Coherent and Thermal Light 365
Sum of N Independent Components of Light 367
Coherent Light Modulated by Gaussian Noise 368
Clipped Digital Correlator 368
7.1.5 Complementary Clipping 369
7.1.6 Randomly Clipped Autocorrelation 371
7.1.7 Scaled Autocorrelation 373
7.1.8 Statistical Accuracy of Estimating the Autocorrelation function 375
Full Autocorrelation Function 375
Normalized Full Correlation 381
Clipped Autocorrelation Function 386
7.1.9 Accuracy of Estimating Unknown Parameters of a Given Spectral Profile 388
7.1.10 Summary 394

7.2 Estimation of Spectral Parameters Based on Measurement of Probability Distributions 395

7.2.1 Probability of Coincidence 395
Multichannel Coincidence........ 396
7.2.2 Time-Interval Probability. Pulse-Separation Technique . 396
The Statistical Accuracy 398
Measurement of Spectral Parameters 399

7.2.3 Single-Photoelectron-Counting Probability 400
7.2.4 Joint-Photoelectron-Counting Probability 403
7.2.5 Instants of Occurrence of a Realization of Photoelectron Events 404

7.3 Single-Photoelectron-Decay Spectroscopy 405

7.3.1 Coherent Light 408
Spectrometer I 408
Spectrometer II 410
7.3.2 Effect of Light Coherence 412

7.4 Estimation of the Spatial Spectrum (the Spatial Coherence Function) 414

7.4.1 The Digital Cross Correlator 417
7.4.2 Statistical Accuracy 417

References.. 421

Subject Index... 433

List of Abbreviations

CAS	Complex analytic signal
cgf	Cumulant generating function
chi_N^2	Chi-square distribution with N degrees of freedom
Cov	Covariance
CRV	Complex random variable
DSP.PP	Doubly stochastic Poisson point process
fmgf	Factorial-moment generating function
JPD	Joint probability distribution
K-L	Karhunen-Loève
MAP	Maximum *a posteriori* probability
mgf	Moment-generating function
ML	Maximum-likelihood
MMSE	Minimum mean-square error
PD	Probability distribution
PDC	Probability distribution of multicoincidence
PP	Point process
P.PP	Poisson point process
Pr	Probability
psf	Point spread function
RV	Random variable
Var	Variance

Functions used in the text[1]

$D_n(x)$	Parabolic cylinder function
$\delta(x)$	Delta function
$Ei(x)$	Exponential-integral function
${}_2F_1(\alpha,\beta;\gamma;z)$	Gaussian hypergeometric function
$\Gamma(x)$	Gamma function, $\Gamma(n) = (n-1)!$
$\Gamma(a,x)$	Incomplete gamma function
$I_n(x)$	Modified Bessel function of first kind
$J_n(x)$	Bessel function of first kind
$J(x)$	$= 2\,J_1(x)/x$
$L_n^m(x)$	Laguerre polynomial
$L_n(x)$	$= L_n^0(x)$
$P_n(x)$	Legendre polynomial
$\Phi(x)$	Probability integral
$\mathrm{sinc}(x)$	$= \sin(\pi x)/\pi x)$
$\mathrm{sh}(x)$	$= \sinh(x)/x$
$W_{n,m}(x)$	Whittaker's function
$\binom{n-1}{m-1}$	$= \dfrac{\Gamma(n)}{\Gamma(m)\,\Gamma(n-m+1)}$

[1] For definitions and properties cf. [184]

1. Introduction

Random fluctuations in physical quantities have always attracted the interest of scientists and engineers. Physicists are usually interested in fluctuations inasmuch as they carry information on the dynamics of the physical phenomena under observation. Engineers, on the other hand, usually treat random fluctuations as undesirable, but unavoidable, effects that limit the reliability of their systems and therefore must be understood.

Random fluctuations of optical fields have always been observed. Slow fluctuations are of course commonly seen in reflections from the surfaces of liquids or in the twinkling of stars. Fluctuations that take place in a much shorter time scale have been indirectly observed, as it has been realized that light waves do not interfere unless they are obtained from the same point source. Light waves from natural sources do not interfere when their random fluctuations are *incoherent* (or uncorrelated). When they are obtained from the same point source, their fluctuations are *coherent* (i.e., correlated or "in harmony"), and they interfere. Only recently were intermediate or *partial* states of *coherence* considered and a very rapid growth of theoretical and experimental studies of random fluctuations in optical radiation began. The success of the new approach to optics in explaining the phenomena of interference, diffraction, image formation, resolution in microscopic images, atmospheric effects on resolution, and speckle, has stimulated continuous interest in the subject.

The experiments of FORRESTER et al. [144] in 1955 and of HANBURY BROWN and TWISS [193, 194] in 1956, on the temporal and spatial correlations of fluctuations of the *intensity* of thermal light, stimulated considerable discussion on photon statistics and pioneered a new approach to optical spectroscopy and interferometry. A new level of sophistication in the statistical description of light was necessary in order to explain the observed intensity-correlation effects.

The introduction of the laser in the sixties contributed in many ways to the present wide interest in coherence and photon statistics. First, observations of laser-light fluctuations constituted an excellent source of information for physicists to test theoretical models of the laser. A great deal of work has been done in this field in many laboratories all over the world.

Second, one of the most exciting applications of the laser in physics, biology, and chemistry is to probe molecular motion. Laser light, which is coherent over considerable space and time is scattered from the molecules; the scattered light fluctuates in accordance with the molecular fluctuations and has the same statistical nature as natural (thermal or chaotic) light, except that the fluctuations are much slower and therefore much easier to measure. The techniques of light-beating and photon-correlation spectroscopy have expanded very rapidly in recent years.

Third, as laser communication and radar systems became a reality, extensive theoretical and experimental work has been done on the effect of random fluctuations in the light beam and the additional fluctuations that are introduced by the detector on the performance of these systems.

Without the continuous developments of the photomultiplier tube as a detector of light, and of fast electronic data-processing circuits, these experiments and systems would have been very difficult to realize.

This book presents the theory of fluctuations of light and of detected photoelectrons, and their applications to spectroscopy and optical communications.

The central theme of the book is that light is described by a stochastic vector field and that photoelectron events are represented by a doubly stochastic Poisson point process. The theories of stochastic fields and of point processes are well-developed branches of mathematical statistics; the book relies heavily on their use.

I have tried to render the book as self-sufficient as possible by including a review, on the engineering level, of the properties of random variables, stochastic processes, and point processes. Although many excellent textbooks cover these subjects, I consider my own review necessary because it provides definitions of the quantities to be used and a collection of the formulas to be referred to in the rest of the book. Moreover, some of the topics that are of particular importance to the theory of photoelectron statistics are not found in conventional textbooks of mathematical statistics (e.g., complex random variables, complex stochastic processes, and the doubly stochastic Poisson point process). Some of the ideas presented in this review were originaly developed in the context of optics, but because of their generality and applicability to other fields, I preferred to present them as part of the mathematical-statistics tools.

The book is divided into three parts: tools, theory, and applications. Part I, which reviews the necessary statistical methods, is divided into two chapters. Chapter 2 covers random variables and stochastic processes, with special emphasis on the properties of circularly symmetric complex random variables and the complex representation of bandpass stochastic processes, using complex analytic signals. Chapter 3 is devoted to point processes and includes a detailed study of the statistical properties of Poisson and doubly stochastic Poisson processes.

Part II, which presents the theory, contains two chapters. Chapter 4 covers the statistical properties of light fields of various origins (e.g., thermal light, laser light, scattered light, modulated light, diffused light). The effect of propagation of light through an optical system on its statistical properties is discussed in general terms. This leads to simple explanations of several important ideas such as intensity interferometry and the phenomenon of speckle. This chapter uses the methods and definitions reviewed in Chapter 2.

The theory of photoelectron statistics is the subject of Chapter 5, which builds on the results of Chapter 4 and uses the methods developed in Chapter 3. This chapter develops a list of single and multifold statistics of photoelectron counts and inter-event times for a large number of important optical fields. The role played by the area of the detector is examined in detail. Nonideal behavior of photodetectors such as the dark current and the dead-time effect are treated at the end of the chapter.

The applications are presented in Part III. The theme here is inversion: determination of the properties of light from measurements of photoelectron statistics. This part is also divided into two chapters. Chapter 6 deals with applications related to optical communication. This includes the optimum design and performance of optical communication systems, based on photoelectron counts or instants of occurrence of the photoelectron events. Many schemes of modulation are considered, and the problem of detecting a radar signal is treated. Error rates in such systems are calculated under different conditions of background radiation or of scintillation effects. Direct and heterodyne detection are examined. Applications to the detection of weak images by using an array of photodetectors are also briefly discussed.

Chapter 7 describes applications to spectroscopy. It outlines methods capable of determining spectra of very narrow bandwidths, 1-10^8 Hz, from measurements taken on the photoelectrons. Detailed emphasis is placed on the very powerful technique of photoelectron-correlation spectroscopy. The technique of single-photoelectron spectroscopy, which is used to measure the decay of fast light pulses, is also discussed. The chapter ends with a discussion of the methods of estimating the spatial spectrum of optical fields.

Because the book is oriented toward applications, the more-rigorous quantum description of electromagnetic field is not used. The electromagnetic field is treated classically, whereas the detection process is treated semi-classically. Photoelectrons, rather than photons, are the subject of interest. This description is adequate for the purpose of this book.

Part I

Tools from Mathematical Statistics

2. Statistical Description of Random Variables and Stochastic Processes

This chapter is a review of the statistical properties of random variables and stochastic processes that are necessary for understanding the optical phenomena described in this book.

In Sec. 2.1, methods of describing the statistics of a random variable are introduced and are generalized in Sec. 2.2 to a set of random variables. Sec. 2.3 discusses the statistical description of complex random variables.

Random functions (stochastic processes) are then discussed in Sec. 2.4 and are generalized to complex random functions in Sec. 2.5. Since bandpass stochastic processes play an important role in optics, emphasis is given to their statistical properties, especially their complex representation, which is treated in Sec. 2.6. A short review of some of the principles of estimation and detection theory is given in Sec. 2.7. This will frequently be referred to in Part III of the book.

Although the material of this chapter is normally covered in standard textbooks on probability theory and stochastic processes (e.g., [35, 105, 106, 115, 135, 138, 139, 250, 358, 364, 461]), the material related to complex random variables, complex stochastic processes and complex representation of bandpass stochastic processes is not conventionally emphasized (see, however, [347]). It is therefore recommended that the reader give these sections special attention.

Several books on subjects like statistical physics, statistical communication theory, control theory, and decision theory give concise and useful reviews of some of the topics covered in this chapter (e.g., [50, 202, 292, 479, 498, 499]).

2.1 Statistical Description of Random Variables

2.1.1 Probability Distribution (PD)

A random variable (RV) is described by a probability distribution $P_x(x)$. If x takes only one of a set of discrete values, then $P_x(x)$ is the probability that the random variable[1] x takes the value x. If x takes a continuum of values, then $P(x)\Delta x$ is the probability that the RV x takes a value between x and $x+\Delta x$, and in this case P(x) is the probability density.

The statistical behaviour of a RV is completely described by its PD.

The *expected value* of a function of the RV x, say f(x), is defined by

$$\begin{aligned} \langle f(x)\rangle &= \int f(x)P(x)dx \quad x \text{ continuous,} \\ &= \sum_j f(x_j)P(x_j) \quad x \text{ discrete.} \end{aligned} \tag{2.1}$$

The moments of a RV x are the expected values of certain algebraic functions of x, as defined in the following.

2.1.2 Moments of a Random Variable

I) Ordinary Moments

The mth order moment of the RV x is defined by

$$\mu_x^{(m)} = \langle x^m \rangle \quad .$$

The first-order moment $\langle x \rangle$ is called the mean, the average, or the expected value of x. We also use the symbol $\bar{x}$ to denote $\langle x \rangle$.

If x takes only nonnegative integer values, then the PD of x can be obtained from the moments by

$$P(x) = \frac{(-1)^x}{x!} \sum_{m=x}^{\infty} \frac{\langle x^m \rangle}{(m-x)!} \quad . \tag{2.2}$$

II) Central Moments

The mth order central moment (central around the mean) is defined by

$$v_x^{(m)} = \langle \Delta x^m \rangle \quad , \quad \text{where} \quad \Delta x = x - \langle x \rangle$$

---

[1]The same symbol x is used to denote the name of a RV and the values it takes. When no ambiguity is likely, we shall write $P_x(x)$ as P(x). Unless otherwise specified, the subscript is the same as the argument.

is the fluctuation of x around its mean. The first-order central moment is obviously zero, and the second-order moment is called the variance,

$$\mathrm{Var}(x) = \langle \Delta x^2 \rangle \quad .$$

The central moments are related to the ordinary moments by

$$\langle \Delta x^m \rangle = \sum_{\ell=0}^{m} \binom{m}{\ell} \langle x^\ell \rangle \, \langle x \rangle^{m-\ell} \quad . \tag{2.3}$$

III) Cumulants

The cumulants (or semi-invariants) $K_x^{(m)}$, m = 1,2, ... of a RV x are defined in terms of the ordinary moments by the identity in t,

$$\exp\left[\sum_{r=1}^{\infty} K_x^{(r)} t^r / r!\right] = \sum_{r=0}^{\infty} \mu_x^{(r)} t^r / r! \quad .$$

The first lower-order cumulants are given by

$$\begin{aligned}
K^{(1)} &= \mu^{(1)} \\
K^{(2)} &= v^{(2)} = \mu^{(2)} - \left[\mu^{(1)}\right]^2 \\
K^{(3)} &= v^{(3)} = \mu^{(3)} - 3\mu^{(1)}\mu^{(2)} + 2\left[\mu^{(1)}\right]^3 \\
K^{(4)} &= v^{(4)} - 3\left[v^{(2)}\right]^2 = \mu^{(4)} - 3\left[\mu^{(2)}\right]^2 - 4\mu^{(1)}\mu^{(3)} + 12\left[\mu^{(1)}\right]^2\mu^{(2)} \\
&\quad - 6\left[\mu^{(1)}\right]^4 \\
&\ldots\ldots ,
\end{aligned} \tag{2.4}$$

and conversely,

$$\begin{aligned}
\mu^{(1)} &= K^{(1)} \\
\mu^{(2)} &= K^{(2)} + \left[K^{(1)}\right]^2 \\
\mu^{(3)} &= K^{(2)} + 3K^{(1)}K^{(2)} + \left[K^{(1)}\right]^3 \\
\mu^{(4)} &= K^{(4)} + 4K^{(1)}K^{(3)} + 3\left[K^{(2)}\right]^2 + 6\left[K^{(1)}\right]^2 K^{(2)} + \left[K^{(1)}\right]^4 \\
&\ldots\ldots .
\end{aligned} \tag{2.5}$$

The first- and second-order cumulants are the mean and the variance, respectively, the most important parameters which describe a RV. The first n cumulants contain just as much information as the first n moments. For a deterministic variable, all cumulants vanish except the first. For a Gaussian RV (cf. Table 2.1), all cumulants except the first and the second vanish. The third-order cumulant displays the asymmetry or "skewness" of the PD around its mean. The coefficient of skewness $K^{(3)} / \left[K^{(2)}\right]^{3/2}$ serves as one indicator of the length of the "tail" of the PD.

IV) Factorial Moments

If x takes only integer values, then the factorial moments,

$$F_x^{(m)} = \left\langle \frac{x!}{(x-m)!} \right\rangle = \langle x(x-1) \ldots (x-m+1) \rangle \quad , \tag{2.6}$$

are also very useful.

2.1.3 Moment-Generating Functions (mgf) of a Random Variable

The study of random variables is greatly facilitated by the introduction of the following moment-generating functions.

I) Ordinary mgf

The mgf is defined by

$$Q_x(s) = \langle e^{-sx} \rangle \quad . \tag{2.7}$$

The coefficients of $(-1)^m s^m/m!$ in a MacLaurin expansion of $\langle e^{-sx} \rangle$ are the ordinary moments $\langle x^m \rangle$

$$Q_x(s) = \sum_{m=0}^{\infty} (-1)^m \frac{s^m}{m!} \langle x^m \rangle \quad , \tag{2.8}$$

$$\langle x^m \rangle = (-1)^m \frac{\partial^m}{\partial s^m} Q_x(s) \Big|_{s=0} \quad . \tag{2.9}$$

Note that if x is nonnegative, then

$$Q_x(s) = \int_0^{\infty} e^{-sx} P(x) dx \quad ,$$

i.e., $Q_x(s)$ is the Laplace transform of P(x), and P(x) can be obtained from $Q_x(s)$ by finding the inverse Laplace transform,

$$P(x) = \frac{1}{2\pi j}\int_{c-j\infty}^{c+j\infty} e^{sx}Q_x(s)ds \quad . \tag{2.10}$$

If s is purely imaginary, the mgf becomes the *characteristic function*,

$$C_x(\xi) = Q_x(j\xi) \quad .$$

II) Cumulant-Generating Function (cgf)

This is defined by

$$Q_x^c(s) = \ln Q_x(s) = \ln \langle e^{-sx}\rangle \quad . \tag{2.11}$$

The cumulants can be obtained from (or are defined by)

$$K_x^{(m)} = (-1)^m \frac{\partial^m}{\partial s^m} Q_x^c(s) \Big|_{s=0} \quad . \tag{2.12}$$

III) Factorial-Moments-Generating Function (fmgf)

If x takes the values (0,1,2, ...), then the function $(1-s)^x$ generates the variables x!/(x-s)! and its expectation generates the factorial moments. Hence

$$Q_x^f(s) = \langle (1-s)^x \rangle \tag{2.13}$$

and

$$F_x^{(m)} = (-1)^m \frac{\partial^m}{\partial s^m} Q_x^f(s) \Big|_{s=0} \quad . \tag{2.14}$$

If we write

$$Q_x^f(s) = \sum_{x=0}^{\infty} (1-s)^x P(x) \quad ,$$

and differentiate m times with respect to s at s = 1, all terms vanish except the one at x = m. We then get

$$P(x) = \frac{(-1)^x}{x!} \frac{\partial^x}{\partial s^x} Q_x^f(s) \Big|_{s=1} \quad . \tag{2.15}$$

This is an important relation, which gives P(x) in terms of $Q_x^f(s)$.

2.1.4 Some Standard Random Variables

Tables 2.1 and 2.2 contain a compilation of the statistical properties of some standard continuous and discrete random variables, which will be used in later chapters.

2.1.5 Transformation of a Random Variable

If y = f(x) is a function of the random variable x, then y itself is also a random variable. If f(x) is a monotonically increasing or decreasing function of x, then the probability distributions of x and y are related by

$$P(y)\left|\frac{dy}{dx}\right| = P(x) \quad . \tag{2.16}$$

This can be seen by observing that if x_0 corresponds to y_0 and $x_0 + \Delta x$ to $y_0 + \Delta y$, then the probability that x lies between x_0 and $x_0 + \Delta x$ should equal the probability that y lies between y_0 and $y_0 + \Delta y$. The reader may like to verify the following transformations (cf. Table 2.1):

I) If x is *normal* with zero mean and $y = x^2$, then y has a *chi_1^2* distribution.

II) If x is *chi_2^2* and $y = \sqrt{x}$, then y is *Rayleigh* distributed.

III) If x is *normal* and $y = e^x$, then y is *log-normal* (see also Fig. 2.1).

2.2 Statistical Description of a Set of Random Variables

2.2.1 Probability Distributions

A set of random variables $x_1, x_2, \dots x_N$ can be described by their *joint* probability distribution (JPD), $P(x_1, x_2, \dots x_N)$. If these RV's are discrete, then $P(x_1, \dots x_N)$ denotes the probability that the RV's $x_1, \dots x_N$ simul-

Table 2.1. Statistical properties of some standard continuous random variables

	$P(x)$	$Q_x(s)$	$\langle x \rangle$	$\langle x^m \rangle$	$K_x^{(m)}/\langle x \rangle^m$
Gaussian (normal	$\frac{1}{\sqrt{2\pi}\,\sigma} \exp[-(x-\mu)^2/2\sigma^2]$	$\exp(-\mu s+\sigma^2 s^2/2)$	μ		σ^2/μ^2 $m=2$ 0 $m>2$
Rayleigh	$\frac{x}{\sigma^2} \exp(-x^2/2\sigma^2)$ $x \geq 0$	$1-s\sqrt{\frac{\pi}{2}}\sigma e^{\sigma^2 s^2/2}\left[1-\Phi(\frac{\sigma s}{\sqrt{2}})\right]$	$\sqrt{\frac{\pi}{2}}\,\sigma$	$\sqrt{\frac{\pi}{2}}\, m!!\sigma^m$ m odd $2^{m/2}\, \frac{m}{2}!!\sigma^m$ m even	
Exponential (chi_2^2)	$\frac{1}{\mu} \exp(-x/\mu)$ $x \geq 0$	$(1+\mu s)^{-1}$	μ	$m!\ \mu^m$	$(m-1)!$
Gaussian-square (chi_1^2)	$\frac{1}{(\pi\mu x)^{1/2}} \exp(-x/\mu)$ $x \geq 0$	$(1+\mu s)^{-1/2}$	$\frac{1}{2}\mu$	$(2m-1)!!\ (\frac{\mu}{2})^m$	$(m-1)!\ 2^{m-1}$
Gamma (chi_{2N}^2)	$\frac{1}{\Gamma(N)}\ \mu^{-N} x^{N-1} \exp(-x/\mu)$ $x \geq 0$	$(1+\mu s)^{-N}$	$N\mu$	$\frac{\Gamma(N+m)}{\Gamma(N)}\ \mu^m$	$(m-1)!/N^{m-1}$
Rician	$\frac{x}{\sigma^2} \exp\left(-\frac{x^2+\mu_0^2}{2\sigma^2}\right) I_0\left(\frac{\mu_0 x}{\sigma^2}\right)$ $x \geq 0$		$\mu_0+\sqrt{\frac{\pi}{2}}\ \sigma$	$\frac{m}{2}!\, L_{\frac{m}{2}}\left(-\frac{\mu_0}{\mu}\right)\mu^{m/2}$ m even	
Rician-square Noncentral chi_2^2 (Bessel)	$\frac{1}{\mu}\exp(-\frac{x+\mu_0}{\mu}) I_0(2\,\frac{\sqrt{x\mu_0}}{\mu})$ $x \geq 0$	$(1+\mu s)^{-1} \exp(\frac{-s\mu_0}{1+\mu s})$	$\mu_0+\mu$	$m!\ L_m(-\frac{\mu_0}{\mu})\mu^m$	$\mu^2+2\mu\mu_0$ $m=2$
Noncentral chi_{2N}^2	$\frac{1}{\mu}(\frac{x}{\mu_0})^{\frac{N-1}{2}} \exp(-\frac{x+\mu_0}{\mu}) I_{N-1}(2\sqrt{x\mu_0}/\mu)$ $x \geq 0$	$(1+\mu s)^{-N} \exp(\frac{-s\mu_0}{1+\mu s})$	$\mu_0+N\mu$	$m!\ L_m^{N-1}(-\frac{\mu_0}{\mu})\mu^m$	
Log-normal	$\frac{1}{\sqrt{2\pi}\,\sigma x} \exp[-(\ln x-\mu)^2/2\sigma^2]$ $x \geq 0$		$e^{\mu+\sigma^2/2}$	$e^{m\mu+m^2\sigma^2/2}$	

Table 2.2. Statistical properties of some standard discrete random variables

	$P(n)$	$Q_n(s)$	$\langle n\rangle$	$\langle\Delta n^2\rangle$	$F_n^m/\langle n\rangle^m$
Poisson	$\mu^n e^{-\mu}/n!$	$\exp[\mu(e^{-s}-1)]$	μ	μ	1
Geometric (Bose-Einstein)	$\frac{\mu^n}{(1+\mu)^{n+1}}$	$(1+\mu-\mu e^{-s})^{-1}$	μ	$\mu + \mu^2$	$m!$
Negative-binomial	$\binom{n+N-1}{n}\frac{\mu^n}{(1+\mu)^{n+N}}$	$(1+\mu-\mu e^{-s})^{-N}$	$N\mu$	$N(\mu + \mu^2)$	$\frac{\Gamma(m+N)}{\Gamma(N)N^m}$
Laguerre	$\frac{\mu^n}{(1+\mu)^{n+1}}\exp\left(-\frac{\mu_o}{1+\mu}\right)L_n\left(-\frac{\mu_o/\mu}{1+\mu}\right)$	$\frac{\exp\left[\frac{\mu_o(e^{-s}-1)}{1+\mu-\mu e^{-s}}\right]}{(1+\mu-\mu e^{-s})}$	$\mu_o+\mu$	$\mu+\mu^2+\mu_o+2\mu\mu_o$	$m!L_m\left(-\frac{\mu_o}{\mu}\right)\left(\frac{\mu}{\mu+\mu_o}\right)^m$
Laguerre order N	$\frac{\mu^n}{(1+\mu)^{n+N}}\exp\left(-\frac{\mu_o}{1+\mu}\right)L_n^{N-1}\left(-\frac{\mu_o/\mu}{1+\mu}\right)$	$\frac{\exp\left[\frac{\mu_o(e^{-s}-1)}{1+\mu-\mu e^{-s}}\right]}{(1+\mu-\mu e^{-s})^N}$	$\mu_o+N\mu$	$N(\mu+\mu^2)+\mu_o+2\mu\mu_o$	$m!L_m^{N-1}\left(-\frac{\mu_o}{\mu}\right)\left(\frac{\mu}{\mu_o+N\mu}\right)^m$

taneously take the values $x_1, x_2, \dots x_N$, respectively. If they are continuous, then $P(x_1, \dots x_N)\Delta x_1 \dots \Delta x_N$ is the probability that $x_1, \dots x_N$ have values between $x_1, x_1 + \Delta x_1, \dots x_N, x_N + \Delta x_N$, respectively. The *expected value* of a function of $x_1, \dots x_N$ is also defined as in the one-dimensional case, namely;

$$\langle f(x_1, \dots x_N)\rangle = \int dx_1 \dots \int dx_N \, f(x_1, \dots x_N)P(x_1, \dots x_N) \quad . \tag{2.17}$$

The *conditional* probability density $P(x_1|x_2)$ denotes the probability density that the RV x_1 takes the value x_1, given that the RV x_2 is known to have the value x_2. This is related to the JPD, $P(x_1,x_2)$ and to the marginal PD, $P(x_2)$ (i.e., the PD of x_2 irrespective of x_1) by the relation,

$$P(x_1|x_2) = P(x_1,x_2) \; / \; P(x_2) \quad , \tag{2.18}$$

that can be regarded as a definition of the conditional probability.

Two RV's are said to be *statistically independent* if their JPD factorizes in the form

$$P(x_1,x_2) = P(x_1)P(x_2) \quad .$$

In this case, the conditional probability of x_1 given x_2, $P(x_1|x_2)$, is simply equal to $P(x_1)$.

2.2.2 Moments and Correlations

The definitions of the previous section can be generalized to cover the joint moments of a set of random variables. In order to economize in notations, let us use the symbol [x] to represent the set $(x_1,x_2, \dots x_N)$, and [m] for $(m_1,m_2, \dots m_N)$, and so on. The set [x] can also be regarded as a column matrix whose elements are $x_1,x_2, \dots x_N$. In addition to the usual rules of matrix algebra (e.g., $[x]'[y] = \sum_i x_i y_i$), we introduce the following notations:

$$[x]^{[y]} = x_1^{y_1} \cdot x_2^{y_2} \dots x_N^{y_N}$$

$$\frac{[x]}{[y]} = \frac{x_1}{y_1} \cdot \frac{x_2}{y_2} \dots \frac{x_N}{y_N}$$

$$([n] - [m])! = (n_1 - m_1)! \, (n_2 - m_2)! \dots (n_N - m_N)!$$

$$\sum_{[m]=[n]}^{[\infty]} = \sum_{m_1=n_1}^{\infty} \sum_{m_2=n_2}^{\infty} \dots \sum_{m_N=n_N}^{\infty}$$

and

$$\frac{\partial^{[m]}}{\partial [s]^{[m]}} = \frac{\partial^{m_1}}{\partial s_1^{m}} \cdot \frac{\partial^{m_2}}{\partial s_2^{m_2}} \cdots \frac{\partial^{m_N}}{\partial s_N^{m_N}} \quad .$$

Let us also assume that [0] and [1] are the sets (0,0,...0) and (1,1 ... 1), respectively, and f([s]) or [f(s)] is the set $[f(s_1), f(s_2), \ldots f(s_N)]$. The generalized moments are defined as follows:

I) Ordinary Moments

$$\mu_{[x]}^{[m]} = \langle [x]^{[m]} \rangle \quad .$$

For example, for N = 2 and [m] = [1],

$$\mu_{x_1,x_2}^{(1,1)} = \langle x_1 x_2 \rangle$$

is the *correlation* between the RV's x_1 and x_2.

II) Central Moments

$$v_{[x]}^{[m]} = \langle [\Delta x]^{[m]} \rangle \quad ,$$

where

$$[\Delta x] = [x] - [\langle x \rangle] \quad .$$

This gives, for N = 2 and [m] = [1],

$$v_{x_1,x_2}^{(1,1)} = \langle \Delta x_1 \Delta x_2 \rangle \quad ,$$

which is known as the *covariance* of x_1 and x_2. This is sometimes written as $\mathrm{Cov}(x_1,x_2)$.

III) Cumulants

The N-fold cumulants will be formally defined in (2.24). We only give here expressions for some low-order cumulants

$$
\begin{aligned}
K^{(1,1)}_{x_1,x_2} &= \langle \Delta x_1 \Delta x_2 \rangle = \text{Covariance} \\
K^{(1,1,1)}_{x_1,x_2,x_3} &= \langle \Delta x_1 \Delta x_2 \Delta x_3 \rangle \\
K^{(1,1,1,1)}_{x_1,x_2,x_3,x_4} &= \langle \Delta x_1 \Delta x_2 \Delta x_3 \Delta x_4 \rangle - \langle \Delta x_1 \Delta x_2 \rangle \langle \Delta x_3 \Delta x_4 \rangle \\
&\quad - \langle \Delta x_1 \Delta x_3 \rangle \langle \Delta x_2 \Delta x_4 \rangle - \langle \Delta x_1 \Delta x_4 \rangle \langle \Delta x_2 \Delta x_3 \rangle \quad .
\end{aligned}
\tag{2.19}
$$

They are defined in such a way that the N-fold cumulant represents the "true" N-fold correlation with all correlations of lower fold subtracted. For example, if x_1 and x_2 are independent from x_3 and x_4, but x_1 and x_2 are not independent, then the 4-fold central moment does not vanish, but the 4-fold cumulant does.

IV) Factorial Moments

For nonnegative integer RV's,

$$
F^{[m]}_{[x]} = \left\langle \frac{[x]!}{([x] - [m])!} \right\rangle \quad . \tag{2.20}
$$

2.2.3 Moment-Generating Functions

I) Ordinary mgf

The N-fold mgf is given by

$$
Q_{[x]}([s]) = \langle e^{-[x]'[s]} \rangle \quad , \tag{2.21}
$$

and the ordinary moments are generated by

$$
\langle [x]^{[m]} \rangle = [-1]^{[m]} \frac{\partial^{[m]}}{\partial [s]^{[m]}} Q_{[x]}([s]) \Big|_{[s] = [0]} \quad . \tag{2.22}
$$

II) Cumulant-Generating Function

The cgf is given by

$$
Q^{c}_{[x]}([s]) = \ln Q_{[x]}([s]) \tag{2.23}
$$

and the cumulants are defined by

$$K^{[m]}_{[x]} = [-1]^{[m]} \frac{\partial^{[m]}}{\partial[s]^{[m]}} Q^{c}_{[x]}([s]) \Big|_{[s] = [0]} \quad . \tag{2.24}$$

III) Factorial-Moment-Generating Function

The fmgf is defined by

$$Q^{f}_{[x]}([s]) = \left\langle ([1] - [s])^{[x]} \right\rangle \quad , \tag{2.25}$$

from which the factorial moments are obtained by

$$F^{[m]}_{[x]} = [-1]^{[s]} \frac{\partial^{[m]}}{\partial[s]^{[m]}} Q^{f}_{[x]}([s]) \Big|_{[s] = [0]} \quad , \tag{2.26}$$

where [x] is a set of nonnegative integer RV's.
Moreover, the JPD of [x] is related to the fmgf by

$$P([x]) = \frac{[-1]^{[x]}}{[x]!} \frac{\partial^{[x]}}{\partial[s]^{[x]}} Q^{f}_{[x]}([s]) \Big|_{[s] = [1]} \quad . \tag{2.27}$$

2.2.4 An Example: A Set of Jointly Gaussian Random Variables

A set of N random variables $[x] = (x_1, x_2, \ldots x_N)$ is said to be jointly Gaussian if their JPD is given by

$$P([x]) = (2\pi)^{-N/2} \left|\mu\right|^{-1/2} \exp(-\tfrac{1}{2}[x]' \mu^{-1} [x]) \quad , \tag{2.28}$$

where [x] represents the column matrix whose elements are $x_1, x_2, \ldots x_N$, $[x]'$ is its transpose, and μ is the matrix whose elements are given by

$$\mu_{ij} = \langle x_i x_j \rangle \tag{2.29}$$

(called the correlation matrix), μ^{-1} being its inverse, and $|\mu|$ its determinant.
Note that it has been implicitly assumed that [x] has zero means. Generalization to a set of random variables with nonzero means is straightforward. The mgf can be obtained by using the definitions (2.21) and 2.17). The result is

$$Q_{[x]}([s]) = \exp(\tfrac{1}{2}[s]^{\sim}\mu[s]) \quad , \tag{2.30}$$

where [s] is the column matrix whose elements are $(s_1, s_2, \dots s_N)$. From (2.23) and (2.30),

$$Q^c_{[x]}([s]) = \tfrac{1}{2}[s]^{\sim}\mu[s] \quad . \tag{2.31}$$

This enables us to determine the cumulants. By use of (2.24), we get

$$\begin{aligned} K^{(1,1)}_{x_1,x_2} &= \langle x_1 x_2 \rangle = \mu_{12} \quad , \\ K^{(1,1,\dots 1)}_{x_1,x_2,\dots x_m} &= 0, \qquad m > 2 \quad . \end{aligned} \tag{2.32}$$

The ordinary moments are given by

$$\begin{aligned} &\langle x_1 x_2 \cdots x_{2r+1} \rangle = 0 \\ &\langle x_1 x_2 \cdots x_{2r} \rangle = \sum_{\pi} \langle x_1 x_{\pi 1} \rangle \langle x_2 x_{\pi 2} \rangle \cdots \langle x_r x_{\pi r} \rangle \quad , \\ &\qquad r = 0,1,2, \dots \quad , \end{aligned} \tag{2.33}$$

where $\sum_{\pi}$ represents summations over the permutations of $(\pi_1, \pi_2, \dots \pi_r)$ among the integers, $(r+1, r+2, \dots 2r)$.

Examples:

$$\begin{aligned} &\langle x_1 x_2 x_3 \rangle = 0 \quad , \\ &\langle x_1 x_2 x_3 x_4 \rangle = \langle x_1 x_2 \rangle \langle x_3 x_4 \rangle + \langle x_1 x_3 \rangle \langle x_2 x_4 \rangle + \langle x_1 x_4 \rangle \langle x_2 x_3 \rangle \quad . \end{aligned} \tag{2.34}$$

2.2.5 Transformation of Random Variables

If $y = f(x_1, x_2)$ is a function of the two RV's x_1 and x_2, then y itself is also a RV. The statistics of y depend on the joint statistics of x_1 and x_2. In order to relate $P(y)$ to $P(x_1, x_2)$, we determine the area in the $x_1 - x_2$ plane that corresponds to values of y between y and $y + \Delta y$, say ΔA, and write

$$P(y)\Delta y = \iint_{\Delta A} P(x_1, x_2) dx_1 dx_2 \quad . \tag{2.35}$$

Another method is to write the mgf

$$Q_y(s) = \langle e^{-sy} \rangle = \langle e^{-sf(x_1,x_2)} \rangle \quad , \tag{2.36}$$

from which P(y) can be determined. For example, if $y = x_1 + x_2$ where x_1 and x_2 are statistically independent, then

$$Q_y(s) = \langle e^{-s(x_1+x_2)} \rangle = \langle e^{-sx_1} \rangle \langle e^{-sx_2} \rangle = Q_{x_1}(s) Q_{x_2}(s) \quad ,$$

from which we get

$$P_y(y) = \int_{-\infty}^{\infty} P_{x_1}(y-x_2) P_{x_2}(x_2) dx_2 = P_{x_1}(y) \circledast P_{x_2}(y) \quad , \tag{2.37}$$

i.e., P(y) is the convolution of $P(x_1)$ and $P(x_2)$.

The above methods can, of course, be directly generalized to a RV which is a function of N RV's. As an exercise, the reader can verify the important transformations sketched in Fig. 2.1 (cf. Table 2.1). This diagram shows that, from a set of statistically independent normal (Gaussian) random vari-

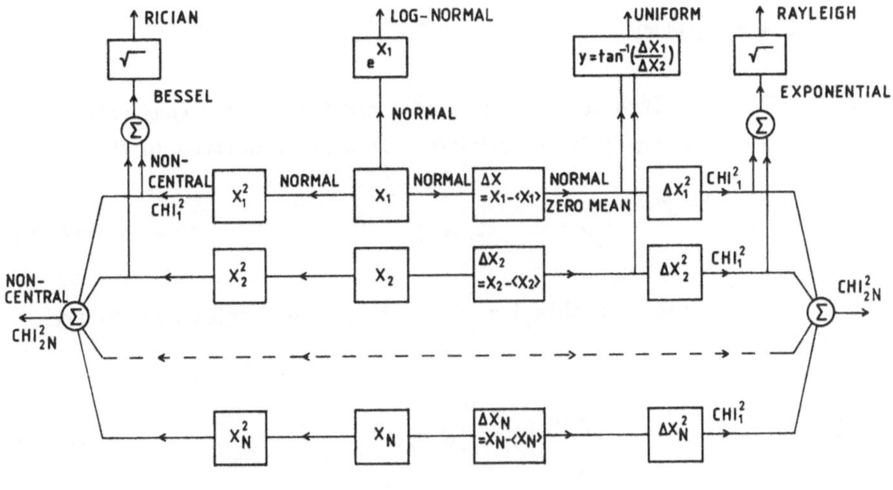

Fig.2.1. Examples of nonlinear transformations of a set of independent identical Gaussian (normal) random variables $x_1, x_2, \ldots x_N$

ables, all the standard distributions of Table 2.1 can be generated by simple nonlinear transformations. Since the Gaussian distribution is the most common in nature, we can appreciate the importance of the RV's defined in Table 2.1.

The importance of the Gaussian distribution can be realized by examining the *central-limit theorem*:

"If x is the sum of N independent RV's, $x_1, \ldots x_N$, then under certain general conditions, P(x) approaches a Gaussian distribution as N increases, regardless of the distributions of $x_1, \ldots x_N$".

The PD, P(x), is the multiconvolution of $P(x_1), \ldots P(x_N)$ see (2.37), and it is a mathematical property that the convolution of a large number of positive functions is Gaussian.

Another reason for the importance of the Gaussian distribution is the fact that any linear transformation of a set of jointly Gaussian RV's is a set of jointly Gaussian RV's.

2.3 Statistical Description of Complex Random Variables

2.3.1 The Complex Random Variable (CRV)

A complex variable, $Z = x + jy$, whose components x and y are random, can be described by the JPD of x and y. The notation,

$$P(Z)d^2Z = P(x,y)dxdy,$$

in which P(Z) is written as a function of Z and its complex conjugate Z^*, is useful. The *moment-generating function* can also be written in the form

$$Q_Z(s) = \left\langle e^{-(S^*Z + SZ^*)} \right\rangle = Q_{x,y}(s_x, s_y) \tag{2.38}$$

where S is the complex variable $S = \frac{1}{2}(s_x + js_y)$. The complex moments are then given by

$$\mu_{Z^*Z}^{(N,M)} = \left\langle Z^{*N} Z^M \right\rangle = (-1)^{N+M} \frac{\partial^N}{\partial S^N} \frac{\partial^M}{\partial S^{*M}} Q_Z(s) \Big|_{S=0} \quad . \tag{2.39}$$

2.3.2 Circularly Symmetric Complex Random Variables

A circularly symmetric CRV is a CRV whose probability density P(Z) is a function of the magnitude $|Z|$ and is independent of the phase $\angle Z$. This means that the phase $\angle Z$ is uniformly distributed over the range $[0,2\pi]$. It therefore follows that the mgf is a function of $|S|$ alone (to prove this, write the expectation in (2.38) as an integration over P(Z) and transform to polar coordinates). Because of this symmetry, the components x and y must have identical distributions. A necessary and sufficient condition for Z to be circularly symmetric is that

$$\langle Z^{*N} Z^{M} \rangle = 0 \quad \forall \; N \neq M \quad . \tag{2.40}$$

The necessity of this condition can be easily shown by writing the expectation as an integral over P(Z) and changing to polar coordinates. To prove its sufficiency, we write the mgf in terms of the moments

$$Q_Z(S) = \sum_N \sum_M \frac{(-S)^N}{N!} \frac{(-S^*)^M}{M!} \langle Z^{*N} Z^M \rangle = \sum_N \frac{|S|^{2N}}{(N!)^2} \langle |Z|^{2N} \rangle \quad .$$

Hence $Q_Z(S)$ is independent of $\angle S$, from which P(Z) is independent of $\angle Z$. If we put N = 0 and M = 2 in (2.40), we get $\langle Z^2 \rangle = 0$, i.e.,

$$\mu_{xx} = \mu_{yy}, \quad \text{and} \quad \mu_{xy} = 0 \quad , \tag{2.41}$$

(where $\mu_{xx} = \langle x^2 \rangle$, $\mu_{yy} = \langle y^2 \rangle$, and $\mu_{xy} = \langle xy \rangle$)
meaning that x and y must be uncorrelated.

The PD of the absolute value $|Z|$ of a circularly symmetric CRV can be obtained from the transformation $P(|Z|) = 2\pi |Z| \, P(Z)$. P(Z) can be obtained by writing the inverse Fourier transform of the characteristic function $C_Z(\xi) = Q_Z(j\xi)$,

$$\begin{aligned} P(Z) &= \frac{1}{4\pi^2} \int\!\!\int_{-\infty}^{\infty} \exp[j(\xi_x x + \xi_y y)] C(|\xi|) d\xi_x d\xi_y \\ &= \frac{1}{4\pi^2} \int_0^{2\pi} d\angle\xi \int_0^{\infty} d|\xi| \,.\, |\xi| C(|\xi|) \exp(j|\xi||Z|\cos(\angle\xi)) \\ &= \frac{1}{2\pi} \int_0^{\infty} |\xi| C(|\xi|) J_0(|\xi||Z|) d|\xi| \quad , \end{aligned}$$

where J_0 is the Bessel function of order zero. Therefore,

$$P(|Z|) = |Z| \int_0^{\infty} uC(u)J_0(u|Z|)du \quad . \tag{2.42}$$

As an example, consider the CRV $Z = A \exp(j\theta)$, where A is a real constant and θ is uniformly distributed. Eq. (2.42) confirms the expected result, $P(|Z|) = \delta(|Z| - A)$, where δ is the delta function.

Another example is the CRV $Z = x + jy$, where x and y are identical independent zero-mean Gaussian variables. Eq. (2.42) gives a Rayleigh distribution for $|Z|$ (cf. Table 2.1 and Fig. 2.1).

2.3.3 The Circularly Symmetric Gaussian CRV

A CRV is said to be Gaussian if its two components are jointly Gaussian.

The mgf of x and y is therefore given by

$$Q_{x,y}(s_x,s_y) = \exp\left[\frac{1}{2}\left(s_x^2\mu_{xx} + 2s_xs_y\mu_{xy} + s_y^2\mu_{yy}\right)\right] \quad , \tag{2.43}$$

where we assumed that the variable has zero mean. Using (2.38), we conclude that

$$Q_Z(S) = \exp\left[\frac{1}{2}\left(S^2\mu_{Z^*Z^*} + 2SS^*\mu_{ZZ^*} + S^{*2}\mu_{ZZ}\right)\right] \quad . \tag{2.44}$$

Thus a Gaussian CRV is completely described by μ_{xx}, μ_{xy} and μ_{yy} or equivalently by μ_{ZZ^*} and μ_{ZZ}. Other moments can be determined as function of these parameters by using (2.33), or by finding the cgf and then its derivatives. This leads to

$$\begin{aligned} K^{(1,1)}_{Z^*,Z} &= \langle Z^*Z \rangle \\ K^{(1,1)}_{Z,Z} &= \langle ZZ \rangle \\ K^{(M,N)}_{Z^*,Z} &= 0 \quad M > 1,\ N > 1 \quad . \end{aligned} \tag{2.45}$$

If the Gaussian CRV is moreover circularly symmetric, then (2.40) is satisfied, and we have

$$Q_Z(S) = \exp(|S|^2 \mu_{Z*Z}) \quad , \tag{2.46}$$

from which

$$P(Z) = \frac{1}{\pi <|Z|^2>} e^{-|Z|^2/<|Z|^2>} = \frac{1}{\sqrt{2\pi\mu_{xx}}} e^{-x^2/2\mu_{xx}} \cdot \frac{1}{\sqrt{2\pi\mu_{yy}}} e^{-y^2/2\mu_{yy}} \quad , \tag{2.47}$$

where the relation

$$\mu_{Z^*Z} = <|Z|^2> = 2\mu_{xx} = 2\mu_{yy}$$

has been used.

The absolute value $|Z|$ has a Rayleigh distribution and its square $|Z|^2$ has a chi_2^2 (chi-square with 2 degrees of freedom) distribution (see Fig. 2.1). The higher-order moments of Z are related to the lower-order moments by

$$<Z^{*N} Z^M> = N! \; <Z^* Z>^N \; \delta_{N,M} \quad . \tag{2.48}$$

From the above equations, it should be fairly easy to show that the necessary and sufficient condition for a Gaussian CRV to be circularly symmetric is that its components be uncorrelated (i.e., statistically independent) and have identical distributions.

2.3.4 A Set of Complex Random Variables

A set of complex random variables $[Z] = (Z_1, Z_2, \ldots Z_N)$ can be described by the 2N-fold joint probability distribution of their real components. Alternatively, we can define the JPD

$$P(Z_1, \ldots Z_N) d^2 Z_1 \ldots d^2 Z_N = P(x_1, y_1; \ldots; x_N, y_N) dx_1 dy_1 \ldots dx_N dy_N$$

and the mgf

$$Q_{Z_1, \ldots Z_N}(S_1, \ldots S_N) = \left\langle \exp\left[-\sum_{j=1}^{N} \left(S_j^* Z_j + S_j Z_j^*\right)\right] \right\rangle \quad .$$

A Set of Complex Gaussian Random Variables

A set of N complex Gaussian random variables $[Z] = (Z_1, \dots Z_N)$ is a set of 2N real Gaussian variables $(x_1,y_1;x_2,y_2; \dots x_N,y_N)$. Their joint PD takes the form

$$P([f]) = (2\pi)^{-N}|\mu|^{-1/2}\exp(-\tfrac{1}{2}[f]^{\sim}\mu^{-1}[f]) \quad , \tag{2.49}$$

where $[f]$ is the vector $(x_1,y_1;x_2,y_2; \dots ;x_N,y_N)$
and $\mu_{ij} = \langle f_i f_j \rangle$.

Not much simplification is gained by using the complex notation in this case.

A Set of Circularly Symmetric Gaussian CRV's

A set of circularly symmetric Gaussian CRV's is a set, $[Z] = (Z_1, \dots Z_N)$, whose JPD is

$$P([Z]) = (2\pi)^{-N}|H|^{-1}\exp(-\tfrac{1}{2}[Z]^{\dagger}H^{-1}[Z]) \tag{2.50}$$

where H is the Hermitian symmetric matrix whose elements are $H_{ij} = \frac{1}{2}\langle Z_i^* Z_j \rangle$, and $[Z]^{\dagger}$ is the complex conjugate of the transpose of $[Z]$.

Note that circular symmetry imposes that

$$\langle Z_i Z_j \rangle = 0, \quad \forall\, i,j \quad .$$

This follows directly from (2.50). Thus every member of the set is a circularly symmetric CRV. Circular symmetry enables us to represent N CRV's (i.e., 2N real RV's) by a complex Hermitian matrix of order NxN. The *moment-generating function* obtained from (2.50) has the form

$$Q_{[Z]}([S]) = \exp(\tfrac{1}{2}[S]^{\dagger}H[S]) \quad . \tag{2.51}$$

The moments of the set [Z] can be determined from (2.51). This gives the factorization rule [395]

$$\langle Z_1^* Z_2^* \dots Z_N^* Z_{N+1} \dots Z_{N+M} \rangle \tag{2.52}$$

$$= \delta_{N,M} \sum_{\pi} \langle Z_1^* Z_{\pi_1} \rangle \langle Z_2^* Z_{\pi_2} \rangle \dots \langle Z_N^* Z_{\pi_N} \rangle$$

where π represents permutations of $\{\pi_1,\pi_2, \ldots \pi_N\}$ over the set of integers $\{N+1,N+2, \ldots 2N\}$ and $\delta_{N,M}$ is the Kronecker delta.

Some special cases of the preceding important rule are

$$<Z_1^*Z_2^*Z_3Z_4> = <Z_1^*Z_3> <Z_2^*Z_4> + <Z_1^*Z_4> <Z_2^*Z_3> \quad , \tag{2.53}$$

$$<(Z_1^*Z_2)^N> = N! <Z_1^*Z_2>^N \quad ,$$

and (2.48).

2.4 Statistical Description of Stochastic Processes

2.4.1 Definitions

A stochastic process $x(t)$ is a random function of an independent variabe t (e.g., time). At every fixed point t, $x(t)$ is a random variable that can be described as in Sec. 2.1. At every fixed set of points $\{t_1,t_2, \ldots t_N\}$, $\{x(t_1),x(t_2), \ldots x(t_N)\}$ is a set of random variables that can be described as in Sec. 2.2. The function $x(t)$ itself is random. This means that it takes one of very many different profiles (called realizations).

A stochastic process can (in most cases) be completely described by the joint probability densities of the sets of random variables $x(t)$ for all t, $\{x(t_1),x(t_2)\}$ for all $\{t_1,t_2\}$ ..., and $\{x(t_1), \ldots x(t_N)\}$ for all $\{t_1, \ldots t_N\}$ and so on for all $N=1,2, \ldots$.

Alternatively, it can be described by the set of moments of the RV's $x(t_1)$, $\ldots x(t_N)$, $N = 1, \ldots \infty$. These moments are functions of $t_1, \ldots t_N$. For example, their moment of order $(1,1, \ldots 1)$ is called the *correlation function* of order N,

$$G_x^{(N)}(t_1,t_2, \ldots t_N) = <x(t_1)x(t_2) \ldots x(t_N)> \quad . \tag{2.54}$$

For $N = 1$, we obtain the mean value at time t, which will always be written $<x(t)>$ instead of $G_x^{(1)}(t)$. For $N = 2$, we have $G_x^{(2)}(t_1,t_2)$. Since this is the most commonly used function, we shall often simply call it $G_x(t_1,t_2)$.

Also, the cumulant functions[2] are just the multifold cumulants of the RV's $x(t_1)$, ... $x(t_N)$

$$K_x^{(N)}(t_1, \dots t_N) = K^{(1,1, \dots 1)}_{x(t_1), x(t_2), \dots x(t_N)}$$

and so on[3].

Note that we need only define the moments of order (1,1, ... 1) because other moments can be easily obtained from them by using the apropriate arguments, e.g.,

$$K_x^{(3)}(t_1,t_1,t_2) = K^{(2,1)}_{x(t_1),x(t_2)} \quad ,$$

$$K_x^{(3)}(t,t,t) = K^{(3)}_{x(t)} \quad .$$

The *moment-generating functions* can be similarly defined,

$$Q_x(s_1,s_2, \dots s_N;t_1,t_2, \dots t_N) = \left\langle \exp\left[-s_1x(t_1)- \dots s_Nx(t_N)\right]\right\rangle , \quad (2.55)$$

and are functions of $t_1,t_2, \dots t_N$ as well as $s_1,s_2, \dots s_N$. From the N-fold mgf, the mgf's of all lower fold can be obtained by setting the appropriate arguments equal to zero. By increasing N, we obtain in the limit, the *moment-generating functional*,

$$L_x(s(t)) = \left\langle \exp\left[-\int s(t)x(t)dt\right]\right\rangle \quad ,$$

which describes the stochastic process x(t) completely.

A random process x(t) is said to be *stationary* (in the strict sense) if its statistical characteristics (probability distributions, moment-generating functions, moments or correlation functions) are invariant to shifts of the time origin. Thus, the correlation functions become functions of time differences,

[2]The cumulant functions defined above are sometimes called correlation functions by other authors.

[3]We shall not always write the superscript N, which indicates the order, if it is clear from the number of arguments of the function. Also, the subscript x may be omitted when no ambiguity is likely.

$$<x(t)> = <x(0)> = \text{constant},$$

$$G_x(t_1,t_2) = G_x(t_2-t_1,0) \quad ,$$

$$\ldots$$

$$G_x^{(N)}(t_1,t_2, \ldots t_N) = G_x^{(N)}(t_2-t_1,t_3-t_1, \ldots t_N-t_1,0) \quad .$$

When no ambiguity is likely, we shall sometimes write $G_x(t_2-t_1,0)$ as $G_x(t_2-t_1)$ or simply $G(t_2-t_1)$.

A random process is said to be stationary of order N if its Nth order moment is time-shift invariant. In particular, stationarity of the second order is called "wide-sense stationarity".

A stochastic process x(t) is called a *Markov process* if, for any finite set of time $t_1<t_2< \ldots <t_N$,

$$P\left[x(t_N)|x(t_1),x(t_2), \ldots x(t_{N-1})\right] = P\left[x(t_N)|x(t_{N-1})\right] \quad . \tag{2.56}$$

This says that if the value of x at t_{N-1} is known, then the probability distribution of the values it will take at t_N is independent of the values it took prior to t_{N-1}. The present state of the process determines its future, irrespective of its past history. A Markov process is therefore completely described by the PD, P[x(t)], and the conditional PD, P[x(t)|x(t')];

$$P\left[x(t_1),x(t_2), \ldots x(t_N)\right] = \left\{\prod_{j=2}^{N} P\left[x(t_j)\middle|x(t_{j-1})\right]\right\} P\left[x(t_1)\right] \quad . \tag{2.57}$$

2.4.2 The Random Spectrum of a Stochastic Process. The Power Spectrum of a Stationary Stochastic Process

The spectrum $F_x(\omega)$ of a process x(t) is defined by the Fourier transform,

$$F_x(\omega) = \int_{-\infty}^{\infty} x(t)e^{-j\omega t}dt \quad , \tag{2.58}$$

and is itself a complex random function of ω. $F_x(\omega)$ should be regarded as a "generalized function", i.e., specified by its values when a linear integral operator is applied to it (like the delta function). Under certain restrictions, we can deal with generalized functions in the same way as the ordinary functions. The complex stochastic process $F_x(\omega)$ can be described by its correlation functions, e.g.,

$$G_x(\omega_1,\omega_2) = \langle F_x^*(\omega_1)F_x(\omega_2)\rangle \quad . \tag{2.59}$$

These functions are called the cross-spectral densities[4] of the process x(t) and are related to the correlation functions of x(t) by the multifold Fourier transform, e.g.,

$$G(\omega_1,\omega_2) = \iint_{-\infty}^{\infty} G(t_1,t_2)e^{j(\omega_1 t_1-\omega_2 t_2)}dt_1dt_2 \quad . \tag{2.60}$$

In Sec. 2.5, the correlation functions of complex stochastic processes will be formally defined and their properties will be presented.

The cross-spectral densities may be singular functions, but they are not more singular than a delta function.

If the process x(t) is stationary (wide-sense), i.e.,

$$G(t_1,t_2) = G(t_2-t_1),$$

then, by substitution in (2.60), we get

$$G(\omega_1,\omega_2) = G(\omega_1)\delta(\omega_1-\omega_2) \quad , \tag{2.61}$$

where

$$G(\omega) = \int_{-\infty}^{\infty} G(\tau)e^{-j\omega\tau}d\tau \tag{2.62}$$

is called the power spectrum of the stationary process x(t). From (2.61), it is concluded that the different frequency components of a stationary process are uncorrelated.

Some special power spectra of interest are shown in Table 2.3. The white process is a mathematical idealization of a rectangular process whose bandwidth is much broader than all other bandwidths of the system.

---

[4]We use the same symbol G to denote both the correlation functions and the cross-spectral densities and identify them by their arguments (t or ω).

Table 2.3. Power spectra and their corresponding correlation functions

	$G(\omega)$	$G(\tau)$
Lorentzian	$\frac{2\Gamma^2}{\Gamma^2+\omega^2}$ 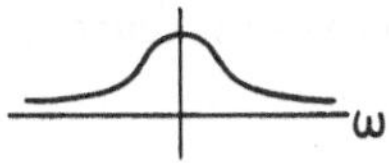	$e^{-\Gamma\lvert\tau\rvert}$ 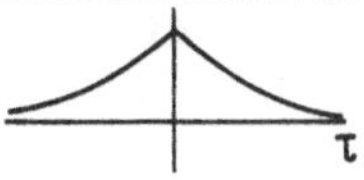
Gaussian	$e^{-\omega^2/2\Gamma^2}$ 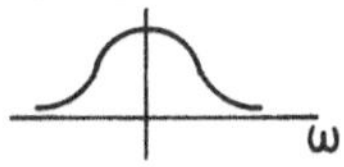	$e^{-\Gamma^2\tau^2/2}$ 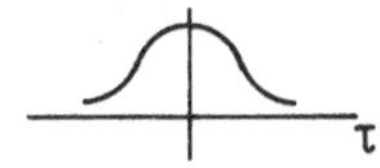
Rectangular	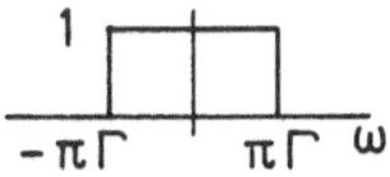	$\pi\Gamma\ \mathrm{sinc}(\Gamma\tau)$ 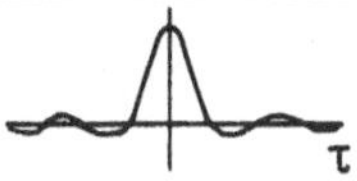
white	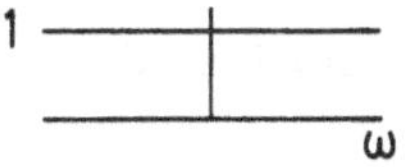	$\delta(\tau)$ 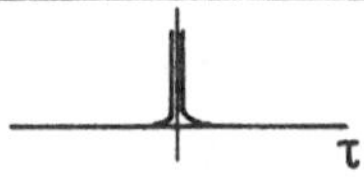
Band-Pass Lorentzian	$\frac{\Gamma^2}{\Gamma^2+(\omega-\omega_0)^2}+\frac{\Gamma^2}{\Gamma^2+(\omega+\omega_0)^2}$ 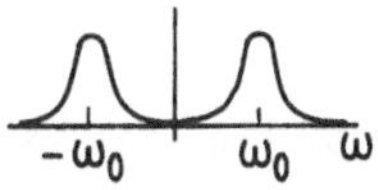	$e^{-\Gamma\lvert\tau\rvert}\cos(\omega_0\tau)$ 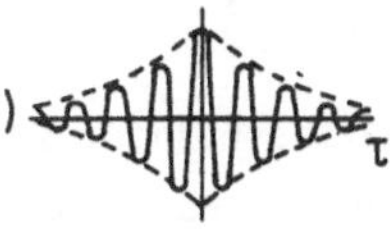

2.4.3 The Gaussian Process

A stochastic process $x(t)$ is said to be Gaussian if, for every finite set of times $\{t_j\}$, the RV's $\{x(t_j)\}$ are jointly Gaussian (Sec. 2.2.4). The statistics of $x(t)$ are then completely determined by the correlation function $G_x(t_1,t_2)$ for all t_1 and t_2. For example, (2.33) gives

$$G^{(2N+1)}(t_1,\ \dots\ t_{2N+1}) = 0 \quad ,$$

$$G^{(2N)}(t_1,\ \dots\ t_N) = \sum_{\pi} G(t_1,t_{\pi 1})G(t_2,t_{\pi 2})\ \dots\ G(t_N,t_{\pi N}) \quad , \tag{2.63}$$

where π is the permutation of $(\pi_1,\pi_2,\ \dots\ \pi_N)$ over $(N+1,n+2,\ \dots\ 2N)$.

It follows that if $x(t)$ is wide-sense stationary and Gaussian, it must also be strict-sense stationary. Two facts give the Gaussian process its prominent place in mathematical statistics: the central limit theorem, and the fact that any linear transformation of a Gaussian process is also a Gaussian process.

2.4.4 Description of a Stochastic Process by the Coefficients of a Karhunen-Loève Expansion

A stochastic function x(t) defined over a time interval [0,T] can be expanded in terms of a set of complete orthogonal deterministic functions $\phi_1(t), \phi_2(t), \ldots$

$$x(t) = \lim_{N\to\infty} \sum_{i=1}^{N} x_i \phi_i(t) \quad , \tag{2.64}$$

where

$$\int_0^T \phi_i(t)\phi_j(t)dt = \delta_{ij} \quad . \tag{2.65}$$

The coefficients $\{x_i\}$ are random variables determined from

$$x_i = \int_0^T x(t)\phi_i(t)dt \tag{2.66}$$

and their joint statistics describe the statistics of x(t) itself. For simplicity, we assume that $\langle x(t)\rangle = 0$. The basis $\{\phi_i(t)\}$ can be chosen in such a way that the coefficients $\{x_i\}$ are uncorrelated, i.e.,

$$\langle x_i x_j \rangle = \mu_i \delta_{ij} \quad . \tag{2.67}$$

Substituting from (2.66) in (2.67) and using (2.65), we can write

$$\int_0^T dt\phi_i(t)\left[\int_0^T G_x(t,t')\phi_j(t')dt' - \phi_j(t)\right] = 0 \quad .$$

A necessary and sufficient condition for this to hold is that

$$\int_0^T G_x(t,t')\phi_j(t')dt' = \mu_j\phi_j(t), \quad 0 \le t \le T \quad . \tag{2.68}$$

This integral equation determines the basis $\{\phi_j(t)\}$ (its eigenfunctions) which ensures the uncorrelatedness of $\{x_i\}$. Its eigenvalues $\{\mu_j\}$ determine the variances of $\{x_j\}$. The basis is called the Karhunen-Loève (K-L) basis.

If x(t) is a Gaussian process then the coefficients $\{x_j\}$ are jointly Gaussian random variables (because (2.66) is a linear transformation). If $\{\phi_j(t)\}$ satisfy (2.68), then the coefficients $\{x_j\}$ are uncorrelated (therefore statistically independent) and are completely determined by their variances $\{\mu_j\}$. The set of real numbers $\{\mu_j\}$ then completely describes the Gaussian process, x(t).

We list here some interesting properties of the Karhunen-Loève expansion:

I) There exists at most a countably infinite set of bounded eigenvalues.

II)

$$\sum_{i=1}^{\infty} \mu_i = \int_0^T G_x(t,t)dt \quad . \tag{2.69}$$

III)

$$\sum_{i=1}^{\infty} \mu_i^2 = \int_0^T\int_0^T dt_1 dt_2 \left[G_x(t_1,t_2)\right]^2 \quad . \tag{2.70}$$

IV)

$$G_x(t_1,t_2) = \sum_{i=1}^{\infty} \mu_i \phi_i(t_1)\phi_i(t_2) \quad , \tag{2.71}$$

(Mercer's theorem).

V) μ_i is a monotonically increasing function of T.

VI) If x(t) is stationary, then in the limit of T much larger than the width of $G_x(\tau)$, we have

$$\mu_n \simeq G_x(\omega_n) \quad , \quad \omega_n = n\frac{2\pi}{T} \quad , \tag{2.72}$$

$$\phi_n(t) \simeq \frac{1}{\sqrt{T}} e^{j\omega_n t} \quad , \quad 0 \leq t \leq T \quad ,$$

i.e., the Karhunen-Loève basis is the harmonic functions and the eigenvalues are samples from the power spectrum.

If we assume that the power spectrum is rectangular with width Γ cycles per second (cf. Table 2.3), then the number of significant eigenvalues (also called modes or degrees of freedom) should be approximately equal to the time-bandwith product $\gamma = \Gamma\, T$.

This illustrates that a stochastic process can practically be described by a finite (and sometimes small) number of random variables. This is why the Karhunen-Loève expansion is indeed very useful.

Now we take some important examples of solutions to (2.68).

The Karhunen-Loève Basis for a Stationary Process with Lorentzian Spectrum

In this case, $G(t_1,t_2) \propto \exp(-\Gamma|t_1-t_2|)$ and

$$\int_0^T \exp(-\Gamma|t-t'|)\phi(t')dt' = \mu\phi(t) \quad 0 \le t \le T \quad . \tag{2.73}$$

A solution of this integral equation can be obtained by changing it into a differential equation. Taking the second derivative with respect to t

$$\left[\frac{\partial^2}{\partial t^2} \exp(-\Gamma|t|) = \left[\Gamma^2 - 2\Gamma\delta(t)\right]\exp(-\Gamma|t|)\right] \; ,$$

we get

$$\frac{d^2\phi}{dt^2} + \omega^2\phi = 0 \; , \quad \omega^2 = 2\Gamma/\mu - \Gamma^2 \quad ,$$

which has a solution, $\phi(t) = A\exp(j\omega t) + B\exp(-j\omega t)$. Substituting in (2.73) and evaluating the integral, we get an identity that has to be satisfied for all t. The condition for this to hold is

$$(\omega^2 - \Gamma^2)\tan(\omega T) = 2\omega\Gamma \quad \mu = 2\Gamma/(\Gamma^2 + \omega^2) \tag{2.74}$$

$$A/B = (\Gamma + j\omega)/(\Gamma - j\omega) \quad .$$

The eigenvalues determined by (2.74) can be separated into two groups that satisfy

$$\begin{aligned} \xi\tan(\xi/2) &= +\gamma \quad & \mu &= 2\gamma T/(\gamma^2 + \xi^2) \\ \bar{\xi}\cot(\bar{\xi}/2) &= -\gamma \quad & \bar{\mu} &= 2\gamma T/(\gamma^2 + \bar{\xi}^2) \end{aligned} \; , \tag{2.75}$$

where

$$\gamma = \Gamma T \quad , \quad \xi = \omega T \quad .$$

The corresponding eigenfunctions are then given by the harmonic functions

$$\phi_j(t) = \left[\frac{1}{2}(T + \mu_j)\right]^{1/2} \cos\left[\xi_j\left(\frac{t}{T} - \frac{1}{2}\right)\right] \quad , \qquad (2.76)$$

$$\bar{\phi}_j(t) = \left[\frac{1}{2}(T + \bar{\mu}_j)\right]^{1/2} \sin\left[\bar{\xi}_j\left(\frac{t}{T} - \frac{1}{2}\right)\right] \quad .$$

Figure 2.2a shows the distribution of the eigenvalues μ_j for several values of $\gamma = \Gamma T$. We see that for very large γ, the number of considerable eigenvalues is equal to γ.

The Karhunen-Loève Basis for a Stationary Process with Rectangular Spectrum

A rectangular power spectrum that vanishes outside the interval $[-\Gamma/2, \Gamma/2]$ cycles per second corresponds to a correlation function

$$G(t_1, t_2) \propto \mathrm{sinc}\,[\Gamma(t_1 - t_2)] \quad , \quad \mathrm{sinc}(x) = \sin(\pi x)/\pi x \quad .$$

The integral equation,

$$\int_{-T/2}^{T/2} \mathrm{sinc}[\Gamma(t-t')]\phi(t')dt' = \mu\phi(t) \quad ,$$

is satisfied by the prolate spheroidal functions [158, 448].

These form a set of band-limited functions that are orthogonal and complete over the finite interval $[-T/2, T/2]$ (and also over the infinite interval). Subroutines for the generation of $\phi_j(t)$ and tables for the eigenvalues μ_j are given in the literature for several values of the parameter $c = (T/2) \cdot \pi\Gamma = (\pi/2)\gamma$, where $\gamma = \Gamma T$ is the time-bandwidth product [143, 448, 449, 462].

In Fig. 2.2b, we give a sample of the eigenvalues μ_j for some values of c and γ.

We observe that μ_j is approximately constant for the first few values of j, then it drops sharply as j exceeds the critical value γ. Therefore the eigenvalues themselves have an approximately rectangular distribution. The approximation is better for large values of γ. Compare these distributions to those in the Lorentzian case (Fig.2.2a).

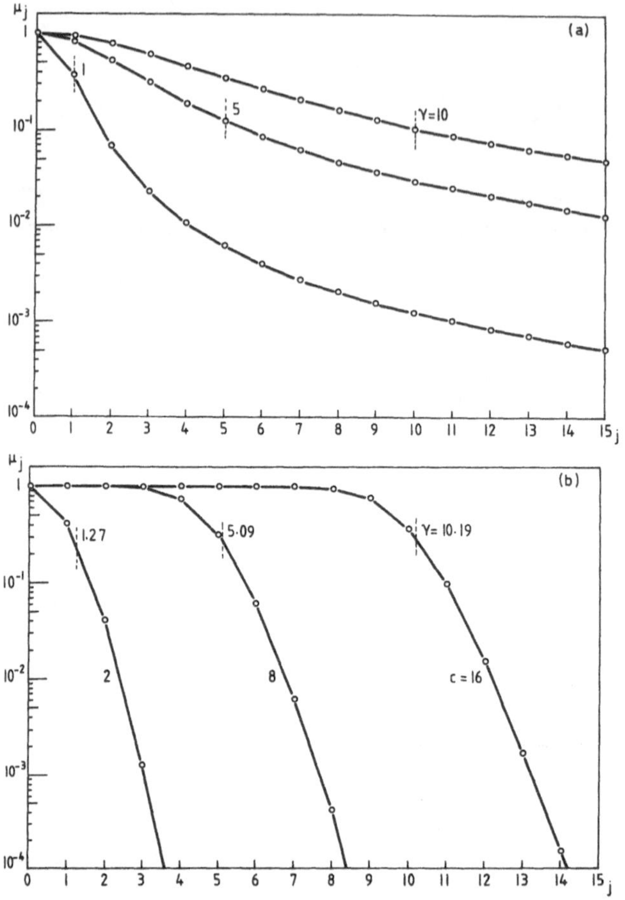

Fig.2.2a and b. Eigenvalues corresponding to: a) a Lorentzian spectrum b) a rectangular spectrum; for the indicated values of time-bandwith product γ

2.4.5 Description of a Stochastic Process by a Differential Equation

A stochastic process can be described by assigning its probability law or its hierarchy of moments. It can also be described by a finite set of random variables (coefficients of its expansion in an orthogonal basis). Alternatively, a stochastic process can be defined by a differential equation that contains some random parameters or another known stochastic process.

For example, the differential equation

$$\frac{dx}{dt} = f(x,t,\omega)$$

defines the stochastic process x(t) in terms of a stochastic process $\omega(t)$ whose probability law is specified.

1) An important class of stochastic processes is generated by a differential equation in which $\omega(t)$ is an additive white Gaussian stochastic process

$$\frac{dx}{dt} = f(x,t) + g(x,t)\omega(t) \quad , \tag{2.77}$$

where

$$G_{\omega}(t_1,t_2) = \langle\omega(t)\omega(t')\rangle = \delta(t-t') \quad .$$

It can be mathematically shown that x(t), as defined by (2.77), is a Markov process (i.e., satisfies (2.56)). This should also be clear from the fact that the rate of change of x(t) depends on x(t) and t but not on the values of x(t) at previous times. The process is therefore completely characterized by the PD, P[x(t)], and the transition probability density, $P[x(t) \mid x(t')]$. If we write these densities in the form P[x(t)] = P(x,t) and $P[x(t)|x(t')] = P(x,t;x',t')$, then both satisfy the differential equation,

$$\frac{\partial P}{\partial t} = -\frac{\partial}{\partial x} J \;, \quad J = Pf - \frac{1}{2}\frac{\partial}{\partial x}(Pg^2) \quad ,$$

$$= L(P) \quad , \quad L(.) = -\frac{\partial}{\partial x}(\cdot f) + \frac{1}{2}\frac{\partial^2}{\partial x^2}(\cdot\, g^2) \quad , \tag{2.78}$$

known as the *Fokker-Planck equation* (or the *Kolmogrov equation*). The differential operator $L(.)$ is known as the forward Kolmogrov operator.

If J is regarded as a "probability current" then (2.78) can be interpreted as an equation of conservation of probability.

2) A set of stochastic processes $[x(t)] = [x_1(t), \ldots x_N(t)]$ can also be described by a set of differential equations,

$$\frac{dx_i}{dt} = f_i([x],t) + g_i([x],t)\omega_i(t) \;, \quad i = 1,2, \ldots N \quad , \tag{2.79}$$

where $\omega_i(t)$ is a set of independent white Gaussian stochastic processes that satisfy $\langle\omega_i(t)\omega_j(t')\rangle = \delta_{ij}\delta(t-t')$. The probability density P([x],t) and the conditional probability density $P([x],t|[x'],t')$ satisfy the Fokker-Planck equation

$$\frac{\partial P}{\partial t} = -\sum_{i=1}^{N} \frac{\partial}{\partial x_i} J_i \quad , \quad J_i = Pf_i - \frac{1}{2}\sum_{j=1}^{N} \frac{\partial}{\partial x_j} (Pg_i g_j) \tag{2.80}$$

3) Now we consider some special cases of (2.77). We start with the case when f and g do not depend explicitly on time

$$\frac{dx}{dt} = f(x) + g(x)\omega(t) \quad . \tag{2.81}$$

This class of equations usually has a steady-state stationary solution. The Fokker-Planck equation can be solved by separation of variables and gives

$$P(x,t) = \psi_0(x) + \sum_{m=1}^{\infty} c_m\psi_m(x) \exp[-\lambda_m(t-t_0)] \quad , \tag{2.82}$$

where $\psi_m(x)$ and λ_m are eigenfunctions and eigenvalues of the differential equation,

$$\frac{d}{dx} f(x)\psi(x) - \frac{1}{2}\frac{d^2}{dx^2} g^2(x)\psi(x) = \lambda\psi(x) \quad ,$$

and c_m are calculated from the initial conditions by use of

$$c_m = \int P(x,t_0) \frac{\psi_m(x)}{\psi_0(x)} dx \quad .$$

As time increases, the stationary solution $P(x,t) = \psi_0(x)$ is obtained.

The transition-probability density can similarly be obtained. When the initial distribution is stationary, it gives the JPD,

$$P[x(t),x(t_0)] = \sum_{m=0}^{\infty} \psi_m[x(t)]\psi_m[x(t_0)]\exp(-\lambda_m|t-t_0|)$$

that corresponds to the correlation function,

$$G_x(t,t+\tau) = \sum_{m=1}^{\infty} h_m^2 \exp(-\lambda_m|\tau|) \quad , \quad h_m = \int x\psi_m(x)dx \quad .$$

4) Another special class of (2.77) is the linear differential equation

$$\frac{dx}{dt} = f(t)x(t) + g(t)w(t) \quad . \tag{2.83}$$

The process x(t) is related to the Gaussian process $w(t)$ by a linear transformation. Therefore, x(t) must be Gaussian. Since it is also Markovian, a process generated by (2.83) is called a *Gauss-Markov process*.

5) An important example of (2.83) is the case of time-independent coefficients f(t) and g(t),

$$\frac{dx}{dt} = -\Gamma x(t) + w(t) \quad . \tag{2.84}$$

The process generated by (2.84) is sometimes called the *Ornstein-Uhlenbeck process*. If $\langle w(t)\rangle = 0$, $G_w(t,t') = 2P_e\Gamma\delta(t-t')$ and $G_x(0,0) = P_i$, then it is easy to show that

$$\langle x(t)\rangle = \langle x(0)\rangle \exp(-\Gamma t) \quad ,$$

$$G_x(t,t') = P_e \exp(-\Gamma|t-t'|) + (P_i-P_e)\exp -\Gamma(t+t') \qquad t,t' \geq 0 \quad , \tag{2.85}$$

and

$$G_x(t,t) = \langle x^2(t)\rangle = P_e + (P_i-P_e)\exp(-2\Gamma t) \quad .$$

The mean values of x(t) decays exponentially from $\langle x(0)\rangle$ to zero with a time constant $1/\Gamma$. The mean square value $G_x(t,t) = \langle x^2(t)\rangle$ decays exponentially from an initial value P_i to an equilibrium value P_e with a time constant $1/2\Gamma$. At equilibrium, the process x(t) has zero mean and a time-invariant exponential correlation function

$$G_x(t,t') = P_e \exp(-\Gamma|t-t'|) \quad .$$

A power spectrum can be defined and is Lorentzian (see Table 2.3).

The Fokker-Planck equation that corresponds to (2.84) gives a Gaussian distribution whose mean and variance are as defined above.

6) A second example of (2.83) is the *Wiener-Lévy* process defined by

$$\frac{dx}{dt} = w(t) \quad .$$

This corresponds to a Fokker-Planck equation,

$$\frac{\partial P}{\partial t} = \frac{1}{2}\frac{\partial^2 P}{\partial x^2} \quad .$$

If $\omega(0) = 0$ with probability 1, this gives the solution

$$P(x,t) = \frac{1}{(2\pi t)^{1/2}} \exp(-x^2/2t) \quad ,$$

a nonstationary Gaussian process whose variance increases with time. The correlation function is given by

$$G_x(t_1,t_2) = \begin{cases} t_2 & t_1 \geq t_2 \\ t_1 & t_1 \leq t_2 \end{cases} \quad .$$

This process has useful applications in the study of Brownian motion.

More examples and a rigorous treatment of stochastic differential equations can be found in the textbooks on stochastic processes mentioned in the begining of this chapter.

2.5 Complex Stochastic Processes

2.5.1 Definitions

The definitions and concepts of the previous section can be easily generalized to apply to complex stochastic processes. A complex stochastic process $Z(t) = x(t) + jy(t)$ is composed of two real stochastic processes and is described by their joint statistics. At an arbitrary set of points $t_1, t_2, \ldots t_N$, we have a set of N complex random variables $Z(t_1), Z(t_2), \ldots Z(t_N)$ whose statistics are described as in Sec. 2.3. The moments are now complex functions of $t_1, t_2, \ldots t_N$. The *correlation functions* are defined by

$$G_Z^{N,M}(t_1, \ldots, t_{N+M}) = \left\langle \prod_{i=1}^{N} Z^*(t_i) \prod_{j=N+1}^{N+M} Z(t_j) \right\rangle \quad . \tag{2.86}$$

These functions satisfy the following properties, which are easily derived from that definition:

I) $$\left[G^{N,M}(t_1, \ldots t_{N+M})\right]^* = G^{M,N}(t_{N+M}, \ldots t_1) \tag{2.87}$$

II) $$G^{N,N}(t_1, \ldots, t_N, t_1, \ldots, t_N) \geq 0 \tag{2.88}$$

III) $$\left|G^{N,N}(t_1, \ldots, t_{2N})\right|^{2N} \leq \prod_{i=1}^{2N} G^{N,N}(t_i, t_i, \ldots t_i) \quad . \tag{2.89}$$

In particular,

$$\left|G^{1,1}(t_1,t_2)\right|^2 \leq G^{1,1}(t_1,t_1)G^{1,1}(t_2,t_2) \quad . \tag{2.90}$$

(This can be proved by using Hölder's inequality).

IV) $$\left|G^{N,N}(t_1, \ldots t_{2N})\right|^2 \leq G^{N,N}(t_1, \ldots t_N; t_1, \ldots t_N) \, G^{N,N}(t_{N+1}, \ldots t_{2N}; t_{N+1}, \ldots t_{2N}) \quad . \tag{2.91}$$

The most commonly used function of the above hierarchy is that of order (1,1). For economy of notations, we shall often call it $G(t_1,t_2)$ and ignore the superscript.
We can also define the *random spectrum* of $Z(t)$,

$$F_Z(\omega) = \int_{-\infty}^{\infty} Z(t)e^{-jwt}dt \quad , \tag{2.92}$$

and the cross-spectral densities

$$G^{N,M}(\omega_1, \ldots \omega_{N+M}) = \left\langle \prod_{i=1}^{N} F_Z^*(\omega_i) \prod_{j=N+1}^{N+M} F_Z(\omega_j) \right\rangle \quad , \tag{2.93}$$

which are the (N+M)-fold Fourier transforms of the correlation functions. If the process is stationary, then $G^{N,M}(t_1, \ldots t_{N+M})$ is time-shift invariant and $G^{N,M}(\omega_1, \ldots \omega_{N+M})$ vanishes unless

$$\sum_{i=1}^{N} \omega_i = \sum_{j=N+1}^{N+M} \omega_j \quad .$$

For example,

$$G_Z(\omega_1,\omega_2) = G_Z(\omega_1)\delta(\omega_1-\omega_2) \quad ,$$

where

$$G_Z(\omega) = \int_{-\infty}^{\infty} G_Z(\tau)e^{-j\omega\tau}d\tau \quad . \tag{2.94}$$

A complex stochastic process is said to be *circularly symmetric* if $\{Z(t_j)\}$ are circularly symmetric for all $\{t_j\}$. In this case,

$$G_Z^{N,M} = 0 \;, \quad N \neq M \quad . \tag{2.95}$$

A complex stochastic process is said to be *circularly symmetric Gaussian* if $\{Z(t_j)\}$ are circularly symmetric Gaussian CRV's. In this case, and from (2.52),

$$G_Z^{N,M}(t_1, \ldots t_{N+M}) = \delta_{N,M}\sum_{\pi} G_Z(t_1,t_{\pi_1}) \ldots G_Z(t_N,t_{\pi_N}) \quad , \tag{2.96}$$

where $\sum_{\pi}$ is the sum of permutations of $(t_{\pi_1}, \ldots t_{\pi_N})$ over $(N+1, \ldots 2N)$.

2.5.2 Karhunen-Loève Expansion of a Complex Stochastic Process

As a generalization of the analysis of Sec. 2.4.4, a complex stochastic process Z(t) can be expanded in the form

$$Z(t) = \lim_{N\to\infty}\sum_{i=1} Z_i\phi_i(t) \;, \quad 0 \le t \le T \quad , \tag{2.97}$$

where $\phi_j(t)$ are complex functions orthonormal over $0 \le t \le T$, i.e.,

$$\int_0^T \phi_i^*(t)\phi_j(t)dt = \delta_{ij} \quad , \tag{2.98}$$

and Z_j are now complex random coefficients given by

$$Z_i = \int_0^T Z(t)\phi_j^*(t)dt \quad .$$

To ensure that these coefficients are uncorrelated, we choose $\{\Phi_j\}$ such that

$$\int_0^T G_Z(t_1,t_2)\Phi_i(t_2)dt_2 = \lambda_i\Phi_i(t_1) \ , \quad 0 \leq t_1 \leq T \quad . \tag{2.99}$$

Equation (2.99) is a generalization of (2.68). Since G_Z is Hermitian (i.e., $G^*(t_1,t_2) = G(t_2,t_1)$), the eigenvalues λ_i must be real. It also follows that

$$\langle Z_i^* Z_j \rangle = \lambda_i \delta_{ij} \quad .$$

Properties similar to those of Sec. 2.4.4 also follow:

I) There is at most a countably infinite set of real bounded eigenvalues.

$$\text{II)} \quad \sum \lambda_i = \int_0^T G_Z(t,t)dt \quad . \tag{2.100}$$

$$\text{III)} \quad \sum \lambda_i^2 = \int_0^T\!\!\int_0^T |G_Z(t_1,t_2)|^2 dt_1 dt_2 \quad . \tag{2.101}$$

$$\text{IV)} \quad G_Z(t_1,t_2) = \sum \lambda_i \Phi_i^*(t_1)\Phi_i(t_2) \quad . \tag{2.102}$$

2.6 Complex Representation of Bandpass Stochastic Processes

2.6.1 Complex Representation of a Real Bandpass Signal

The Complex Analytic Signal. A real function (signal) $x(t)$ can be represented by its Fourier transform $F_x(\omega)$,

$$F_x(\omega) = \int_{-\infty}^{\infty} x(t)e^{-j\omega t}dt \quad ,$$

$$x(t) = \int_{-\infty}^{\infty} F_x(\omega)e^{j\omega t}d\omega/2\pi \quad .$$

Because x(t) is real, $F_x(-\omega) = F_x^*(\omega)$. It is therefore sufficient to know the complex function $F_x(\omega)$ in the frequency range $\omega \geq 0$ in order to describe completely the function x(t). If we define the complex signal,

$$X(t) = \int_0^{\infty} F_x(\omega)e^{j\omega t}d\omega/2\pi \quad , \tag{2.103}$$

we can easily see that

$$x(t) = 2\,\mathrm{Re}\{X(t)\} \quad , \tag{2.104}$$

which means that x(t) can be easily reconstructed from X(t). In many situations, working with X(t) is much simpler than with x(t). A familiar example is the signal $x(t) = \cos(\omega_0 t)$, which corresponds to $X(t) = \exp(j\omega_0 t)$. (X(t) is a rotating phasor and x(t) is its projection on the horizontal axis). The function X(t) is called the *complex analytic signal*[5] (CAS) corresponding to x(t). The CAS itself can be represented by its Fourier transform,

$$F_X(\omega) = \int_{-\infty}^{\infty} X(t)e^{-j\omega t}dt \quad . \tag{2.105}$$

Note that

$$F_X(\omega) = 2F_x(\omega) \ , \quad \omega \geq 0 \ ,$$
$$= 0 \ , \quad \omega < 0 \ .$$

Complex Envelope (Amplitude) of a Bandpass Signal

Let x(t) be a real bandpass signal, i.e., its Fourier transform $F_x(\omega)$ is centered around a central frequency ω_0 and $-\omega_0$ ($F_X(\omega)$ is centered around ω_0). Now we can write the CAS in the form

$$X(t) = X(t)e^{j\omega_0 t} \tag{2.106}$$

---

[5]It can be shown that the real and imaginary parts of X(t) as a function of a complex t are both analytic and regular in the lower half of the complex t plane. From the analytic properties, it follows that the real and imaginary parts of X(t) form a Hilbert-transform pair.

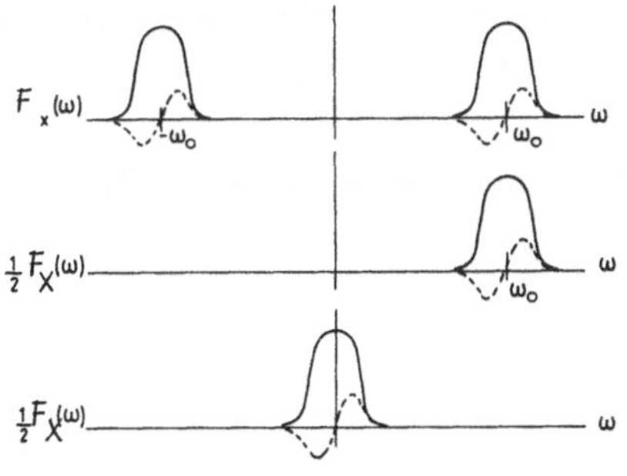

Fig.2.3. The Fourier transform of a bandpass signal x, its complex analytic signal X, and its complex amplitude $\mathcal{X}$: —— magnitude; ---- phase

and call $\mathcal{X}(t)$ the complex envelope (or the complex amplitude).[6] The Fourier transform of $\mathcal{X}(t)$ is centered around zero frequency (Multiplication by $\exp(j\omega_0 t)$ in the time domain corresponds to shifting of the Fourier spectrum by ω_0). Thus, the complex envelope contains only low frequencies (see Fig. 2.3).

In terms of its real and imaginary parts, $\mathcal{X}(t)$ can be written as

$$\mathcal{X}(t) = x_c(t) - jx_s(t) \quad , \tag{2.107}$$

where both $x_c(t)$ and $x_s(t)$ are real low-pass signals (i.e., contain only low frequencies). Using (2.106) and (2.104), we can write our original band-pass real function $x(t)$ in terms of the two low-frequency real functions $x_c(t)$ and $x_s(t)$ in the form

$$x(t) = 2x_c(t)\cos(\omega_0 t) + 2x_s(t)\sin(\omega_0 t) \quad . \tag{2.108}$$

The functions $x_c(t)$, $x_s(t)$ and $\mathcal{X}(t)$ can be obtained from $x(t)$ by the operations

$$\begin{aligned} x_c(t) &= [x(t)\cos(\omega_0 t)]_{LP} \\ x_s(t) &= [x(t)\sin(\omega_0 t)]_{LP} \\ \mathcal{X}(t) &= [x(t)\exp(j\omega_0 t)]_{LP} \quad , \end{aligned} \tag{2.109}$$

---

[6] $\mathcal{X}(t)$ is not an analytic signal and its real and imaginary parts do not form a Hilbert-transform pair.

where $[.]_{LP}$ denotes the operation of passing the argument through an ideal low-pass filter with a bandwidth equal to that of the original bandpass signal.

The simplest example that illustrates these concepts is again the function, $x(t) = \cos(\omega_0 t)$, which has

$$X(t) = e^{j\omega_0 t}$$

$$X(t) = 1 \ , \quad x_c(t) = 1 \ , \quad x_s(t) = 0$$

$$F_x(\omega) = \frac{1}{2}[\delta(\omega-\omega_0) + \delta(\omega+\omega_0)]$$

$$F_X(\omega) = \delta(\omega-\omega_0)$$

$$F_X(\omega) = \delta(\omega) \quad .$$

2.6.2 Complex Representation of a Real Bandpass Stochastic Process

Let our real signal $x(t)$ be a realization of a real stochastic process. Then we can still define the complex analytic signal $X(t)$ and the complex envelope $X(t)$ as before but they are now complex stochastic processes. The components of $X(t)$ ($x_c(t)$ and $x_s(t)$), are low-frequency real stochastic processes.

The statistical properties of $x(t)$ can be described by the set of real correlation functions of $x(t)$, or the complex correlation functions of $X(t)$ or $X(t)$, or alternatively by the correlations and cross correlations of the real processes $x_c(t)$ and $x_s(t)$. To illustrate this, we write the second-order correlation functions:

I) $x(t)$, $$G_x(t_1,t_2) = \langle x(t_1)x(t_2)\rangle \quad . \tag{2.110}$$

II) $X(t)$, $$G_X(t_1,t_2) = \langle X^*(t_1)X(t_2)\rangle \ , \tag{2.111}$$

$$G_X^{0,2}(t_1,t_2) = \langle X(t_1)X(t_2)\rangle \quad . \tag{2.112}$$

III) $X(t)$, $$G_X(t_1,t_2) = \langle X^*(t_1)X(t_2)\rangle \quad , \tag{2.113}$$

$$G_X^{0,2}(t_1,t_2) = \langle X(t_1)X(t_2)\rangle \quad . \tag{2.114}$$

IV) $x_c(t), x_s(t)$, $\quad G_{x_c}(t_1,t_2) = \langle x_c(t_1)x_c(t_2)\rangle$, (2.115)

$$G_{x_s}(t_1,t_2) = \langle x_s(t_1)x_s(t_2)\rangle \quad , \tag{2.116}$$

$$G_{x_c x_s}(t_1,t_2) = \langle x_c(t_1)x_s(t_2)\rangle \quad . \tag{2.117}$$

These functions are related by the following equations which are easily derived from the definitions:

$$G_X(t_1,t_2) = G_X(t_1,t_2)\exp[-j\omega_0(t_1-t_2)] \tag{2.118}$$

$$G_X^{0,2}(t_1,t_2) = G_X^{0,2}(t_1,t_2)\exp[j\omega_0(t_1+t_2)] \tag{2.119}$$

$$G_x = 2\,\mathrm{Re}\left\{G_X + G_X^{0,2}\right\} \tag{2.120}$$

$$G_X = \left[G_{x_c} + G_{x_s}\right] + j\left[G_{x_s x_c} - G_{x_c x_s}\right] \tag{2.121}$$

$$G_X^{0,2} = \left[G_{x_c} - G_{x_s}\right] - j\left[G_{x_s x_c} + G_{x_c x_s}\right] \quad . \tag{2.122}$$

Higher-order correlations can be also defined, e.g.,

$$\begin{aligned} G_X^{N,M}(t_1, \ldots t_{N+M}) &= \left\langle \prod_{i=1}^{N} X^*(t_i) \prod_{j=N+1}^{N+M} X(t_j)\right\rangle \\ &= G_X^{N,M}(t_1, \ldots t_{N+M})\exp\left[-j\omega_0\left(\sum_{i=1}^{N} t_i - \sum_{j=N+1}^{N+M} t_j\right)\right] \quad , \end{aligned} \tag{2.123}$$

and can similarly be related to the higher-order correlation functions of x(t). They obviously satisfy the properties of (2.87-91). The cross-spectral density is defined by

$$G_X^{N,M}(\ _1, \ldots \ _{N+M}) = \left\langle \prod_{i=1}^{N} F_X^*(\omega_i) \prod_{j=N+1}^{N+M} F_X(\omega_j)\right\rangle \quad .$$

It should vanish if any one of the arguments is negative.

2.6.3 Complex Representation of a Stationary Real Bandpass Stochastic Process

Let x(t) be a real stochastic process that is stationary (in the second-order), i.e., $G_x(t_1,t_2) = G_x(t_2-t_1)$. Then its corresponding CAS (or complex envelope) has a time-shift-invariant correlation function, i.e.,

$$G_X(t_1,t_2) = G_X(t_2-t_1) \quad .$$

Moreover,

$$G_X^{2,0}(t_1,t_2) = G_X^{0,2}(t_1,t_2) = 0 \quad . \tag{2.124}$$

This can be proved as follows: Because x(t) is stationary, then from (2.61),

$$G_x(\omega,\omega') = G_x(\omega)\delta(\omega-\omega') \quad . \tag{2.125}$$

By using the definition of X(t), (2.103), and substituting in the definition of G_X, (2.111), we get

$$G_X(t_1,t_2) = \int_0^\infty\int_0^\infty G_x(\omega,\omega')\exp[-j(\omega t_1-\omega' t_2)]d\omega d\omega' \quad .$$

Substituting from (2.125), we get

$$G_X(t_1,t_2) = \int_0^\infty G_x(\omega)\exp[j\omega(t_2-t_1)]d\omega = G_X(t_2-t_1)$$

i.e.,

$$G_X(\tau) = \int_0^\infty G_x(\omega)\exp(-j\omega\tau)d\omega \quad . \tag{2.126}$$

Equation (2.124) can be proved by substituting from (2.103) in the definition of $G_X^{0,2}$, (2.112), and getting

$$G_X^{0,2}(t_1,t_2) = \int_0^\infty\int_0^\infty \langle F_x(\omega)F_x(\omega')\rangle\exp[j(\omega' t_2+\omega t_1)]d\omega d\omega'$$

$$= \iint\limits_{0\;-\infty}^{\infty\;0} \langle F_x(\omega)F_x(-\omega')\rangle \exp[j(\omega' t_2-\omega t_1)]d\omega d\omega'$$

$$= \iint\limits_{0\;-\infty}^{\infty} \langle F_x(\omega)F_x^*(\omega')\rangle \exp[j(\omega' t_2-\omega t_1)]d\omega d\omega' \quad .$$

Substituting from (2.125), we can easily see that $G_X^{0,2}$ vanishes.

Another way of seeing this is to consider (2.119). Because of the factor $\exp[j\omega(t_1+t_2)]$, $G_X^{0,2}$ cannot be time-shift invariant, so it must vanish.

From (2.124) and (2.118), we conclude that

$$G_X(\tau) = G_X(\tau)\exp(j\omega\tau) \quad ;$$

and from (2.120) we conclude that

$$G_x(\tau) = 2\,\mathrm{Re}\{G_X(\tau)\} \quad ,$$

$$G_{x_c}(\tau) = \tfrac{1}{2}\,\mathrm{Re}\{G_X(\tau)\} \;, \quad G_{x_s}(\tau) = \tfrac{1}{2}\,\mathrm{Im}\{G_X(\tau)\} \quad , \tag{2.127}$$

$$G_{x_c x_s}(\tau) = 0 \quad .$$

It can also be shown that the necessary and sufficient conditions for a bandpass real process $x(t)$ to be stationary (second order) are that the quadrature components $x_c(t)$ and $x_s(t)$ of its CAS should be stationary (second order) and should satisfy the relations

$$G_{x_c}(\tau) = G_{x_s}(\tau)$$

$$G_{x_c x_s}(\tau) = -\,G_{x_s x_c}(\tau) \quad . \tag{2.128}$$

In other words,

$$G_X^{0,2}(\tau) = 0 \quad .$$

Necessity can be shown as follows. If $x(t)$ is stationary, then from (2.124), $G_X^{0,2} = 0$; and from (2.122), (2.128) follows.

Substituting in (2.121), we get

$$G_X = 2G_{x_c} + 2jG_{x_s x_c} \quad .$$

Because G_X is time-shift invariant, its real and imaginary parts have to be time-shift invariant also from which it follows that $x_c(t)$ and $x_s(t)$ are stationary.

Sufficiency can be shown by noting that if x_c and x_s are stationary, then G_X is time-shift invariant; and if $G_X^{0,2} = 0$, then from (2.120), G_x has to be time-shift invariant, i.e., x(t) is stationary.

Note that the condition (2.128) is satisfied if X(t) is a circularly symmetric complex random variable (because then $x_c(t)$ and $x_s(t)$ are identical and statistically independent). Yet this alone is not sufficient for x(t) to be stationary. The real processes $x_c(t)$ and $x_s(t)$ themselves have to be stationary.

Now let us consider the effect of stationarity on correlation functions of higher order. If x(t) is stationary, then X(t) is also stationary. As we noted before (cf. (2.93)), its cross-spectral density $G_X^{N,M}(\omega_1, \ldots \omega_{N+M})$ vanishes unless

$$\sum_{i=1}^{N} \omega_i = \sum_{j=N+1}^{N+M} \omega_j \quad . \tag{2.129}$$

By definition, X(t) has only positive frequencies. This puts severe restrictions on $G_X^{N,M}$. For example, if M = 0, then (2.129) cannot be satisfied and $G^{N,0} = 0$ for all N. A special case of this is $G^{2,0} = 0$, which we have already considered. In general, if x(t) has a narrow spectrum with bandwidth $\Delta\omega$, then the condition that $G^{N,M}$ does not vanish is

$$\frac{|N - M|}{N + M} \le \frac{1}{2}\frac{\Delta\omega}{\omega_0} \quad . \tag{2.130}$$

If $\Delta\omega \ll \omega_0$ (which is often the case in optics), then the correlation functions of order $N \neq M$ vanish unless N and M are very large numbers. Moreover, all functions of order $N \gg M$ or $M \ll N$ must vanish.

A special case is the function $\langle x^*(t)^N X(t)^M \rangle$ which should vanish for all moderately large and unequal values of N and M. By using the arguments of

Sec. 2.3.2, we see that the complex random variable $X(t)$ has to be approximately circularly symmetric (i.e., with uniformly distributed phase).

2.6.4 Processes with Stationary Quadrature Components. Quasi-stationary Processes

A real bandpass stochastic process $x(t)$ is said to be quasi-stationary if its complex envelope $X(t)$ is stationary (or its quadrature components $x_c(t)$ and $x_s(t)$ are stationary). As we already know, the stationarity of $X(t)$ (or $x_c(t)$ and $x_s(t)$) is not sufficient to imply that the main process $x(t)$ is itself stationary (or that the CAS is stationary). The conditions (2.128) must also be satisfied.

A quasi-stationary process has a CAS with time-shift invariant $G_X(t_1,t_2)$, but with a time-shift variant and not necessarily vanishing $G_X^{2,0}(t_1,t_2)$.

Example

Consider the process $x(t) = x_c(t)\cos(\omega_0 t)$ where $x_c(t)$ is a low-frequency stationary real process. It is obvious that $x(t)$ is not stationary because

$$G_x(t_1,t_2) = G_{x_c}(t_2-t_1)\cos(\omega_0 t_1)\cos(\omega_0 t_2) \quad .$$

The CAS is $X(t) = x_c(t)\exp(j\omega_0 t)$ and the complex envelope (which is real in this case) $X(t) = x_c(t)$ is stationary.

The conditions (2.128) are obviously not satisfied. The correlation functions of $X(t)$ are given by

$$G_X(t_1,t_2) = G_{x_c}(t_2-t_1)\exp[j\omega_0(t_2-t_1)]$$

$$G_X^{0,2}(t_1,t_2) = G_{x_c}(t_2-t_1)\exp[j\omega_0(t_1+t_2)] \quad .$$

This shows that $G_X(t_1,t_2)$ is time-shift invariant; this is why $x(t)$ is called quasi-stationary. Higher-order correlation functions of order $N=M$ are all time-shift invariant, but those of order $N \neq M$ are not necessarily so.

2.6.5 Complex Representation of Gaussian Bandpass Stochastic Processes

Let $x(t)$ be a *Gaussian* process. Because of the invariance of a Gaussian distribution under linear transformations, it follows that the components

$x_s(t)$ and $x_c(t)$ of the complex envelope of $x(t)$ are jointly Gaussian. Hence $X(t)$ and $X(t)$ are complex Gaussian processes.

If $x(t)$ is also *stationary*, then

I) The quadrature components $x_c(t)$ and $x_s(t)$ are stationary, jointly Gaussian, and satisfy

$$G_{x_c}(\tau) = G_{x_s}(\tau) \quad , \quad G_{x_c x_s}(\tau) = -G_{x_s x_c}(\tau) \quad . \tag{2.131}$$

II) The complex analytic signal $x(t)$ and the complex envelope $X(t)$ are stationary, complex Gaussian, and satisfy

$$G_X^{2,0} = G_X^{0,2} = 0 \quad .$$

If $x(t)$ is *Gaussian, stationary* and has a *symmetric power spectrum* (symmetric around the central frequency ω_0), then

I) The quadrature components are stationary, statistically independent Gaussian processes (because symmetry implies that $G_{x_c x_s} = 0$).

II) The complex analytic signal and the complex envelope are circularly symmetric complex Gaussian processes.

However, if $x(t)$ is *Gaussian* and has stationary quadrature components, i.e., *quasi-stationary*, then

I) The quadrature components are jointly Gaussian, but they do not necessarily satisfy (2.131).

II) The complex envelope is a stationary complex Gaussian process.

III) The complex analytic signal is Gaussian but not strict-sense stationary.

2.6.6 Karhunen-Loève Expansion of the Complex Envelope of a Bandpass Stochastic Process

As a complex stochastic process, the complex envelope can also be expanded in a Karhunen-Loève series as discussed in Sec. 2.5.2. We now find the relationship between the K-L expansion of a real bandpass process $x(t)$ and the K-L expansion of its complex envelope $X(t)$.

Let $x(t)$ be a stationary process expanded in an orthonormal basis

$$x(t) = \sum x_i \phi_i(t) \quad ,$$

where

$$\int_0^T G_x(t_1,t_2)\phi_i(t_1)dt_1 = \mu_i\phi_i(t_2) \; ; \tag{2.132}$$

and let its complex envelope $X(t)$ be expanded in an orthonormal basis

$$x(t) = \sum x_i\Phi_i(t) \quad ,$$

where

$$\int_0^T G_X(t_1,t_2)\Phi_i(t_i)dt_1 = \lambda_i\Phi_i(t_2) \quad . \tag{2.133}$$

If we put

$$\mu_i = 2\lambda_i \quad ,$$

$$\phi_i(t) = 2\mathrm{Re}\{\Phi_i(t)\}\exp(j\omega_0 t+\theta) \quad ,$$

and use (2.127) in (2.132), we find that (2.133) is automatically satisfied for all values of θ. Thus, each of the eigenvalues μ_i has a degenerate eigenspace of dimension 2. This means that every eigenvalue λ_j corresponds to two equal eigenvalues μ_j each having two different eigenfunctions. The coefficients of the expansion are then related by

$$\begin{array}{lll} x_1 = \mathrm{Re}\{x_1\} & \langle x_1^2\rangle = \mu_1 & \langle|x_1|^2\rangle = \lambda_1 = 2\mu_1 \\ x_2 = \mathrm{Im}\{x_1\} & \langle x_2^2\rangle = \mu_1 & \\ x_3 = \mathrm{Re}\{x_2\} & \langle x_3^2\rangle = \mu_2 & \langle|x_2|^2\rangle = \lambda_2 = 2\mu_2 \\ x_4 = \mathrm{Im}\{x_2\} & \langle x_4^2\rangle = \mu_2 & \end{array} \tag{2.134}$$

and so on.

As an example, we take again a bandpass process with Lorentzian spectrum centered around a frequency ω_0, i.e., $G_X(\tau) = \exp(-\Gamma|\tau|)$. We already have a solution for (2.133) (Sec. 2.4.4).

Were we to work with the real function directly, and not with its complex representation, we would then have to solve the integral equation (2.132) with

$$G_x(\tau) = 2\exp(-\Gamma|\tau|)\cos(\omega_0\tau) \quad .$$

This is another example that shows the simplification realized by the complex representation.

2.7 A Short Review of Some Principles of Estimation and Detection Theory

2.7.1 Test of Hypotheses

Consider a system that can assume only two states, denoted by hypothesis $H^{(1)}$ and hypothesis $H^{(2)}$. The outcome of a measurement taken on the system is $[x] = (x_1, x_2, \ldots x_N)$, a set of random variables with conditional probability distributions $P[[x]|H^{(1)}]$ and $P[[x]|H^{(2)}]$. Based on the observation x, it is desired to determine the state of the system. The decision is a choice between $H^{(1)}$ and $H^{(2)}$.

A decision strategy is a division of the space of all outcomes [x] into two regions $D^{(1)}$ and $D^{(2)}$ such that $H^{(k)}$ is selected whenever [x] falls in $D^{(k)}$. The performance of a decision strategy is determined by the probability of error. Q_1 is the probability of selecting $H^{(2)}$ when $H^{(1)}$ is correct and Q_2 is the probability of selecting $H^{(1)}$ when $H^{(2)}$ is correct. They are given by

$$Q_1 = \int_{D^{(2)}} P\left[[x]|H^{(1)}\right] d[x] \quad , \qquad Q_2 = \int_{D^{(1)}} P\left[[x]|H^{(2)}\right] d[x] \quad . \tag{2.135}$$

The Maximum-Likelihood (ML) Strategy

According to the ML strategy, $H^{(2)}$ is chosen if $\phi^{(2)} > \phi^{(1)}$, where

$$\phi^{(k)} = P\left[[x]|H^{(k)}\right] \tag{2.136}$$

are the likelihood functions.

This amounts to choosing the hypothesis that makes the observation more likely to occur. It is interesting to know that the ML strategy is that which makes the sum of the probabilities of error $Q_1 + Q_2$ a minimum.

Bayes Strategy

According to Bayes strategy, prior probabilities P_1 and P_2 are assigned to $H^{(1)}$ and $H^{(2)}$. The probability of error is then

$$P_e = P_1Q_1 + P_2Q_2 \quad . \tag{2.137}$$

The two kinds of error are weighted by different costs C_1 and C_2 representing their relative importance. The average cost is

$$\bar{C} = P_1Q_1C_1 + P_2Q_2C_2 \quad . \tag{2.138}$$

Bayes strategy is based on minimizing the average cost. The solution to this optimization problem is the decision rule

$$\Lambda([x]) \underset{1}{\overset{2}{\gtrless}} \eta \tag{2.139}$$

i.e., choose $H^{(2)}$, when $\Lambda(x) > \eta$. Here

$$\Lambda([x]) = \phi^{(2)}/\phi^{(1)} \tag{2.140}$$

is called the likelihood ratio, and

$$\eta = P_1C_1/P_2C_2 \tag{2.141}$$

is called the decision level.

Note that if $C_1 = C_2$ and $P_1 = P_2$ we obtain $\eta = 1$ and the ML strategy is reproduced.

The above principles can be generalized to the test of multiple hypotheses. Here, $H^{(k)}$, $k = 1, \ldots M$ hypotheses are possible. The conditional-probability distributions $P[[x]|H^{(k)}]$ are known. [x] is measured and a choice among the hypotheses is sought.

Again the ML strategy is based on comparing the likelihood functions $\phi^{(k)} = P[[x]|H^{(k)}]$, $k = 1, \ldots M$. $H^{(q)}$ is chosen if $\phi^{(q)} > \phi^{(k)}$ for all $k \neq q$.

Bayes strategy can also be formulated in terms of prior probabilities and error costs. But the solution to the optimization problem is not as simple as in (2.139).

2.7.2 Parameter Estimation

The state of a system depends on the value of a parameter θ. An observation is made and the random variable [x] is measured. The probability distribution of the observed variable $P([x]|\theta)$ is a known function of θ. Given the result of the measurement [x], we wish to find an estimate $\hat{\theta}$ for θ.

Since $\hat{\theta}$ is determined from x , it is a random variable. An important measure of the quality of the estimation $\hat{\theta}$ is the bias

$$B = \langle\hat{\theta}\rangle - \theta \quad . \tag{2.142}$$

If B = 0, the estimator is called unbiased and, on the average, the estimator $\hat{\theta}$ gives the correct value of θ.

The quality of the estimator $\hat{\theta}$ is also described by the mean square error,

$$\langle(\hat{\theta} - \theta^2)\rangle = \mathrm{Var}(\hat{\theta}) + B^2 \quad . \tag{2.143}$$

If B = 0, then the width of the distribution of $\hat{\theta}$ around its mean θ is an obvious measure of the quality of estimation.

Maximum-Likelihood (ML) Estimation

The ML estimator is the value of θ which makes $P([x]|\theta)$ a maximum, i.e., the value which makes the observation most likely to occur. We call $P([x]|\theta)$ the likelihood function and determine $\hat{\theta}$ from

$$\frac{\partial}{\partial\theta} P([x]|\theta) \Big|_{\theta=\hat{\theta}} = 0 \, , \quad \text{or} \quad \frac{\partial}{\partial\theta} \ln P([x]|\theta) \Big|_{\theta=\hat{\theta}} = 0 \quad . \tag{2.144}$$

Bayes Estimation

The Bayes criterion assigns to the parameter θ an *a priori* probability distribution $P(\theta)$. To each combination of parameter θ and parameter estimator $\hat{\theta}$ a cost of error $C(\hat{\theta},\theta)$ is assigned. The estimator $\hat{\theta}$ is chosen so that the expected total cost

$$C(\hat{\theta},\theta) = \iint C\left[\hat{\theta}([x]),\theta\right] P([x]|\theta)P(\theta)d[x]d\theta \tag{2.145}$$

is minimized.

If $C(\hat{\theta},\theta) = (\hat{\theta} - \theta)^2$, we obtain the *minimum mean square error (MMSE)* estimator. The minimization problem gives

$$\hat{\theta}([x]) = \langle\theta|[x]\rangle \quad ,$$

the *a posteriori* conditional mean. This can be obtained from the *a posteriori* probability distribution,

$$P(\theta|[x]) = P([x]|\theta)P(\theta)/\int P([x]|\theta)P(\theta)d\theta \quad ,$$

by using

$$\langle\theta|[x]\rangle = \int \theta P(\theta)|[x])d\theta \quad . \tag{2.146}$$

If $C(\hat{\theta},\theta) = |\hat{\theta} - \theta|$, then the value of θ which minimizes the expected cost is that which maximizes the *a posteriori* PD, $P(\theta|[x])$. This is called the *maximum a posteriori* (MAP) estimator and it can be obtained from

$$\frac{\partial}{\partial\theta} P(\theta|[x]) \Big|_{\theta=\hat{\theta}} = 0 \quad , \quad \text{or} \quad \frac{\partial}{\partial\theta}\Big[\ln P([x]|\theta) + \ln P(\theta)\Big]\Big|_{\theta=\hat{\theta}} = 0. \tag{2.147}$$

Note that the ML estimator requires the knowledge of $P([x]|\theta)$; and both the MMSE and the MAP estimator require the knowledge of $P([x]|\theta)$ and $P(\theta)$. Sometimes these probability distributions are not available. The above criteria cannot then be used. An example of an alternative approach is given below.

MMSE Nonlinear Fitting

Assume that the observation $[x] = x_1,x_2, \ldots x_L$ is a set of points on a curve. The mean value $\langle x_\ell\rangle = \bar{x}_\ell(\theta)$ is a known function of an unknown parameter θ. The probability distribution $P([x]|\theta)$ is unknown, but the variances and covariances of $[x]$, $\Lambda_{\ell k} = \langle\Delta x_\ell \Delta x_k\rangle$, are known and are, in general, functions of θ.

We write the cost as

$$C = \sum_{\ell=1}^{L} \xi_\ell [x_\ell - \bar{x}_\ell(\theta)]^2 \quad , \tag{2.148}$$

where ξ_ℓ are weighting factors, and find the value θ which minimizes this cost. We obtain a nonlinear equation

$$\sum_\ell \xi_\ell [x_\ell - \bar{x}_\ell(\theta)] \frac{\partial}{\partial\theta} \bar{x}_\ell(\theta) = 0 \quad , \tag{2.149}$$

which can be solved for $\hat{\theta}$ in terms [x]. The estimator $\hat{\theta}$ is therefore a random variable whose variance can be related to $\Lambda_{\ell k}$ by linearizing (2.149). This gives

$$\mathrm{Var}(\hat{\theta}) = \sum_{\ell,k} \xi_\ell \xi_k \Lambda_{\ell k} \frac{\partial}{\partial\theta} \bar{x}_\ell \frac{\partial}{\partial\theta} \bar{x}_k \Big/ \left[\sum_\ell \xi_\ell \left(\frac{\partial}{\partial\theta} \bar{x}_\ell\right)^2\right]^2 \quad . \tag{2.150}$$

The weighting factors ξ_ℓ can then be chosen so as to minimize $\mathrm{Var}(\hat{\theta})$. When $(x_1, x_2, \ldots x_L)$ are uncorrelated this gives

$$\xi_\ell \propto 1/\Lambda_{\ell\ell} = 1/\mathrm{Var}(x_\ell) \quad , \tag{2.151}$$

and

$$\mathrm{Var}(\hat{\theta}) = \left[\sum_\ell \left(\frac{\partial}{\partial\theta} \bar{x}_\ell\right)^2 / \mathrm{Var}(x_\ell)\right]^{-1} \quad . \tag{2.152}$$

Note that if we choose equal weights, we obtain a higher estimation error

$$\mathrm{Var}(\hat{\theta}) = \sum_\ell \left(\frac{\partial}{\partial\theta} \bar{x}_\ell\right)^2 \mathrm{Var}(x_\ell) \Big/ \left[\sum_\ell \left(\frac{\partial}{\partial\theta} \bar{x}_\ell\right)^2\right]^2 \quad . \tag{2.153}$$

The above analysis can also be generalized to the estimation of more than one parameter.

3. Point Processes

When weak light is observed by a photodetector, a series of individual photoelectrons is emitted at random points of time. An observer who wishes to collect the full information carried by the detected electrons must record these random times. Useful information can be extracted from this very large set of times only by a statistical analysis in which smoothed statistical averages are utilized.

In this chapter, we introduce some tools from mathematical statistics that are necessary for the statistical description of random points. The theory of point processes is an advanced branch of statistics to which several books are totally devoted (e.g., [103, 104, 453, 457]).

Many books on probability theory and statistics devote one or more chapters to point processes (e.g., Chap. 13 in [138], Chap. 6 in [461], Chap. 9 in [105], Chap. 16 in [358]). Several articles give a sufficiently general treatment of the subject (e.g., [56, 112, 302, 327, 328]).

3.1 One-Dimensional Point Processes

A point process (PP) is a stochastic process whose realizations are series of point events. The events occur in a continuous one-dimensional time and they are distinguished only by their instants of occurrence. In the study of point processes, it is generally assumed that the probability for two or more events to occur simultaneously is zero. The time origin is not in general a special point, and it can be taken at any arbitrary point on the time axis.

A realization of a PP can be described by one of several ways (as illustrated in Fig. 3.1).

I) By specifying the times of occurrence of the events ($\dots, t_1, t_2, \dots t_q$, ...). The times of occurrence measured from an arbitrary time t_0 (t_{10},

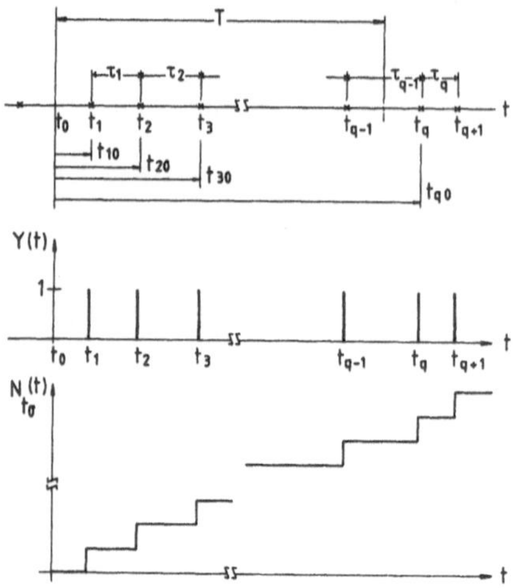

Fig.3.1. Events of a point process and their description by the functions Y(t) and $N_{t_0}(t)$

t_{20}, ... t_{q0} ...) are called forward recurrence times[7], residual lifetimes or waiting times. Alternatively, the intervals between the events (...., τ_1, τ_2, ... τ_q, ...) (also called lifetimes or inter-event times) may be specified.

II) By the discontinuous function Y(t) that takes the value 1.0 if an event occurs in an interval of width Δt surrounding t, and the value zero otherwise.

III) By the discontinuous nondecreasing step function with unit jumps, $N_{t_0}(t)$ = number of events in the semi-closed interval $(t_0, t_0+t]$.

The statistics of a PP can be completely described by specifying either of the stochastic processes Y(t) or $N_{t_0}(t)$, using any of the methods mentioned in Chap. 2. The discontinuous nature of these functions makes such a description difficult. Alternatively, the statistics of a PP can be described by assigning the probability functionals of one of the sets of random variables $(t_1, t_2, \ldots t_q)$, $(t_{10}, t_{20}, \ldots t_{q0})$ or $(\tau_1, \tau_2, \ldots \tau_q)$. This will be done in the following section. But let us here only define the density of events of a PP, $\lambda(t)$. This function can be defined as

---

[7]The backward recurrence times can similarly be defined as the times from an arbitrary time point to the previous events.

$$\lambda(t) = \lim_{\Delta t \to 0^+} \frac{1}{\Delta t} \Pr[N_t(t+\Delta t) \geq 1]$$
$$= \lim_{\Delta t \to 0^+} \frac{1}{\Delta t} \langle N_t(t+\Delta t) \rangle \qquad (3.1)$$
$$= \lim_{\Delta t \to 0^+} \frac{1}{\Delta t} \Pr[Y(t) = 1]$$

and is assumed to exist and to be continuous. It is assumed that

$$\lim_{\Delta t \to 0^+} \frac{1}{\Delta t} \Pr[N_t(t+\Delta t) \geq 2] = 0.$$

This means that in an infinitesimal interval Δt, the random variable N_t $(t+\Delta t)$ can only be one with probability $\lambda(t)\Delta t$; or zero with probability $[1-\lambda(t)\Delta t]$. Under this assumption of "regularity", $Y(t) = N_t(t+\Delta t)$.

Although the density of events function $\lambda(t)$ is a very important parameter, it does not, in general, provide a complete description of the PP. For a stationary PP, the probability distribution of $N_{t_0}(t)$ should be independent of t_0 hence $\lambda(t) = \lambda$ is a constant, and the process is then called homogenous.

3.2 Statistics of Times of Occurrence

Several functions may be used to describe the statistics of the times of occurrence of the events of a point process.

3.2.1 Joint Probability Density of Multicoincidence

The probability density of multicoincidence (PDC) (also called the distribution function or the product density), $\lambda(t_1, t_2, \ldots t_q)$, is the JPD that events occur simultaneously in q given distinct instants, $t_1, t_2, \ldots t_q$. Mathematically, the PDC can be written in the form

$$\lambda(t_1, t_2, \ldots t_q) = \lim_{\{\Delta t_i\} \to 0^+} \frac{1}{\prod_{i=1}^{q} \Delta t_i} \Pr[Y(t_1)=1, Y(t_2) = 1, \ldots, Y(t_1)=1] , \qquad (3.2)$$

or in the form

$$\lambda(t_1,t_2, \dots t_q) = \lim_{\{\Delta t_i\}\to 0^+} \frac{1}{\prod_{i=1}^{q} \Delta t_i} \langle \prod_{i=1}^{q} Y(t_i) \rangle \quad . \tag{3.3}$$

For q = 1, the density of events $\lambda(t)$ is obtained as defined in the previous section.

Because a realization of a point process is specified by the instances of occurrence on the time axis, i.e., by the points at which Y(t) equals one, the PDC for all $q = 1,2, \dots \infty$ provides a complete description of the point process. Another useful function is the *conditional probability density of coincidence*, an example of which is

$$\lambda_c(t_2|t_1) = \lambda(t_1,t_2)/\lambda(t_1) \quad , \tag{3.4}$$

which is the conditional probability that an event occurs at time t_2 provided one has already occurred at t_1.

3.2.2 Joint Probability Density of Forward Recurrence Times

The probability density $P_f(t_{10},t_{20}, \dots t_{q0}:t_0)$ is the q-fold JPD that if observation of the process begins at a time t_0, the first q events occur at times $t_1 = t_0+t_{20}$, ..., and $t_q = t_0+t_{q0}$. Of course, if the process is stationary, then this probability density is independent of t_0. The special case q = 1 gives $P_f(t_{10})$, the PD of the *forward recurrence time* (or the *time of arrival of the first event*).

The PD of *the time of arrival of the* q^{th} *event* $P_f(t_{q0}:t_0)$ can be obtained from $P_f(t_{10}, \dots, t_{q0}:t_0)$ by integrating over the possible values of t_{10}, $t_{20}, \dots t_{q0}$,

$$P_f(t_{q0}:t_0) = \frac{1}{(q-1)!}\int_{t_0}^{t_q} dt_{10} \dots \int_{t_0}^{t_q} dt_{q-1,0} P_f(t_{10},t_{20}, \dots t_{q0}:t_0). \tag{3.5}$$

The function $P_f(t_{10}, \dots t_{q0}:t_0)$ can be obtained from the PDC, $\lambda(t_1, \dots t_q)$ for all values of $q = 1,2, \dots \infty$, by using the relation

$$P_f(t_{10}, \ldots, t_{q0}:t_0) = \sum_{r=0}^{\infty} \frac{(-1)^r}{r!} \int_{t_0}^{t_q} d\theta_1 \ldots \int_{t_0}^{t_q} d\theta_r \lambda(t_1, \ldots t_q; \theta_1, \ldots, \theta_r) \,,$$

$$t_i = t_0 + t_{i0} \,, \qquad (3.6)$$

a derivation of which can be found in STRATONOVICH [461].

3.2.3 Joint Probability Density of Intervals Between Events

The distribution $P_i(\tau_1, \tau_2, \ldots \tau_{q-1}:t_1)$ represents the joint probability density that the time intervals between q consecutive events, beginning from the event that occurs at t_1, are $\tau_1, \tau_2, \ldots \tau_{q-1}$. This is the same as the conditional joint probability density that, beginning from time t_1, the forward recurrence times are $0, t_{21}, \ldots t_{q1}$ $(t_{j1} = t_j - t_1)$, provided that an event does occur at t_1. This means that P_i and P_f are related by

$$P_i(\tau_1, \tau_2, \ldots \tau_{q-1}:t_1) = P_f(0, t_{21}, t_{31}, \ldots t_{q1}:t_1)/\lambda(t_1), \quad q > 1,$$

$$t_{j1} = t_j - t_1 = \tau_1 + \tau_2 + \ldots \tau_{j-1} \,. \qquad (3.7)$$

For example, for $q = 2$, $P_i(\tau_1:t_1)$, the PD of the time between the event that occurs at t_1 and the following one is given by

$$P_i(\tau_1:t_1) = P_f(0, \tau_1:t_1)/\lambda(t_1) \,. \qquad (3.8)$$

By using (3.6), the relation between P_f and λ, together with (3.7), the following relation between P_i and λ is obtained

$$P_i(\tau_1, \tau_2, \ldots, \tau_{q-1}:t_1) = \frac{1}{\lambda(t_1)} \sum_{r=0}^{\infty} \frac{(-1)^r}{r!} \int_{t_1}^{t_q} d\theta_1 \ldots \int_{t_1}^{t_q} d\theta_r \lambda(t_1, \ldots t_q;$$

$$\theta_1, \ldots \theta_r) \quad t_j = t_1 + \tau_1 + \ldots \tau_{j-1} \,. \qquad (3.9)$$

3.2.4 Joint Probability Density of the Number of Events and Their Instants of Occurrence in a Closed Interval

Given a time interval $[t_0,t_0+T]$ in the domain of the point process, and during which the process is observed, we define

$$P_c(q;t_1,t_2,\ \dots\ t_q:[t_0,t_0+\tau])$$

as the JPD that the total number of events that occur in this interval is q, and that their instants are $t_1,t_2,\ \dots\ t_q$.

This JPD is related to the probability of multicoincidence by the relation [276]

$$P_c(q,t_1,\ \dots\ t_q:[t_0,t_0+T])$$

$$=\sum_{r=0}^{\infty}\frac{(-1)^r}{r!}\int_{t_0}^{t_0+T}d\theta_1\ \dots\int_{t_0}^{t_0+T}d\theta_r\lambda(t_1,\ \dots\ t_q,\theta_1,\ \dots\ \theta_r)\quad, \tag{3.10}$$

whose inverse is

$$\lambda(t_1,\ \dots\ t_q)=\sum_{r=0}^{\infty}\frac{1}{r!}\int_{t_0}^{t_0+T}d\theta_1\ \dots\int_{t_0}^{t_0+T}d\theta_r P_c(q+r;t_1,\ \dots,\ t_q,\ \theta_1,\ \dots\ \theta_r:[t_0,t_0+T])\quad. \tag{3.11}$$

3.2.5 Generating Functional

As before, q is the number of occurrences in $[t_0,t_0+T]$ and $t_1,\ \dots\ t_q$ are their instants of occurrence. Then the generating functional is defined by

$$L[v(t)]=\left\langle\prod_{j=1}^{q}[1+v(t_j)]\right\rangle\ . \tag{3.12}$$

The generating functional is related to the distribution function by

$$L[v(t)] = 1+\sum_{r=1}^{\infty}\frac{1}{r!}\int_{t_0}^{t_0+T}dt_1\ \dots\int_{t_0}^{t_0+T}dt_r\lambda(t_1,\ \dots\ t_r)v(t_1)\ \dots\ v(t_r) \tag{3.13}$$

(which follows from (3.12) by use of the definition of the expectation and by expanding the products). The inverse of (3.13) is

$$\lambda(t_1, \dots t_q) = \frac{\delta^q L[v(t)]}{\delta v(t_1) \dots \delta v(t_q)} \Bigg|_{v(t) \equiv 0} , \tag{3.14}$$

where δ/δ stands for the functional derivative.

Other generating functions can be obtained from $L[v(t)]$. The simplest example is the moment-generating function of the random variable q. By putting $v(t) = e^{-t}-1$, we obtain (cf. Sec. 2.1.3)

$$L[e^{-s}-1] = \langle e^{-sq} \rangle = Q_q(s) \quad . \tag{3.15}$$

Likewise, the factorial moment generating function $Q_q^f(s) = \langle (1-s)^q \rangle$ is given by

$$Q_q^f(s) = L[-s] \quad . \tag{3.16}$$

3.3 Counting Statistics

Instead of observing the instants of occurrence of the events of a point process or the function $N_t(\tau)$ at all times, it is sometimes simpler to divide the available time into a set of intervals (preferably disjoint) and to observe the number of occurrence (also called counts) in each interval. It is obvious that the observed counts carry less information than the instances of occurrence themselves. From the instants, we can calculate the number of counts in an interval, whereas when only the number of occurrences in an interval are recorded, information about their positions in time is lost.

In a counting experiment, the observed data are the number of counts $(n_1,n_2, \dots n_j, \dots n_q)$ in intervals (t_j,t_j+T_j) $j = 1, \dots q$. They form a set of nonnegative integers which can be described by their joint probability, $P(n_1,n_2, \dots n_q)$. For $q = 1$, this reduces to $P(n)$, the probability of observing n counts in an interval. Again, $P(n_1,n_2, \dots n_q)$ completely describes the random variables $(n_1, \dots n_q)$ but it does not completely describe the PP itself.

The PD, $P(n)$, can be obtained from $P_c(n,t_1, \dots t_n:[t_0,t_0+T]$ simply by integrating over $t_1, \dots t_n$ and dividing by n!, the number of permutations of n points falling in $[t_0,t_0+T]$. Thus,

$$P(n) = \frac{1}{n!}\int_{t_0}^{t_0+T} dt_1 \dots \int_{t_0}^{t_0+T} dt_n Pc(n;t_1, \dots t_n:[t_0,t_0+T]) \quad . \tag{3.17}$$

The set of variables $[n] = (n_1,n_2, \dots n_q)$ can also be described by their moments $\mu_{[n]}^{[m]}$, their central moments $v_{[n]}^{[m]}$, their cumulants $K_{[n]}^{[m]}$, or their factorial moments $F_{[n]}^{[m]}$. They can be further described by the moment-generating function $Q_{[n]}^{c}([s])$, the cumulant-generating function $Q_{[n]}^{c}([s])$, or the factorial moment-generating function, $Q_{[n]}^{f}([s])$, all defined in Sec. 2.1.3. Expressions for these moments and moment-generating functions cannot be given until the statistical nature of the underlying PP is specified.

The factorial moments of [n] can be obtained from the PDC by using the relations [302]

$$F_n^{(m)} = \int_t^{t+T} d\theta_1 \dots \int_t^{t+T} d\theta_m \lambda(\theta_1,\theta_2, \dots \theta_m) \quad , \tag{3.18}$$

and

$$F_{[n]}^{[m]} = \left(\prod_{i=1}^{q} \int_{I_i} d\theta_1^i \int_{I_i} d\theta_2^i \dots \int_{I_i} d\theta_{m_i}^i\right)\lambda(\theta_1^1, \dots,\theta_{m_1}^1;\theta_1^2, \dots, \theta_{m_2}^2; \dots;\theta_1^q, \dots, \theta_{m_q}^q) \tag{3.19}$$

$$I_i = [t_i,t_i+T_i] \quad .$$

The ordinary moments $\mu_{[n]}^{[m]}$ can be obtained from the factorial moments $F_{[n]}^{[m]}$. Moreover, the counting probability distribution $P([n])$ can be written as a q-fold summation over $\mu_{[n]}^{[m]}$ $m_1,m_2, \dots m_q = 1,2, \dots$. Consequently, $P([n])$ can be determined from the distribution function $\lambda(\theta_1,\theta_2, \dots)$.

Triggered Counting

Another scheme of counting can be devised by using the occurrence of one event to trigger a counting interval. The counting distribution obtained can be described by the conditional probability $P(n|t)$ that n events occur in the interval [t,t+T], provided that an event has already occurred at t, using the law of conditional probability,

$$P(n|t) = P(n)/\lambda(t) \quad , \tag{3.20}$$

where P(n) is the probability that n counts occur in [t,t+T].

3.4 The Poisson Process

3.4.1 Definition

A Poisson point process (P.PP) is defined as a point process (PP) whose PDC (or distribution function) fatorizes in the form

$$\lambda(t_1,t_2, \ldots, t_q) = \prod_{i=1}^{q} \lambda(t_i) \quad . \tag{3.21}$$

This means that the events of such a process are *statistically independent.*

An Alternative Definition

A P.PP is a PP with a generating functional given by

$$L[v(t)] = \exp\left[\int_t^{t+T} \lambda(\theta)v(\theta)d\theta\right] \quad . \tag{3.22}$$

That the second definition (3.22) follows from the first definition (3.21) can be demonstrated by using (3.21) in (3.13). Also (3.21) can be shown to follow from (3.22) by using (3.22) in (3.14). Other definitions based on different combinations of the properties listed below can, of course, be used.

It should be noted that a P.PP is completely describable by only one function, namely $\lambda(t)$, the density of events. If $\lambda(t)$ is constant, the process is called a *homogeneous Poisson PP* (HP.PP). By far, the P.PP and its generalizations are the most important point processes encountered in practice.

3.4.2 Statistics of Times

By use of the above definitions of a P.PP, expressions for the probability distributions that describe a PP (defined in Sec. 3.2) will be found.

1)

$$P_f(t_{10}, \ldots t_{q0}) = \exp\left[-\int_{t_0}^{t_q} \lambda(\theta)d\theta\right] \prod_{j=1}^{q} \lambda(t_j) \quad ,$$

$$t_j = t_0 + t_{j0} \quad . \tag{3.23}$$

This can be shown by using the factorization property (3.21), together with (3.6), and finding the resultant summation. For q = 1,

$$P_f(t_{10}) = \lambda(t_1)\exp -\left[\int_{t_0}^{t_1} \lambda(\theta)d\theta\right] \quad . \tag{3.24}$$

For a HP.PP, $\lambda(t) = \lambda$ and t_{10} has an exponential distribution,

$$P_f(t_{10}) = \lambda e^{-\lambda(t_1-t_0)} = \lambda e^{-\lambda t_{10}} \quad , \tag{3.25}$$

which is, as expected, independent of t_0.

The time of arrival of the qth event has a distribution

$$P_f(t_{q0}) = \lambda(t_q) \frac{\left[\int_{t_0}^{t_q} \lambda(\theta)d\theta\right]^{q-1}}{(q-1)!} \exp -\left[\int_{t_0}^{t_q} \lambda(\theta)d\theta\right] \quad . \tag{3.26}$$

This can be obtained by realizing that $t_{q0} = t_{10}+\tau_1+\tau_2+ \ldots \tau_{q-1}$ is the sum of q independent random variables with distributions given by (3.24). For a HP.PP,

$$P_f(t_{q0}) = \frac{\lambda^q}{(q-1)!} t_{q0}^{q-1} e^{-\lambda t_{q0}} \tag{3.27}$$

This is the distribution of the sum of q independent, exponentially distributed, random variables.

2)

$$P_i(\tau_1,\tau_2, \ldots \tau_{q-1}:t_1) = \exp\left[-\int_{t_1}^{t_q}\lambda(\theta)d\theta\right]\prod_{j=2}^{q}\lambda(t_j)$$

$$t_j = t_1 + \tau_1 + \tau_2 + \ldots \tau_{j-1} \quad . \tag{3.28}$$

This relation is obtained by using (3.23) and (3.7). For $q = 2$

$$P_i(\tau_1:t_1) = \frac{1}{\lambda(t_1)}\lambda(t_1)\lambda(t_2)\exp -\left[\int_{t_1}^{t_2}\lambda(\theta)d\theta\right] = \lambda(t_1+\tau_1)$$

$$\exp -\left[\int_{t_1}^{t_1+\tau_1}\lambda(\theta)d\theta\right] \quad . \tag{3.29}$$

For a HP.PP,

$$P_i(\tau_1:t_1) = \lambda\exp(-\lambda\tau_1) \tag{3.30}$$

which is an exponential distribution, independent of t_1 (as expected).

Note that both the forward recurrence time and the interval between occurrences have identical distribution. This is a remarkable property of the P.PP not shared by any other point process.

3)

$$P_c(q;t_1,t_2, \ldots t_q:[t_0,t_0+T]) = \exp\left[-\int_{t_0}^{t_0+T}\lambda(\theta)d\theta\right]\prod_{i=1}^{q}\lambda(t_i) \quad . \tag{3.31}$$

This is directly obtained from (3.10) by substituting from the factorization definition of a P.PP, (3.21).

For a HP.PP,

$$P(q,t_1,t_2, \ldots t_q:[t_0,t_0+T]) = \exp(-\lambda\tau)\lambda^q \tag{3.32}$$

This shows that the q events that occur in $[t_0,t_0+T]$ are equally likely to occur any where within this interval.

4)

$$\lambda_c(t_2|t_1) = \lambda(t_1,t_2)/\lambda(t_1) = \lambda(t_2) \quad (3.33)$$

showing that the occurrence of an event at t_1 does not in any way affect the likelihood of an occurrence at t_2.

3.4.3 Counting Statistics

Probabilities

The number of counts n in a single interval $[t_0,t_0+T]$ has a Poisson distribution

$$P(n) = \frac{W^n}{n!} e^{-W} \quad , \quad (3.34)$$

where W is the rate of events integrated over the counting interval, i.e.,

$$W = \int_{t_0}^{t_0+T} \lambda(\theta)d\theta \quad . \quad (3.35)$$

Eq. (3.34) can be proved in many ways. One way is to use the relation between P(n) and $P_c(n,t_1, \dots t_n:[t_0,t_0+T])$, (3.17), and substitute from (3.31).

For a HP.PP,

$$P(n) = \frac{(\lambda T)^n}{n!} e^{-\lambda T} \quad . \quad (3.36)$$

The numbers of counts in several intervals $n_1,n_2, \dots n_N$ are statistically independent (as a direct consequence of the definition of a P.PP). Hence their joint probability is simply the product of the individual probabilities, i.e.,

$$P(n_1,n_2, \dots n_q) = \prod_{j=1}^{q} P(n_j) = \prod_{j=1}^{q}\left(\frac{W_j^{n_j}}{n_j!} e^{-W_j}\right) \quad , \quad (3.37)$$

or

$$P([n]) = \frac{[W]^{[n]}}{[n]!} e^{-[W]} \quad ,$$

where

$$W_j = \int_{t_j}^{t_j+T} \lambda(\theta)d\theta \quad . \tag{3.38}$$

The triggered counting PD is given by

$$P(n|t) = P(n)/\lambda(t) = \frac{W^n}{n!} e^{-W}/\lambda(t) \quad . \tag{3.39}$$

For a HP.PP,

$$P(n|t) = T \frac{(\lambda T)^{n-1}}{n!} e^{-\lambda T} \quad . \tag{3.40}$$

Moment-Generating Functions

The integer random variable n can be described by one of its moment-generating functions.

I) Ordinary mgf

$$Q_n(s) = \langle e^{-ns} \rangle = \sum_{n=0}^{\infty} e^{-ns} P(n) \quad .$$

By using (3.34), this summation can be shown to give

$$Q_n(s) = e^{W(e^{-s}-1)} \quad . \tag{3.41}$$

This same expression can be obtained by using the relation between $Q_n(s)$ and $L[v(t)]$, (3.15), together with the definition (3.22).

II) Cumulant gf

$$Q_n^c(s) = \ln Q_n(s) = W(e^{-s}-1) \quad . \tag{3.42}$$

III) Factorial mgf

$$Q_n^f(s) = \langle(1-s)^n\rangle = \sum_{n=0}^{\infty} (1-s)^n P(n) = e^{-sW} \quad . \tag{3.43}$$

This can also be obtained by directly using (3.16) and (3.22).

Equations (3.41 - 43) can be generalized to the multidimensional case. By use of the notations of Sec. 2.2.2,

$$Q_{[n]}([s]) = \exp([W]^{\sim}[e^{-s}-1]) \quad , \tag{3.44}$$

$$Q_{[n]}^c([s]) = [W]^{\sim}[e^{-s}-1] \quad , \tag{3.45}$$

$$Q_{[n]}^f([s]) = \exp(-[W]^{\sim}[s]) \quad . \tag{3.46}$$

Moments

I) Ordinary Moments

The ordinary moments of n can be obtained by direct use of the mgf given by (3.41) and (3.44) and the rules of Sec. 2.1.3 and 2.2.3. The low-order moments thus obtained are listed below and will be used later.

$$\begin{aligned} \langle n\rangle &= W \qquad \text{(mean)} \\ \langle n^2\rangle &= W^2 + W \\ \langle n^3\rangle &= W^3 + 3W^2 + W \\ \langle n^4\rangle &= W^4 + 6W^3 + 7W^2 + W \\ &\ldots\ldots\ldots\ldots\ldots \end{aligned} \tag{3.47}$$

$$\begin{aligned} \langle n_1 n_2\rangle &= W_1 W_2 \quad \text{(correlation)} \\ \langle n_1 n_2 \ldots n_q\rangle &= W_1 W_2 \ldots W_q \\ &\ldots\ldots\ldots\ldots\ldots \end{aligned} \tag{3.48}$$

$$\begin{aligned} \langle n_1^2 n_2\rangle &= (W_1^2+W_1)\,(W_2) \\ \langle n_1^2 n_2^2\rangle &= (W_1^2+W_1)\,(W_2^2+W_2) \\ \langle n_1^3 n_2\rangle &= (W_1^3+3W_1^2+W_1)\,(W_2) \\ \langle n_1^3 n_2^2\rangle &= (W_1^3+3W_1^2+W_1)\,(W_2^2+W_2) \end{aligned} \tag{3.49}$$

....

and so on.

II) Central Moments

The central moments $\nu_n^{(m)}$ can be obtained from $\mu_n^{(m)}$ by employing the relations in Sec. 2.1.3. The lower-order central moments are given by

$$<\Delta n> = 0$$

$$<(\Delta n)^2> = W = <n> \qquad \text{(variance)} \tag{3.50}$$

$$<\Delta n_1 \Delta n_2> = 0 \qquad \text{(covariance)}$$

etc.

III) Factorial Moments

Using (3.43) and (3.46), together with (2.14) and (2.26), we obtain

$$F_n^{(m)} = \left\langle \frac{n!}{(n-m)!} \right\rangle = W^m \tag{3.51}$$

and

$$F_{[n]}^{[m]} = [W]^{[m]} \quad . \tag{3.52}$$

The simplicity and generality of this expression accounts for the importance of the factorial moments in the study of the Poisson process and its generalizations.

IV) Cumulants

Similarly, by using (3.42) and (3.45) together with (2.12) and (2.24), we get

$$K_n^{(m)} = W \tag{3.53}$$

$$K_{[n]}^{[m]} = [W]^{[1]} \quad , \tag{3.54}$$

e.g.,

$$K_{n_1,n_2}^{m_1,m_2} = W_1 W_2 \quad . \tag{3.54}$$

Cumulants of a given fold are all equal. Departure from this property indicates a non-Poissonian nature.

3.5 Doubly Stochastic Poisson Point Processes (DSP.PP)

3.5.1 Definitions

It has been shown that a P.PP is statistically completely determined by its rate density function $\lambda(t)$.

A doubly stochastic (or compound) Poisson point process (DSP.PP) is a P.PP whose rate density function $\lambda(t)$ is itself a stochastic process. For each realization of $\lambda(t)$, the resulting PP is a Poisson point process. Therefore, the statistical properties of a DSP.PP are completely specified if the statistical properties of the $\lambda(t)$ are given. The moments and mgf's can be determined by averaging the corresponding moments and mgf's over the different realizations of $\lambda(t)$. This will be done in the following sections.

3.5.2 Counting Statistics

Moment-Generating Functions

The moment-generating functions of a DSP.PP can be obtained from those of a P.PP by averaging over the realizations of the rate λ or its integration over the counting time W. Thus from (3.41 - 43) we obtain

$$Q_n(s) = \langle e^{W(e^{-s}-1)} \rangle = Q_W(1-e^{-s}) \tag{3.55}$$

$$Q_n^c(s) = \ln Q_n(s) = \ln Q_W(1-e^{-s}) = Q_W^c(1-e^{-s}) \tag{3.56}$$

$$Q_n^f(s) = \langle e^{-sW} \rangle = Q_W(s) \quad , \tag{3.57}$$

i.e., the mgf's of the process n are simply related to those of the process W. Conversely,

$$Q_W(s) = Q_n[\ln (1+s)] = \exp\left\{Q_n^c[\ln (1+s)]\right\} = Q_n^f(s) \quad . \tag{3.58}$$

The simplest of the above relations is (3.57), which relates the factorial mgf of n to the mgf of W.

Generalization of the preceding equations to the multivariate case results in

$$Q_{[n]}([s]) = Q_{[W]}([1 - e^{-s}]) \tag{3.59}$$

$$Q^c_{[n]}([s]) = \ln Q_{[n]}([s]) = Q^c_{[W]}([1 - e^{-s}]) \tag{3.60}$$

$$Q^f_{[n]}([s]) = Q_{[W]}([s]) \quad . \tag{3.61}$$

Expressions for $Q_{[W]}$ cannot be given unless the statistics of the process W are specified. This will be done in later chapters.

Moments

In this section, the relations between the various moments of [n] and those of [W] are listed. These relations can be obtained directly from the moment-generating function (3.55), or by using the expressions of the moments of a P.PP, (3.47), and averaging over the realizations of W.

I) Ordinary Moments

$$\begin{aligned} \langle n\rangle &= \langle W\rangle \\ \langle n^2\rangle &= \langle W^2\rangle + \langle W\rangle \\ \langle n^3\rangle &= \langle W^3\rangle + 3\langle W^2\rangle + \langle W\rangle \\ \langle n^4\rangle &= \langle W^4\rangle + 6\langle W^3\rangle + 7\langle W^2\rangle + \langle W\rangle \end{aligned} \tag{3.62}$$

....

$$\begin{aligned} \langle W\rangle &= \langle n\rangle \\ \langle W^2\rangle &= \langle n^2\rangle - \langle n\rangle \\ \langle W^3\rangle &= \langle n^3\rangle - 3\langle n^2\rangle + 2\langle n\rangle \end{aligned} \tag{3.63}$$

etc.

Also,

$$\begin{aligned} \langle n_1 n_2\rangle &= \langle W_1 W_2\rangle \\ \langle n_1 n_2 \cdots n_q\rangle &= \langle W_1 W_2 \cdots W_q\rangle \end{aligned} \tag{3.64}$$

$$\begin{aligned}
\langle n_1^2 n_2\rangle &= \langle W_1^2 W_2\rangle + \langle W_1 W_2\rangle \\
\langle n_1^2 n_2^2\rangle &= \langle W_1^2 W_2^2\rangle + \langle W_1^2 W_2\rangle + \langle W_1 W_2^2\rangle + \langle W_1 W_2\rangle \\
\langle n_1^3 n_2\rangle &= \langle W_1^3 W_2\rangle + 3\langle W_1^2 W_2\rangle + \langle W_1 W_2\rangle \\
\langle n_1^2 n_2 n_3\rangle &= \langle W_1^2 W_2 W_3\rangle + \langle W_1 W_2 W_3\rangle
\end{aligned} \tag{3.65}$$

....

and so on.

In the set of equations (3.62), because we note that $\langle W^r\rangle \geq \langle W\rangle^r$ and that the coefficients of expansion of $\langle n^m\rangle$ in terms of $\langle W^r\rangle$, $r = 1, \ldots m$ are always larger than or equal to one, we immediately see that the moment $\langle n^m\rangle$ for a DSP.PP is always larger than or equal to the corresponding moment for a pure P.PP.

II) Central Moments

$$\langle \Delta n^2\rangle = \langle \Delta W^2\rangle + \langle W\rangle \quad . \tag{3.66}$$

Comparison with (3.50) shows that the variance of n for a DSP.PP is always larger than that of a P.PP of the same mean. This excess variance, or "excess noise", is equal to the variance of W. Also,

$$\langle \Delta n_1 \ldots \Delta n_q\rangle = \langle \Delta W_1 \ldots \Delta W_q\rangle \quad . \tag{3.67}$$

III) Factorial Moments

The factorial moments of n are obtained from the factorial mgf of n by using (2.14). By using the relation between the mgf of n and that of W, (3.57), we get

$$F_n^{(m)} = (-1)^m \frac{\partial^m}{\partial s^m} Q_W(s)\Big|_{s=0} \quad , \tag{3.68}$$

or

$$F_n^{(m)} = \mu_W^{(m)} \quad , \quad \left\langle \frac{n!}{(n-m)!} \right\rangle = \langle W^m\rangle \quad ; \tag{3.69}$$

and in general

$$F^{[m]}_{[n]} = \mu^{[m]}_{[W]} \quad . \tag{3.70}$$

Note that (3.69) and (3.70) follow directly from (3.18) and (3.19).

IV) Cumulants

By expansion of both sides of (3.56) and equating the coefficients of s, the following relations can be established:

$$\begin{aligned}
K_n^{(1)} &= K_W^{(1)} \\
K_n^{(2)} &= K_W^{(2)} + K_W^{(1)} \\
K_n^{(3)} &= K_W^{(3)} + 3\,K_W^{(2)} + K_W^{(1)} \\
K_n^{(4)} &= K_W^{(4)} + 6\,K_W^{(3)} + 7\,K_W^{(2)} + K_W^{(1)}
\end{aligned} \tag{3.71}$$

....

$$K^{[1]}_{[n]} = K^{[1]}_{[W]} \quad . \tag{3.72}$$

Also,

$$\begin{aligned}
K_W^{(1)} &= K_n^{(1)} \\
K_W^{(2)} &= K_n^{(2)} - K_n^{(1)} \\
K_W^{(3)} &= K_n^{(3)} - 3\,K_n^{(2)} + 2\,K_n^{(1)} \\
K_W^{(4)} &= K_n^{(4)} - 6\,K_n^{(3)} + 11\,K_n^{(2)} - 6\,K_n^{(1)} \quad ,
\end{aligned} \tag{3.73}$$

and so on.

Relation Between the Moments of W and Those of λ(t)

We have seen that the moments of the number of counts n are related to the moments of the integrated density of events W. Because

$$W = \int_{t_0}^{t_0+T} \lambda(\theta)d\theta \quad ,$$

the moments of W can be related to those of the density of events by

$$\langle W^m \rangle = \int_{t_0}^{t_0+T} d\theta_1 \ldots \int_{t_0}^{t_0+T} d\theta_n G_\lambda^{(m)}(\theta_1,\theta_2, \ldots \theta_m) \quad , \tag{3.74}$$

where

$$G_\lambda^{(m)}(t_1,t_2, \ldots t_m) = \langle \lambda(t_1)\lambda(t_2) \ldots \lambda(tm) \rangle \tag{3.75}$$

is the correlation function of the stochastic process $\lambda(t)$. Similarly, for the cumulants,

$$K_W^m = \int_{t_0}^{t_0+T} d\theta_1 \ldots \int_{t_0}^{t_0+T} d\theta_m \, K_\lambda^{(m)}(\theta_1,\theta_2, \ldots \theta_m) \tag{3.76}$$

(see Sec. 2.4.1). Also,

$$\langle W_1 W_2 \ldots W_q \rangle = \int_{t_1}^{t_1+T_1} d\theta_1 \ldots \int_{t_q}^{t_q+T_q} d\theta_q \, G_\lambda^{(q)}(\theta_1, \ldots \theta_q) \quad ; \tag{3.77}$$

and conversely,

$$\langle \lambda(t_1+T_1)\lambda(t_2+T_2) \ldots \lambda(t_q+T_q) \rangle = \frac{\partial^q}{\partial T_1 \partial T_2 \ldots \partial T_q} \langle W_1 W_2 \ldots W_q \rangle \quad . \tag{3.78}$$

Probability Distribution

When (3.34) and (3.37) are averaged over the realizations of W, we obtain

$$P(n) = \left\langle \frac{W^n}{n!} e^{-W} \right\rangle \quad , \tag{3.79}$$

and its generalization,

$$P([n]) = \left\langle \frac{[W]^{[n]} e^{-[W]}}{[n]!} \right\rangle \quad . \tag{3.80}$$

$P(n)$ can be obtained from $Q_W(s)$ by differentiation at $s = 1$. This can be shown simply by writing

$$P(n) = \left\langle \frac{W^n}{n!} e^{-W} \right\rangle = \left\langle \frac{(-1)^n}{n!} \frac{\partial^n}{\partial s^n} e^{-Ws} \Big|_{s=1} \right\rangle = \frac{(-1)^n}{n!} \frac{\partial^n}{\partial s^n} \left\langle e^{-Ws} \right\rangle \Big|_{s=1}$$

i.e.,

$$P(n) = \frac{(-1)^n}{n!} \frac{\partial^n}{\partial s^n} Q_W(s) \Big|_{s=1} \quad . \tag{3.81}$$

Similarly, by expanding $Q_W(s)$ around $s = 1$ in a MacLaurin expansion and using (3.81), we can write $Q_W(s)$ in terms of $P(n)$ in the form

$$Q_W(s) = \sum_{n=0}^{\infty} (-1)^n P(n)(s-1)^n \quad . \tag{3.82}$$

The averages in (3.79) and (3.80) will be determined for several special distributions of W in later chapters. Some general properties can be shown if we write the expectation in (3.79) in the form,

$$P(n) = \int_0^{\infty} dW P(W) \frac{W^n}{n!} e^{-W} \quad , \tag{3.83}$$

and note that $P(n)$ and $P(W)$ are related by the so-called Poisson transform, properties of which are given in Sec. 3.6.

Triggered Counting

By using (3.39) and averaging over variations in λ and W, we get

$$P(n|t) = \left\langle \frac{W^n}{n!} e^{-W} \right\rangle / \langle \lambda(t) \rangle \quad . \tag{3.84}$$

3.5.3 Statistics of Times

The statistics of times for a DSP.PP can be determined by averaging the expressions of Sec. 3.4.2 over the distribution of $\lambda(\theta)$. This leads to

$$\text{I)} \quad \lambda(t_1, t_2, \dots t_q) = \left\langle \prod_{i=1}^{q} \lambda(t_i) \right\rangle = G_\lambda^{(q)}(t_1, t_2, \dots t_q) \quad , \tag{3.85}$$

i.e., the distribution function of the events is equal to the correlation function of the stochastic rate density. This can also be regarded as a definition of the DSP.PP. Also,

$$\lambda_c(t_2|t_1) = \langle\lambda(t_1)\lambda(t_2)\rangle/\langle\lambda(t_1)\rangle$$
$$= G_\lambda(t_1,t_2)/\langle\lambda(t_1)\rangle \quad . \tag{3.86}$$

II)

$$P_f(t_{10}, \ldots t_{q0}) = \left\langle \exp\left[-\int_{t_0}^{t_q}\lambda(\theta)d\theta\right]\prod_{j=1}^{q}\lambda(t_j)\right\rangle \tag{3.87}$$
$$t_j = t_0 + t_{j_0} \quad ,$$

$$P_f(t_{10}) = \left\langle \lambda(t_1)\exp\left[-\int_{t_0}^{t_1}\lambda(\theta)d\theta\right]\right\rangle \quad , \tag{3.88}$$

$$P_f(t_{q0}) = \frac{1}{(q-1)!}\left\langle \lambda(t_q)\left[\int_{t_0}^{t_q}\lambda(\theta)d\theta\right]^{q-1}\exp\left[-\int_{t_0}^{t_q}\lambda(\theta)d\theta\right]\right\rangle \quad . \tag{3.89}$$

III)

$$P_i(\tau:t) = \left\langle \lambda(t)\lambda(t+\tau)\exp\left[-\int_{t}^{t+\tau}\lambda(\theta)d\theta\right]\right\rangle /\langle\lambda(t)\rangle \quad , \tag{3.90}$$

$$P_i(\tau_1,\tau_2, \ldots \tau_{q-1}:t) = \left\langle \exp\left[-\int_{t_1}^{t_q}\lambda(\theta)d\theta\right]\prod_{j=1}^{q}\lambda(t_j)\right\rangle/\langle\lambda(t_1)\rangle \quad . \tag{3.91}$$

Note that in deriving (3.90) and (3.91) from (3.29) and (3.28), we resorted to the original definition of the conditional probability as a ratio between two probabilities and averaged each probability separately.

IV)

$$P_c(q;t_1, \ldots t_q:[t_0,t_0+T]) = \left\langle \exp\left[-\int_{t_0}^{t_0+T}\lambda(\theta)d\theta\right]\prod_{i=1}^{q}\lambda(t_i)\right\rangle \quad . \tag{3.92}$$

Once the nature of the stochastic process $\lambda(t)$ is specified, the expectations in (3.86-92) cam be calculated. It is, however, useful to note that some time statistics can be directly related to the mgf of the integrated density rate $Q_W(s)$.

By comparing

$$Q_W(s) = \left\langle \exp\left[- s \int_{t_0}^{t_0+T} \lambda(\theta)d\theta\right]\right\rangle$$

to (3.88) and (3.90), we obtain [173]

$$P_f(t_{10}) = - \frac{\partial}{\partial T} Q_W(1) \quad , \tag{3.93}$$

$$P_i(\tau:t_0) = \frac{1}{\langle\lambda(t_0)\rangle} \frac{\partial^2}{\partial T^2} Q_W(1) \quad . \tag{3.94}$$

Note that $P_f(0) = \langle\lambda\rangle$ and $P_i(0) = \langle\lambda^2\rangle/\langle\lambda\rangle$. Hence, $P_i(0) \geq P_f(0)$; the equality holds only when λ is deterministic. Also the probability density of the time of arrival of the qth event can be obtained from [48]

$$P_f(t_{q0}) = \frac{1}{\langle\lambda(t_0)\rangle} \frac{(-1)^q}{q!} \frac{\partial^q}{\partial s^q}\left[\frac{1}{s^2} \frac{\partial^2}{\partial T^2} Q_W(s)\right]\Bigg|_{s=1} \quad , \tag{3.95}$$

or from [417]

$$P_f(t_{q0}) = - \frac{\partial}{\partial T}\sum_{j=0}^{q-1} \frac{(-1)^j}{j!} \frac{\partial^j}{\partial s^j} Q_W(s) \Big|_{s=1} \quad . \tag{3.96}$$

Eq. (3.95) and (3.96) can be easily shown to verify (3.89).

The important role played by the mgf $Q_W(s)$ can be realized by noting its usefulness in determining the counting mgf (3.55-57), the counting factorial moments (3.68), the counting probability density (3.81) and some times statistics (3.93-96).

3.6 Appendix: The Poisson Transform

3.6.1 Definition and Properties

The Poisson transform of a function f(x) is a function P(n) defined by the linear transformation,

$$P(n) = \mathcal{P}[f(x)] \triangleq \int_0^\infty dx\ f(x)\ \frac{x^n e^{-x}}{n!} \quad . \tag{3.97}$$

A list of the Poisson transforms of some important functions (all of them are probability density functions) is given in Table 3.1.

An examination of (3.97) shows that the Poisson transform of a function is always broader and smoother than the function itself. This is due to the factors x^n and e^{-x} which damp out any irregular behaviour of f(x) in the regions $x < 1$ and $x > 1$, respectively. The list of examples in Table 3.1 demonstrates this point clearly.

Some properties of the Poisson transform are listed below (for proofs see [210]).

I) The Poisson transform P(n) and the Laplace transform F(s) of a function f(x)

$$\left(F(s) = \mathcal{L}[f(x)] = \int_0^\infty f(x) e^{-sx} dx\right)$$

are related by

$$P(n) = \frac{(-1)^n}{n!} \frac{\partial^n}{\partial s^n} F(s)\Big|_{s=1} \tag{3.98}$$

and

$$F(s) = \sum_{n=0}^{\infty} \frac{(-1)^n}{n!} P(n) s^n \quad . \tag{3.99}$$

Notice that if f(x) represents the probability density of a nonnegative random variable x, then F(s) represents its moment generating function $Q_x(s)$. This property could be important in extracting the inverse Poisson transform by using the formula

$$f(x) = \mathcal{P}^{-1}[P(n)] = \mathcal{L}^{-1}[F(s)] = \mathcal{L}^{-1}\left[\sum_{n=0}^{\infty} \frac{(-1)^n}{n!} P(n) s^n\right] \quad , \tag{3.100}$$

where $\mathcal{L}^{-1}$ and $\mathcal{P}^{-1}$ represent the inverse Laplace and Poisson transforms, respectively.

Table 3.1. Poisson transforms of some probability distribution functions

	f(x)	P(n)	
Constant	$\delta(x - \mu)$	$\frac{\mu^n e^{-\mu}}{n!}$	Poisson
Exponential (chi_2^2)	$\frac{1}{\mu} e^{-x/\mu}$	$\frac{\mu^n}{(1+\mu)^{n+1}}$	Bose-Einstein (geometric)
Gaussian-square (chi_1^2)	$\frac{1}{\sqrt{\pi\mu x}} e^{-x/\mu}$	$\frac{(2n-1)!!}{n!\ 2^n} \frac{\mu^n}{(1+\mu)^{n+1/2}}$	
Chi_{2N}^2	$\frac{1}{\Gamma(N)} \mu^{-N} x^{N-1} e^{-x/\mu}$	$\binom{n+N-1}{n} \frac{\mu^n}{(1+\mu)^{n+N}}$	Negative-binomial
Bessel (noncentral chi_2^2)	$\frac{1}{\mu} \exp\left(-\frac{x+\mu_0}{\mu}\right) I_0 (2\sqrt{\mu_o x}/\mu)$	$\frac{\mu^n}{(1+\mu)^{n+1}} \exp\left(-\frac{\mu_0}{\mu+1}\right) L_n\left(\frac{-\mu_0/\mu}{1+\mu}\right)$.	Laguerre
Noncentral chi_{2N}^2	$\frac{1}{\mu}\left(\frac{x}{\mu_0}\right)^{(N-1)/2} \exp\left(-\frac{x+\mu_0}{\mu}\right) I_{N-1}\left(\frac{2\sqrt{\mu_0 x}}{\mu}\right)$	$\frac{\mu^n}{(1+\mu)^{n+N}} \exp\left(-\frac{\mu_0}{1+\mu}\right) L_n^{N-1}\left(\frac{-\mu_0/\mu}{1+\mu}\right)$	

II)

$$P[f(\tfrac{x}{a})] = \frac{(-1)^n}{n!} a^{n+1} \frac{\partial^n}{\partial s^n} F(s) \Big|_{s=a} \quad . \tag{3.101}$$

III)

$$P[f(x-a)] = e^{-a} \sum_{\ell=0}^{n} (-a)^{n-\ell} \binom{n}{\ell} \frac{\partial^\ell}{\partial s^\ell} F(s) \Big|_{s=1} \quad . \tag{3.102}$$

IV)

$$P[e^{-ax} f(x)] = \frac{(-1)^n}{n!} \frac{\partial^n}{\partial s^n} F(s) \Big|_{s=a+1} \quad . \tag{3.103}$$

V)

$$P\{f(x)[\ln(x)]^\ell\} = \frac{1}{n!} \frac{d^\ell}{dn^\ell} \{n!\ P[f(x)]\} \quad . \tag{3.104}$$

VI)

$$P[F(x)] = \int_0^\infty \frac{f(t)}{(t+1)^{n+1}} dt \quad . \tag{3.105}$$

VII) If

$$P_1(n) = P[f(x)], \quad \text{and} \quad P_2(n) = P[f_2(x)] \quad ,$$

then

$$P[f_1(x) \circledast f_2(x)] = P_1(n) \circledast P_2(n) \quad , \tag{3.106}$$

where $\circledast$ represents the convolutions,

$$f_1(x) \circledast f_2(x) = \int_0^\infty f_1(y) f_2(x-y) dy \quad ,$$

and

$$P_1(n) \circledast P_2(n) = \sum_{m=0}^{n} P_1(m) P_2(n-m) \quad .$$

This property is important when x and n represent random variables whose probability distributions f(x) and P(n) are related by a Poisson transform. In this case, if x is the sum of two statistically independent random variables with distributions $f_1(x)$ and $f_2(x)$, then $f(x) = f_1(x) \circledast f_2(x)$ and P(n) =

$P_1(n) \circledast P_2(n)$ where $P_1(n)$ and $P_2(n)$ are the Poisson transforms of $f_1(x)$ and $f_2(x)$, respectively. This can, of course, be generalized to the sum of several random variables. Other properties of the Poisson transform can be found in [211-213].

3.6.2 Inversion of the Poisson Transform

It is of great interest to develop methods for determining the function f(x) whose Poisson transform P(n) is known. Such methods would permit determination of the probability density of the integrated density of events of a DSP.PP, P(W), if the probability distribution of the number of counts P(n) is known.

Method I [515]

From (3.99), we can simply write

$$f(x) = \mathcal{L}^{-1}\left[\sum_{n=0}^{\infty} \frac{(-1)^n}{n!} P(n)s^n\right] \quad . \tag{3.107}$$

This shows that the Poisson transform can, in principle, be inverted. The usefulness of (3.107) is obviously limited to cases when a closed expression is available for P(n). For example, if $P(n) = \bar{n}^{-n}/(1+\bar{n})^{n+1}$, the sum in (3.107) gives $(\bar{n}+1+s\bar{n})^{-1}$ whose inverse Laplace transform is the exponential function, $f(x) = \bar{n}e^{-x/\bar{n}}$ (cf. Table 3.1).

Method II [39, 385]

This method is based on expanding f(x) in an infinite series of orthogonal Laguerre functions

$$f(x) = \sum_{m=0}^{\infty} a_m L_m(x) \quad , \tag{3.108}$$

where

$$L_m(x) = \sum_{\ell=0}^{m} \binom{m}{\ell} \frac{(-x)^\ell}{\ell!} \quad , \quad \int_0^\infty e^{-x} L_n(x) L_m(x) dx = \delta_{n,m} \tag{3.109}$$

and the coefficients a_m are given by the inner product,

$$a_m = \int_0^\infty e^{-x} f(x) L_m(x) \quad . \tag{3.110}$$

Substituting (3.109) in (3.110) and using the definition of the Poisson transform, we get

$$a_m = \sum_{\ell=0}^{m} (-1)^\ell \binom{m}{\ell} P(\ell) \tag{3.111}$$

which, when substituted in (3.108) gives the desired inversion formula

$$f(x) = \sum_{m=0}^{\infty} \sum_{\ell=0}^{m} (-1)^\ell \binom{m}{\ell} L_m(x) P(\ell) \quad . \tag{3.112}$$

If the distributions P(n) of Table 3.1 are substituted in (3.112), the correct f(x)'s are obtained. But the question remains as to how fast the summations in this formula converge and how much is the error in the calculated f(x) that results from possible inaccuracies in P(n).

Method III

This method depends on the equality between the factorial moments of P(n) and the ordinary moments of f(x) (cf. (3.69)). Given P(n), $F_n^{(m)}$ can be calculated from which $\langle x^m \rangle$ can be estimated. There remains the problem of determining f(x) from its moments $\langle x^m \rangle$. This is a conventional problem in statistics; many methods exist for its solution.

Part II

Theory

4. The Optical Field: A Stochastic Vector Field or, Classical Theory of Optical Coherence

Light is an electromagnetic phenomenon. According to *classical electromagnetic theory*, an electromagnetic field consists of an electric vector field and a magnetic vector field, both being functions of space and time. These fields have to satisfy a set of partial differential equations, Maxwell's equations, which govern their generation by accelerated electric charges as well as their propagation through media that have different electric and magnetic properties. All optical phenomena can be explained by solving Maxwell's equations under the proper boundary conditions.

Manipulation of Maxwell's equations in a charge-free space leads directly to the wave equation, which any of the components of the electric and magnetic fields in any direction must satisfy.

Visible light is an electromagnetic field that propagates in waves that have frequencies in the range from 10^{14} to 10^{15} Hz, corresponding to wavelengths from 4000 to 8000 Å. The frequency of a light wave determines its color. A wave that contains only one frequency is called monochromatic. Normally we deal with a sum of waves that have a relatively narrow distribution of frequencies (quasi-monochromatic light).

Although the classical electromagnetic theory assumes that the instantaneous field vibrations at all positions could in principle be followed and measured, such a fast phenomenon (which takes place in intervals of the order of 10^{-15}s) has of course never been directly observed.

Yet the wave behaviour of light, which is a consequence of its electromagnetic nature, very simply accounts for the basic optical phenomena of diffraction and interference. Moreover, geometrical optics can be deduced from the wave theory by taking the limit of very short wavelengths and replacing plane waves by rays.

The wealth of optical phenomena that can be understood by use of the simple Maxwell's equations can be appreciated by examination of the textbook of BORN and WOLF [65] which is completely based on solutions of those equations.

At the end of the last century and after the long controversy between the corpuscular and the wave theories of light seemed to have been finally settled in favor of the latter, it was thought that an answer had been found to the old (and everlasting) question, "What is light?" Unfortunately (or rather fortunately) around the turn of the century, the whole classical physics (mechanics and electromagnetic theory) faced many difficulties when it was found incapable of explaining several phenomena. This led to a fundamental shake-up of the basic concepts of physics and to the emergence of the quantum theory.

Today it is believed that the *quantum electrodynamic theory* explains all known phenomena related to the generation, propagation, and detection of light and seems to be limited only by the difficulties of performing the required mathematical calculations. Yet the emergence of the quantum theory has not diminished the role of the classical theory. In fact, more optical phenomena can be understood in terms of classical electromagnetic theory than necessitate quantum electrodynamics.

On the other hand, many phenomena related to the interaction of radiation with matter can be understood on the basis of a *semiclassical theory*, in which the radiation is described classically and the matter is described quantum theoretically. The photoelectric effect is one such phenomenon. Although this effect played an important role in the discovery of quantum theory, it can be understood without quantization of the electromagnetic field and even without invoking the concept of the "photon".

As is mentioned in the Introduction, the subject of this book is the study of light as a tool in spectroscopic investigations and as a carrier of information in optical communication systems. Therefore, we shall not be dealing with phenomena in which a complete quantum description is necessary. Hence, throughout this book, light is described classically.

The description of light in terms of monochromatic plane (or spherical) waves or in terms of other mathematical functions of space-time is clearly an idealization that, though sometimes useful, is in many cases completely inadequate. Light can be described only statistically. The following arguments justify this:

I) Because the electromagnetic field in the optical region is always generated by a very large number of radiators, which may radiate independently, a detailed rigorous solution of Maxwell's equations under the proper boundary conditions is not only very difficult but is useless because it would mask the important physical effects. This point of view is analogous to that used to justify the introduction of statistical mechanics to describe the dynamics of a large number of molecules.

II) A description in terms of statistical averages is dictated by the fact that no instrument is yet available (or because it is fundamentally impossible) to measure the very fast instantaneous variations of an optical field. On the other hand, instruments are available (as we shall see later) for the measurement of statistical averages that vary at a much slower rate. The statistical description of light (the theory of coherence) aims at describing the behaviour of the statistical averages of appropriate functions of the optical fields.

III) The statistical approach is indeed capable of explaining many phenomena under the most general conditions (e.g., diffraction, interference, image formation, resolution in microscopic images, etc.).

IV) The randomness of light may originate from (or be supplemented by) random fluctuations in the medium through which it propagates (e.g., due to turbulence in the atmosphere, scattering from suspended objects in random motion, or even scattering from rough surfaces). Sometimes light is used to probe a fluctuating target. In this case, this randomness carries the useful information.

Moreover, the process of light detection itself introduces some randomness, due to its quantum nature, which must be treated statistically.

The theory of optical coherence has been treated in several books [51, 65, 147, 268, 353, 359, 369, 487, 501]. Quite a number of articles review this subject, together with the theory of light detection and photoelectron statistics [9, 10, 15, 53, 171 - 176, 296, 309, 318, 332, 336, 381, 410, 431, 488, 491, 510 - 512]. MANDEL and WOLF [319] reprinted a selection of original papers on the subject, together with an extensive bibliography on the whole field up to 1966.

In Sec. 4.1, the general theoretical aspects of the classical statistical description of light are introduced. Section 4.1.1 is a short outline of the equations that classically describe an optical field and its propagation. In Sec. 4.1.2, the randomness of the optical field is introduced and the various "smoothed" averages that describe the field are defined and their tempo-spatial properties discussed. The rules of propagation of these averages are briefly outlined in Sec. 4.1.3. In Sec. 4.2, several statistical models of light fields of special practical importance are defined and studied in detail.

4.1 Classical Statistical Description of Light

4.1.1 The Classical Description of an Electromagnetic Field

An electromagnetic field is described by the electric field $\underline{E}(\underline{x})$ and the magnetic induction $\underline{B}(\underline{x})$, which are vector functions of the space- time point $\underline{x} = \underline{r},t$. These fields satisfy Maxwell's equations, which, for a charge-free, homogeneous, and isotropic medium take the form,

$$\nabla \cdot \underline{E} = 0 \quad , \qquad \nabla \cdot \underline{B} = 0 \quad ,$$

$$\nabla \times \underline{E} = -\frac{\partial \underline{B}}{\partial t} \quad , \qquad \nabla \times \underline{B} = \frac{1}{c^2}\frac{\partial \underline{E}}{\partial t} \quad , \tag{4.1}$$

where c is the velocity of light in the medium. It follows from (4.1) that all components of $\underline{E}$ and $\underline{B}$ have to satisfy the wave equation

$$\left[\nabla^2 - \frac{1}{c^2}\frac{\partial^2}{\partial t^2}\right] v(\underline{x}) = 0 \quad , \tag{4.2}$$

where $v(\underline{x})$ represents any such component.

Equations (4.1) and (4.2) do not include the radiation source. Throughout this book, we shall not need to solve radiation problems. Normally we shall simply assume the existence of optical fields that have certain properties, study their propagation in some systems and concentrate most of our efforts on studying the process of their detection.

The complex analytic signal $V(\underline{x})$ (cf. Sec. 2.6.1) that corresponds to $v(\underline{x})$ also satifies a wave equation

$$\left(\nabla^2 - \frac{1}{c^2}\frac{\partial^2}{\partial t^2}\right) V(\underline{x}) = 0 \tag{4.3}$$

(this can be shown by using the definition of the CAS Sec. 2.6.1); its Fourier transform $V(\underline{r},\omega)$ satisfies the Helmholtz equation

$$\left[\nabla^2 + \left(\frac{\omega}{c}\right)^2\right] V(\underline{r},\omega) = 0 \quad . \tag{4.4}$$

One solution of (4.4) is the plane wave

$$V(\underline{r},\omega) \propto \exp(-j\underline{K}\cdot\underline{r}) \quad , \qquad |\underline{K}| = \frac{\omega}{c} \quad ,$$

which propagates in the direction of $\underline{K}$ with a wave length $\lambda = 2\pi/|\underline{K}|$.

Another more realistic solution is the Gaussian beam that propagates in the z direction (BOYD and GORDON [66])

$$V(\underline{r},\omega) \propto \frac{b}{(b^2+z^2)^{\frac{1}{2}}} \exp\left(-\frac{\omega}{c}\frac{b}{2}\frac{x^2+y^2}{b^2+z^2}\right) \exp\left\{-j\frac{\omega}{c}\left[b+z+\frac{z(x^2+y^2)}{2(b^2+z^2)}\right] -j\tan^{-1}\left(\frac{b-z}{b+z}\right)\right\} , \qquad (4.5)$$

where $\sqrt{bc/\omega}$ is the width of the beam at its waist (i.e. at $z = 0$).

A general solution of (4.3) is a weighted sum of a continuum of plane waves that have all possible frequencies and wave vectors such that $\omega = c|\underline{K}|$ (see, e.g., [440])

$$V(\underline{r},t) = \int V(\underline{r},\omega)\exp(j\omega t)d\omega$$

$$V(\underline{r},\omega) = \iint V(\ell_x,\ell_y,\omega)\exp\left[-j\frac{\omega}{c}(\ell_x x+\ell_y y+\ell_z z)\right] d\ell_x d\ell_y$$

$$\ell_z = (1-\ell_x^2-\ell_y^2)^{\frac{1}{2}} , \quad \ell_x^2+\ell_y^2 \le 1$$

$$= j(\ell_x^2+\ell_y^2-1)^{\frac{1}{2}} , \quad \ell_x^2+\ell_y^2 > 1 \qquad (4.6)$$

This angular-spectrum representation has found many useful applications in optics.

In a general optical system, we are usually interested in relating the optical field distribution at some plane (the image plane) to its distribution at another plane (the object plane). Between the two planes we may have apertures, lenses, filters, or other optical components. If we are interested in only one polarization, then we have to solve the wave equation (4.4) taking into account the proper boundary conditions in the system. If we are interested in the two polarizations, then two such equations have to be solved. In any case, because of the linearity of (4.4), the field at the image plane (coordinates $\underline{r}$) is related to the field at the object plane (coordinates $\underline{r}'$) by a general linear transformation

$$V(\underline{r},\omega) = \int h(\underline{r},\underline{r}';\omega)\, V'(\underline{r}';\omega)d\underline{r}' \quad . \qquad (4.7)$$

The function $h(\underline{r},\underline{r}';\omega)$ represents the image of a point object at position $\underline{r}'$ and is called the system's impulse response, point spread function psf, or Green's function at the frequency ω. The most important example of an optical system is the plane wave of light that illuminates an aperture in a screen (Fig. 4.1). The image plane is a far plane facing the screen.

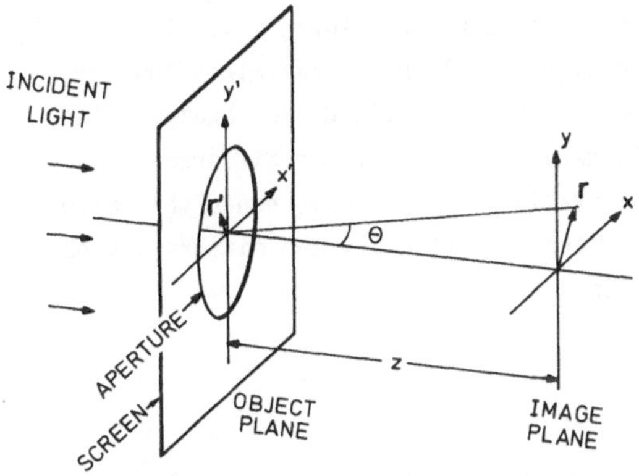

Fig.4.1. Geometry of a diffraction system

The Kirchhoff's theorem of diffraction is simply a solution to (4.4) under the boundary condition that the field vanishes at all points on the back surface of the screen except at the aperture where it equals the incident field. This gives in the far field under the Fresnel approximation,

$$h(\underline{r},\underline{r}',\omega) \propto \frac{\exp(-j\,\frac{\omega}{c}\,z)}{j\lambda z}\exp\left(-j\frac{\omega}{2cz}\,|\underline{r}-\underline{r}'|^2\right)\;, \tag{4.8}$$

and under the Fraunhofer approximation,

$$h(\underline{r},\underline{r}',\omega) \propto \frac{\exp(-j\,\frac{\omega}{c}\,z)}{j\lambda z}\exp\left(-j\frac{\omega}{2cz}\,r^2\right)\exp\left(j\,\frac{\omega}{cz}\,\underline{r}\cdot\underline{r}'\right)\;, \tag{4.9}$$

where z is the distance between the aperture and the image planes. Other examples of the impulse responses of more complicated systems can be found, for example, in the textbooks [180, 356, 359].

Systems that involve polarization can be described by the general linear transformation

$$V_i(\underline{r},\omega) = \sum_{j=1}^{2} \int h_{ij}(\underline{r},\underline{r}',\omega)V_j'(\underline{r}',\omega)d\underline{r}', \quad i = 1,2 \quad , \tag{4.10}$$

where h_{ij} is a 2 x 2 matrix function determined by the polarization properties of the components in the system (polarizers, rotators, wave plates, etc.). The impulse response matrix function $h_{ij}(\underline{r},\underline{r}',\omega)$ in general mixes the temporal, spatial, and polarization properties of the field. For example, the spatial response of the system is in general different for different temporal frequencies and different polarizations. If the components that alter the polarization in an optical system treat identically the different spectral components at all positions, then

$$h_{ij}(\underline{r},\underline{r}',\omega) = T_{ij}h(\underline{r},\underline{r}',\omega) \quad ,$$

where T_{ij} is a 2 x 2 matrix of complex numbers. This means that polarization effects are separate and independent of the spatio-temporal behaviour of the system.

For example, a system composed of a perfect polarizer is represented by the matrix

$$\begin{bmatrix} 1 & 0 \\ 0 & 0 \end{bmatrix} .$$

A retarder is represented by

$$\begin{bmatrix} 1 & 0 \\ 0 & \exp(-j\Delta) \end{bmatrix} ,$$

where Δ is the differential retardation and so on. Details on this subject can be found in [98, 356, 441].

Summarizing, an optical field is described by a complex analytic signal $V_j(\underline{r},t)$, j = 1,2 which has to satisfy the wave equations (4.3,4), and whose propagation through linear optical systems can most generally be described by the linear transformations (4.7,10), which describe the change of its temporal, spatial, and polarization dependence.

Now we turn to the statistical description of this signal.

4.1.2 The Statistical Description of Light

The statistical description of an electromagnetic field starts by declaring $\underline{E}(\underline{x})$ and $\underline{B}(\underline{x})$ to be vector stochastic fields, i.e., all their components $V_j(\underline{x})$ are stochastic functions of space and time.

The components of $\underline{E}$ and $\underline{B}$ are stochastic functions of time with a relatively very narrow band centered around a large central frequency. Therefore, the use of their complex analytic signal representation provides the simplifications discussed in Sec. 2.6 (see also [312]). Thus, from now on, $\underline{E}(\underline{x})$ and $\underline{B}(\underline{x})$ represent the complex analytic signals (CAS) of the real electric and magnetic fields, and $V(\underline{x})$ represents the CAS of any of their components. Hence, throughout the present chapter we shall be involved with one problem - the description of one or two statistically fluctuating complex analytic functions of space and time. This should therefore cover the temporal, spatial, and polarization fluctuations and their interdependence.

We have already discussed the different tools for describing complex stochastic functions (in Chap. 2); we should therefore be well prepared to use these tools straightforwardly.

The Joint Probability Densities and the Correlation Functions (Coherence Function)

Under the most general conditions, the components of $\underline{E}$ and $\underline{B}$ (say $\{V_j\}$) are described by the joint probability densities

$$P\left\{V_{j_1}(\underline{x}_1), V_{j_2}(\underline{x}_2), \ldots, V_{j_N}(\underline{x}_N)\right\} \quad N = 1,2, \ldots ,$$

and the set of correlations and cross-correlations between these components at different space-time points

$$G^{N,M}_{j_1, \ldots j_{N+M}}(\underline{x}_1, \ldots \underline{x}_{N+M}) = \left\langle V^*_{j_1}(\underline{x}_1) \ldots V^*_{j_N}(\underline{x}_N) \right. \\ \left. V_{j_{N+1}}(\underline{x}_{N+1}) \ldots V_{j_{N+M}}(\underline{x}_{N+M}) \right\rangle \quad . \tag{4.11}$$

We shall normally be interested in only one field (say $\underline{E}$) and in its two transverse components. Hence j in the above definitions is an index $j = 1,2$ that represents these two components.

Under the scalar approximation [186, 322], i.e., for optical phenomena that can approximately be described by one scalar function (or one polarization of the field), we drop the subscript j and have the probability densities,

$$P\{V(\underline{x}_1), \ldots V(\underline{x}_N)\} \quad N = 1,2, \ldots$$

and the correlation function,

$$G^{N,M}(\underline{x}_1, \ldots \underline{x}_{N+M}) = \langle V^*(\underline{x}_1) \ldots V^*(\underline{x}_N)V(\underline{x}_{N+1}) \ldots V(\underline{x}_{N+M})\rangle \quad . \tag{4.12}$$

If only one space point $\underline{r}$ is of interest, then we can also drop the position dependence and get

$$P\{V(t_1), \ldots V(t_N)\} \quad N = 1,2, \ldots$$

and

$$G^{N,M}(t_1, \ldots t_{N+M}) = \langle V^*(t_1) \ldots V^*(t_N)V(t_{N+1}) \ldots V(t_{N+M})\rangle \quad . \tag{4.13}$$

Similarly, if the statistics of the field at different positions at a certain time are of interest, we can drop the time dependence.

These probability densities and correlation functions satisfy the general properties discussed in Sec. 2.5.1. The stochastic process $V(\underline{x})$ can also be described by its probability functional [50, 518, 519]. We do not concern ourselves with the functional formulation in this book.

Most important of the above-defined correlation functions is the one whose order is (1,1),

$$G(\underline{x}_1,\underline{x}_2) = G^{1,1}(\underline{x}_1,\underline{x}_2) \quad .$$

This is known as the *mutual coherence function* [508]; as we shall see, it is of great importance in understanding many phenomena such as diffraction, interference, and imagery.

An optical field is said to be *stationary* if its CAS is a stationary stochastic process, i.e., if the probability distribution and correlation functions are time-shift invariant. In this case, the mutual coherence functions depends on only the time difference

$$G(\underline{x}_1,\underline{x}_2) = G(\underline{r}_1,\underline{r}_2,t_2-t_1) \quad .$$

On the other hand, the field is called *homogeneous* if the probability distributions and correlation functions are invariant to shifts in the space-coordinate system. Although homogeneity is, in general, less likely to occur, it may be satisfied over a finite region of space.

Description in the Temporal-Frequency Domain

In order to make use of the full machinary of stochastic theory, we consider the spectral density of the optical field, $V(\underline{r},\omega)$. This is also a complex stochastic function (or generalized function), which is described by a set of correlation functions (cross-spectral densities) (cf. [316])

$$G^{N,M}(\underline{r}_1,\omega_1; \; \dots \; ; \underline{r}_{N+M},\omega_{N+M}) = \langle V^*(\underline{r}_1,\omega_1) \; \dots \; V(\underline{r}_{N+M},\omega_{N+M})\rangle \quad . \tag{4.14}$$

As we remember from Sec. 2.5.1, if the field is stationary, the cross-spectral densities vanish unless

$$\sum_{i=1}^{N} \omega_i = \sum_{j=N+1}^{N+M} \omega_j \quad .$$

In particular,

$$G(\underline{r}_1,\omega_1; \underline{r}_2,\omega_2) = G(\underline{r}_1, \underline{r}_2,\omega_1)\delta(\omega_1-\omega_2) \quad , \tag{4.15}$$

where $G(\underline{r}_1,\underline{r}_2,\omega)$ is the cross-power spectrum related to the mutual coherence function by the Fourier transform

$$G(\underline{r}_1,\underline{r}_2,t_2-t_1) = \int_0^\infty G(\underline{r}_1,\underline{r}_2,\omega)e^{j\omega(t_2-t_1)}d\omega \quad . \tag{4.16}$$

The function $G(\underline{r},\underline{r},\omega)$ represents the optical power spectrum of light at the point $\underline{r}$ and is the quantity of fundamental importance in spectroscopy.

It has also proved useful in some applications to define correlation functions in the *angular-frequency domain*. In this case, emphasis is given to the stochastic process $V(\ell_x,\ell_y,\omega)$ and its correlations [323].

Normalization of the Coherence Functions. The Degree of Coherence of Light

The normalized mutual coherence function

$$g(\underline{x}_1,\underline{x}_2) = \frac{G(\underline{x}_1,\underline{x}_2)}{[G(\underline{x}_1,\underline{x}_1)]^{\frac{1}{2}}[G(\underline{x}_2,\underline{x}_2)]^{\frac{1}{2}}} \qquad (4.17)$$

gives a measure of the degree of coherence or correlatedness of light fluctuations at $\underline{x}_1$ and $\underline{x}_2$.
This can be seen by noting that if $V(\underline{x}_1)$ and $V(\underline{x}_2)$ are uncorrelated, then

$$|g(\underline{x}_1,\underline{x}_2)| = 0 \quad .$$

On the other hand, if $V(\underline{x}_1)$ and $V(\underline{x}_2)$ are deterministic, we get

$$|g(\underline{x}_1,\underline{x}_2)| = 1 \quad .$$

If $V(x) = c\, f(x)$, where $f(x)$ is deterministic but c is a complex random variable, then $|g(\underline{x}_1,\underline{x}_2)| = 1$.
Also, if $V(\underline{x}_1)$ and $V(\underline{x}_2)$ are random, but they are deterministically related by a linear relation (e.g., $V(\underline{x}_2) = aV(\underline{x}_1)$, where a is deterministic), then $|g(\underline{x}_1,\underline{x}_2)| = 1$.

Moreover, by using (2.90) it follows that

$$0 \le |g(\underline{x}_1,\underline{x}_2)| \le 1 \quad . \qquad (4.18)$$

Because of the above considerations, the function $g(\underline{x}_1,\underline{x}_2)$ is called the *"complex degree of coherence"* [65, 318].

The cross-spectral densities can also be similarly normalized, e.g.,

$$g(\underline{r}_1,\underline{r}_2,\omega) = \frac{G(\underline{r}_1,\underline{r}_2,\omega)}{[G(\underline{x}_1,\underline{x}_1)]^{\frac{1}{2}}[G(\underline{x}_2,\underline{x}_2)]^{\frac{1}{2}}} \qquad (4.19)$$

is the *normalized cross-power spectrum* and is related to the complex degree of coherence by a Fourier transform

$$g(\underline{r}_1,\underline{r}_2,\tau) = \int_0^{\infty} g(\underline{r}_1,\underline{r}_2,\omega)e^{j\omega\tau}d\omega \quad . \tag{4.20}$$

Note that $|g(\underline{r}_1,\underline{r}_2,\omega)|$ is not bounded between zero and one.

In the temporal-frequency domain, if we limit ourselves to the component of light that has a frequency ω then the degree of cross-correlation between fluctuations at $\underline{r}_1$ and at $\underline{r}_2$ is not given by $g(\underline{r}_1,\underline{r}_2,\omega)$ as defined in (4.19) but rather by

$$g_s(\underline{r}_1,\underline{r}_2,\omega) = \frac{G(\underline{r}_1,\underline{r}_2,\omega)}{[G(\underline{r}_1,\underline{r}_1,\omega)]^{\frac{1}{2}}[G(\underline{r}_2,\underline{r}_2,\omega)]^{\frac{1}{2}}} \quad , \tag{4.21}$$

which is called the *complex degree of spectral coherence* [321]. Note that $g_s(\underline{r}_1,\underline{r}_2,\omega)$ has an absolute value that is bounded between zero and one. Note also that $g_s(\underline{r}_1,\underline{r}_2,\omega)$ is not related to $g(\underline{r}_1,\underline{r}_2,\tau)$ by a Fourier transform. From (4.21) and (4.19) it follows that

$$g_s(\underline{r}_1,\underline{r}_2,\omega) = \frac{g(\underline{r}_1,\underline{r}_2,\omega)}{[g(\underline{r}_1,\underline{r}_1,\omega)]^{\frac{1}{2}}[g(\underline{r}_2,\underline{r}_2,\omega)]^{\frac{1}{2}}} \quad . \tag{4.22}$$

The concept of spectral coherence is useful in understanding problems of radiometry [85, 513, 514] and can be of use in understanding light-scattering phenomena, which are described most naturally in the $\underline{r}$-ω domain.

Complete Coherence

An optical field is said to be completely coherent (in the second order) if its coherence function (of order 1,1) factorizes in the form

$$G(\underline{x}_1,\underline{x}_2) = \overset{*}{U}(\underline{x}_1)U(\underline{x}_2) \quad , \quad \forall \underline{x}_1,\underline{x}_2 \quad . \tag{4.23}$$

It follows from this factorization that the complex degree of coherence attains its maximum value

$$|g(\underline{x}_1,\underline{x}_2)| = 1 \quad \forall \underline{x}_1,\underline{x}_2 \quad . \tag{4.24}$$

It can also be shown that (4.24) is a sufficient condition for (4.23) to hold. Therefore (4.24) can also be taken as a definition of second-order complete coherence.

Complete Coherence and Monochromaticity

A deterministic field is completely coherent (it satisfies (4.23) and (4.24) but it is not necessarily monochromatic. An example is a plane wave whose amplitude is pulsed in time (such a field is obviously not monochromatic, because it contains a central frequency and sidebands, but it is completely coherent because it is deterministic).

However, *if a completely coherent field is stationary it must be monochromatic.*

Formally, if a stationary field satisfies the condition $|g(\underline{r}_1,\underline{r}_2,\tau)| = 1$ for all points $\underline{r}_1,\underline{r}_2$ of a finite region $\mathcal{D}$ and for all time delays, and if $\underline{r}_0$ does not necessarily belong to $\mathcal{D}$, then

$$G(\underline{r}_1,\underline{r}_2,\tau) = \overset{*}{U}(\underline{r}_1)U(\underline{r}_2)\exp(j\omega_0\tau)$$

$$g(\underline{r}_1,\underline{r}_2,\tau) = \exp(j\omega_0\tau) \ , \quad g(\underline{r}_1,\underline{r}_2,\omega) = \delta(\omega-\omega_0) \tag{4.25}$$

$$g(\underline{r}_1,r_0,\tau) = g(\underline{r}_1,\underline{r}_0,0)\exp(j\omega_0\tau) \ , \quad g(\underline{r}_1,\underline{r}_0,\omega) = g(\underline{r}_1,\underline{r}_0,0)\delta(\omega-\omega_0) \ ,$$

i.e., the field is monochromatic. This follows from the fact that the only solution of the equation

$$G(t_1-t_2) = \overset{*}{U}(t_1)U(t_2)$$

is the harmonic function. For a formal mathematical proof see [334] (cf. also [339]).

Complete Incoherence

We can of course take the other extreme and call the field that satisfies

$$|G(\underline{r}_1,\underline{r}_2,\tau)| = 0 \qquad \forall\underline{r}_1,\underline{r}_2,\tau$$

a completely incoherent field. But such a field cannot exist in free space. Moreover, if $|G| = 0$ for all pairs of points on some continuous closed surface, the surface does not radiate. This may become obvious by noting that this implies that

$$G(\underline{r},\underline{r},0) = \langle|V(\underline{r},0)|^2\rangle = 0 \quad ,$$

i.e., that the average radiance of the field at all points in space is zero.

We should therefore exclude the point $\underline{r}_1 = \underline{r}_2$ and $\tau = 0$ from the definition and define a field that is completely incoherent in space by

$$|g(\underline{r}_1,\underline{r}_2,\tau)| = |g(\underline{r}_1,\underline{r}_1,\tau)|\delta(\underline{r}_1-\underline{r}_2) \quad ,$$

where δ is a very narrow function that vanishes unless its argument is zero (a delta function) (cf. BERAN and PARRENT [51] Sec. 4.4).

Normalization of Higher-Order Coherence Functions.
Degree of Higher-Order Coherence

The coherence function of order (N,N) can be normalized in the form [333, 372]

$$g^{N,N}(\underline{x}_1, \dots \underline{x}_{2N}) = \frac{G^{N,N}(\underline{x}_1, \dots \underline{x}_{2N})}{\left[\prod_{j=1}^{2N} G^{N,N}(\underline{x}_j, \dots \underline{x}_j)\right]^{1/2N}} \quad . \tag{4.26}$$

Other normalization schemes are certainly possible [168, 170, 172, 268] and all normalizations reproduce (4.17) for the order (1,1).

The above normalization compares the (N,N) correlation of the variables $V(\underline{x}_1)$, ... $V(\underline{x}_{2N})$ to a geometrical average of the (N,N) correlations of these variables when each is taken alone. Therefore,

$$g^{N,N}(\underline{x},\underline{x}, \dots \underline{x}) = 1 \quad .$$

The normalization (4.26) guarantees that

$$0 \le |g^{N,N}| \le 1 \tag{4.27}$$

(see Sec. 2.5.1).

Complete Coherence of Higher Order

Complete coherence of order (N,N) can be defined by the factorization property

$$G^{N,N}(\underline{x}_1, \dots \underline{x}_{2N}) = \prod_{j=1}^{N} U^*(\underline{x}_j) \prod_{i=N+1}^{2N} U(\underline{x}_i) \quad , \tag{4.28}$$

which is a necessary and sufficient condition for

$$|g^{N,N}(\underline{x}_1, \dots \underline{x}_{2N})| = 1 \quad . \tag{4.29}$$

The necessity of (4.28) is obvious. Its sufficiency was proved by MEHTA [334] who also showed that complete coherence of any order (N,N) implies complete coherence of all orders (M,M)!

Factorization of the Temporal and Spatial Dependence of the Coherence Function: Cross-Spectral Purity

An optical field is said to be cross-spectrally pure [308] if the power spectral profiles at any pair of points $\underline{r}_1$ and $\underline{r}_2$ are identical and if the cross-power spectral profile at $\underline{r}_1$ and $\underline{r}_2$ is also identical to them except for a phase factor that is proportional to the frequency ω. Mathematically, this can be put in the form

$$g(\underline{r}_1, \underline{r}_1, \omega) = g(\underline{r}_2, \underline{r}_2, \omega) = g(\omega) \tag{4.30}$$

and

$$g(\underline{r}_1, \underline{r}_2, \omega) = g(\underline{r}_1, \underline{r}_2, \tau_0) g(\omega) \exp(-j\omega\tau_0) \quad , \tag{4.31}$$

where τ_0 is arbitrary.

By using (4.22), we see that the definition (4.31) implies that

$$g_s(\underline{r}_1, \underline{r}_2, \omega) = g(\underline{r}_1, \underline{r}_2, \tau_0) \exp(-j\omega\tau_0) \quad , \tag{4.32}$$

i.e., the absolute value of the degree of spectral coherence is the same for all frequency components. All frequency components have the same degree of cross-correlation between fluctuations at $\underline{r}_1$ and those at $\underline{r}_2$ [321]. By taking the Fourier transform of (4.30) and (4.31), we obtain

$$g(\underline{r}_1,\underline{r}_1,\tau) = g(\underline{r}_2,\underline{r}_2,\tau) = g(\tau) \quad , \tag{4.33}$$

$$g(\underline{r}_1,\underline{r}_2,\tau) = g(\underline{r}_1,\underline{r}_2,\tau_0)g(\tau-\tau_0) \quad , \tag{4.34}$$

where $g(\tau)$ and $g(\omega)$ are Fourier-transform pair. Eqs. (4.33) and (4.34) show that, for cross-spectrally pure light, the complex degree of coherence factorizes into the product of a function of $\underline{r}_1$ and $\underline{r}_2$ and another of the time delay τ. This can be written in the form

$$g(\underline{r}_1,\underline{r}_2,\tau) = g(\underline{r}_1,\underline{r}_2)g(\tau) \quad .$$

It will be shown later, (Sec. 4.1.3) that as cross-spectrally pure light propagates, it retains its power-spectral profile under certain geometrical conditions. It will also be shown that cross-spectral purity of a propagating field is preserved under some geometrical limits. An example of a cross-spectrally pure field is light in the vicinity of a radiating surface whose elements independently radiate light waves with identical spectra. As light propagates, its temporal and spatial dependence are mixed and it may not continue to be spectrally pure. We shall discuss this point further in Sec. 4.1.3.

Coherence Time and Coherence Area

Consider a stationary and cross-spectrally pure optical field whose complex degree of coherence is

$$g(\underline{r}_1,\underline{r}_2,\tau) = g(\underline{r}_1,\underline{r}_2)g(\tau) \quad .$$

The *coherence time* is the width of the function $g(\tau)$, i.e., the time delay τ_c at which $g(\tau)$ drops significantly (Fig.4.2). There is no unique way of defining the width of a function. We shall arbitrarily choose to define

$$\tau_c = \int_{-\infty}^{\infty} |g(\tau)|^2 d\tau \quad . \tag{4.35}$$

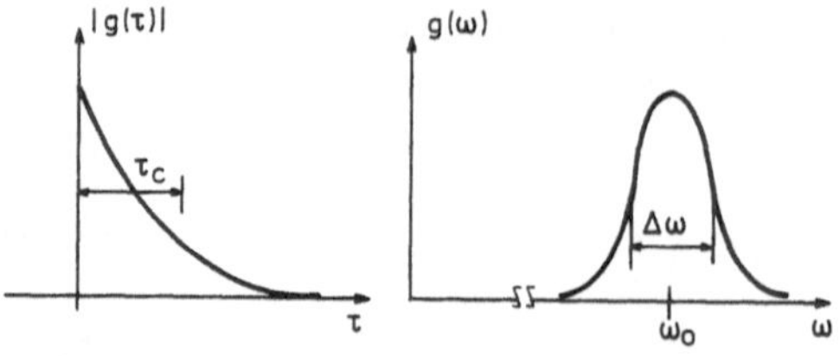

Fig.4.2. Coherence time τ_0 and bandwidth $\Delta\nu = \Delta\omega/2\pi$ of an optical signal

For example, for an exponential correlation function

$$g(\tau) = \exp(-\Gamma|\tau|) \quad ,$$

we get

$$\tau_c = 1/\Gamma \quad .$$

In the frequency domain, the *bandwidth* is the width of the spectral distribution which is centered around the central frequency,

$$\Delta\nu = \int_{-\infty}^{\infty} |g(\omega)|^2 \frac{d\omega}{2\pi} \quad , \qquad \Delta\omega = 2\pi\Delta\nu \quad .$$

Because $g(\omega)$ and $g(\tau)$ are Fourier-transform pairs, it follows from Parceval's theorem that $\Delta\nu^{-1} = \tau_c$.

If the field is not cross-spectrally pure, then the coherence time would be a function of position. The *coherence area* can be defined by fixing $\underline{r}_1$ and observing $g(\underline{r}_1,\underline{r}_2)$ as a function of $\underline{r}_2$ in a given plane. The area within which the function $g(\underline{r}_1,\underline{r}_2)$ is significant and outside of which it is small is the area of coherence. Mathematically, we can use an expression similar to (4.35)

$$A_c - \iint_{-\infty}^{\infty} |g(\underline{r}_1,\underline{r}_2)|^2 d\underline{r}_2 \quad . \tag{4.36}$$

The area of coherence is then a function of $\underline{r}_1$ unless the field is homogeneous.

The Light Intensity: A Real Stochastic Process

The electromagnetic flux density at a space-time point $\underline{x}$ is given by the Poynting vector $\underline{E}(\underline{x}) \times \underline{H}(\underline{x})$ where $\underline{H}$ is the magnetic field (proportional to the magnetic induction). In a nonconducting medium H is proportional to E; the energy flux density is then proportional to $|\underline{E}|^2$. The energy flux density carried by one polarization (component of $\underline{E}$) is given by $v^2(t)$ where v represents this component.

For an optical field, $v(\underline{r},t)$ oscillates at the frequency of light and v^2 oscillates at double that frequency. Because no detector can follow such a fast rate, we must find the time average over a time interval longer than the period of oscillation, but much shorter than the time within which the envelope of v(t) itself varies (Fig.4.3). We shall see in Sec. 5.1 that a photodetector detects this average, automatically. The result of the averaging is what we shall call the light intensity (or instantaneous light intensity).

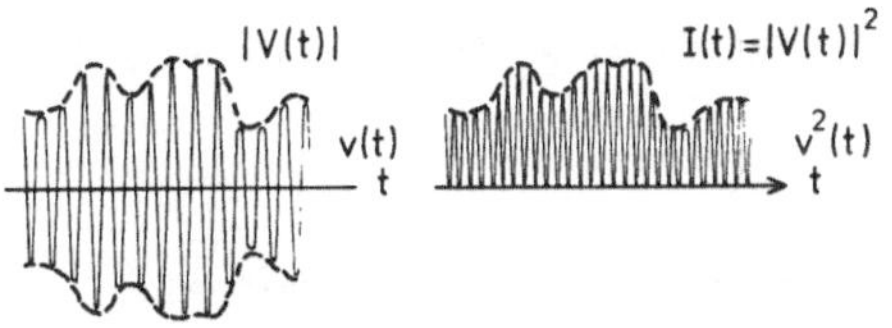

Fig.4.3. Instantaneous intensity

$$I(\underline{x}) = |V(\underline{x})|^2 = |\mathcal{V}(\underline{x})|^2 \quad , \tag{4.37}$$

where V is the CAS and $\mathcal{V}$ is the complex envelope of v (see Sec. 2.6.1).

The light intensity is a real and nonnegative stochastic process. It can therefore be described by the probability densities

$$P\{I(\underline{x}_1), \ldots I(\underline{x}_N)\} = P\{|V(\underline{x}_1)|^2, \ldots, |V(\underline{x}_N)|^2\} \quad ,$$

which can be determined from $P\{V(\underline{x}_1), \ldots V(\underline{x}_N)\}$ by using the techniques of transformation of random variables. The intensity correlation functions are defined by

$$G_I^{(N)}(\underline{x}_1, \ldots \underline{x}_N) = \langle I(\underline{x}_1) \ldots I(\underline{x}_N)\rangle \tag{4.38}$$

and are related to the field correlation functions by

$$G_I^{(N)}(\underline{x}_1, \ldots \underline{x}_N) = G_V^{N,N}(\underline{x}_1, \ldots, \underline{x}_N, \underline{x}_1, \ldots \underline{x}_N) \quad . \tag{4.39}$$

The intensity correlation function can also be normalized. We shall adopt the normalization

$$g_I^{(N)}(\underline{x}_1, \ldots \underline{x}_N) = \frac{G_I^{(N)}(\underline{x}_1, \ldots \underline{x}_N)}{\langle I(\underline{x}_1)\rangle\langle I(\underline{x}_2)\rangle \ldots \langle I(\underline{x}_N)\rangle} , \tag{4.40}$$

which is different from (4.26). It satisfies however the inequality

$$0 \leq |g_I^{(N)}| \leq 1 \quad . \tag{4.41}$$

The most important function in the hierarchy (4.38) is the average intensity $G_I^{(1)}(\underline{x}) = \langle I(\underline{x})\rangle$. Then comes the second-order intensity-correlation function $G_I^{(2)}(\underline{x}_1,\underline{x}_2)$ (which we shall also write as $G_I(\underline{x}_1,\underline{x}_2)$ or $G^{(2)}(\underline{x}_1,\underline{x}_2)$). This can be measured by correlating the intensity at $\underline{x}_1$ to the intensity at $\underline{x}_2$.

4.1.3 Statistical Description of Light Propagation in a Linear Optical System

So far, our statistical description of light has been based on making the field's CAS, $V(\underline{x})$, and its intensity $I(\underline{x}) = |V(\underline{x})|^2$ stochastic functions of time and space and defining their probability densities and a hierarchy of statistical averages. But every statistical realization of $V(\underline{x})$ must satisfy the wave equation and the boundary conditons set by the optical components in the system, as indicated in Sec. 4.1.1. It also has to satisfy the linear transformation (4.7). We may ask, how would that affect the correlation functions that we defined at one or a set of space-time points?

Let us start by considering the effect of the wave equation on the *mutual coherence function*. By multiplying the wave equation (4.3) by $V^*(\underline{x}_2,t_2)$ and taking the ensemble average, we get

$$\left(\nabla_j^2 - \frac{1}{c^2}\frac{\partial^2}{\partial t_j^2}\right) G(\underline{r}_1,t_1;\underline{r}_2,t_2) = 0 , \qquad j = 1,2, \tag{4.42}$$

where ∇_j^2 is the Laplacian operator with respect to $\underline{r}_j$. For a stationary field, we simply have

$$\left(\nabla_j^2 - \frac{1}{c^2}\frac{\partial^2}{\partial \tau^2}\right) G(\underline{r}_1,\underline{r}_2,\tau) = 0 , \quad j = 1,2 \quad . \tag{4.43}$$

This means that the mutual-coherence function itself has to satisfy two partial differential equations that govern its time-space variation.

Only in few cases has (4.43) been directly solved. One solution is

$$G(\underline{r}_1,\underline{r}_2,\tau) \propto J_0\left\{\frac{\left[(x_2-x_1)^2+(y_2-y_1)^2\right]^{\frac{1}{2}}}{\alpha}\right\}\exp\left\{-j\left[\left(\frac{\omega_0}{c}\right)^2-\frac{1}{\alpha^2}\right]^{\frac{1}{2}}(z_2-z_1)+j\omega_0\tau\right\}. \quad (4.44)$$

This is a homogeneous plane wave that propagates in the z direction and that has a coherence area $A_c = \alpha^2$ in the transverse plane.

A much more direct approach to the propagation problem is to solve first the wave equation in terms of one realization of the fluctuating field and then to form the statistical averages. For example, we know from (4.6) that the field

$$V(\underline{r},t) = \int d\omega \exp(j\omega t) \iint d\ell_x d\ell_y \, V(\ell_x,\ell_y,\omega)\exp\left[-j\,\frac{\omega}{c}(\ell_x x+\ell_y y+\ell_z z)\right]$$

$$\ell_z = (1-\ell_x^2-\ell_y^2)^{\frac{1}{2}} \qquad \ell_x^2+\ell_y^2 \le 1 \qquad (4.45)$$

$$= j(\ell_x^2+\ell_y^2-1)^{\frac{1}{2}} \qquad \ell_x^2+\ell_y^2 > 1$$

satisfies the wave equation. By forming the correlation function, we get

$$G(\underline{r}_1,\underline{r}_2,\tau) = \int d\omega \exp(j\omega\tau) \int\cdots\int d\ell_x\, d\ell_y\, d\ell_x'\, d\ell_y'\, G(\ell_x,\ell_y;\ell_x',\ell_y',\omega)$$
$$\exp\left[j\,\frac{\omega}{c}(\ell_x x_1+\ell_y y_1+\ell_z z_1-\ell_x' x_2-\ell_y' y_2-\ell_z' z_2)\right] \quad , \qquad (4.46)$$

which automatically satisfies (4.43) for all $\underline{r}_1,\underline{r}_2,\tau$. For a general optical system described by (4.7) the cross-power spectra at the image and object planes are related by the transformation

$$G_{IM}(\underline{r}_1,\underline{r}_2,\omega) = \iint G_{OB}(\underline{r}_1',\underline{r}_2',\omega)h^*(\underline{r}_1,\underline{r}_1',\omega)h(\underline{r}_2,\underline{r}_2',\omega)d\underline{r}_1'd\underline{r}_2' \quad . \qquad (4.47)$$

Some special cases of (4.47) are of importance.

Diffraction of Spatially Incoherent Light

For example if the object is spatially incoherent, i.e.,

$$G_{OB}(\underline{r}_1',\underline{r}_2',\omega) = G_{OB}(\underline{r}_1',\underline{r}_1',\omega)\delta(\underline{r}_1'-\underline{r}_2') \quad ,$$

then (4.47) reduces to

$$G_{IM}(\underline{r}_1,\underline{r}_2,\omega) = \int G_{OB}(\underline{r}',\underline{r}',\omega)h^*(\underline{r}_1,\underline{r}',\omega)h(\underline{r}_2,\underline{r}',\omega)d\underline{r}' \quad . \tag{4.48}$$

It is therefore apparent that the light in the image plane is not completely incoherent. The field must have gained some coherence by the mere fact of propagation. This can be visualized by noting that each point of the object contributes to light in a spot in the image plane (as large as the width of the psf of the system). Therefore, a portion of the light which illuminates two points within such a spot comes from the same point in the object. Hence, it is not surprising that these two points are partially correlated.

Pursuing the mathematical formulas, we also see that if *the object is a point source*

$$G_{OB}(\underline{r}',\underline{r}',\omega) \propto \delta(\underline{r}'-\underline{r}_0')$$

the mutual coherence function in the image plane (4.47) factorizes and the image light is also spatially completely coherent. Light radiated by a point source is spatially coherent. On the other hand, if *the incoherent object has a uniform intensity distribution* within an area A, then

$$G_{IM}(\underline{r}_1,\underline{r}_2,\omega) \propto \int_A h^*(\underline{r}_1,\underline{r}',\omega)h(\underline{r}_2,\underline{r}',\omega)d^2\underline{r}' \quad . \tag{4.49}$$

Let us illustrate this by an example, *diffraction of incoherent homogeneous light that illuminates a circular aperture* of radius a. Substituting from the Fraunhofer formula (4.9) in (4.49) and normalizing, we get

$$g_{IM}(\underline{r}_1,\underline{r}_2,\omega) = J\left(\frac{\omega a}{cz}\,|\underline{r}_1-\underline{r}_2|\right)\exp\left[j\,\frac{\omega}{2cz}\left(r_1^2-r_2^2\right)\right] \quad ,$$

$$J(\theta) = 2J_1(\theta)/\theta \quad . \tag{4.50}$$

This gives for $\Delta\omega|\tau| \ll 1$ (i.e., for light with a small bandwidth),

$$g_{IM}(\underline{r}_1,\underline{r}_2,\tau) = J\left(\frac{\omega_0 a}{cz}|\underline{r}_1-\underline{r}_2|\right)\exp\left[j\ \frac{\omega_0}{2cz}\left(r_1^2-r_2^2\right)+j\omega_0\tau\right] \quad . \tag{4.51}$$

This equation is known as the *Van Cittert-Zernike theorem*. The behaviour of $|g_{IM}|$ is shown in Fig.4.4, which gives us a good example of a spatial co-

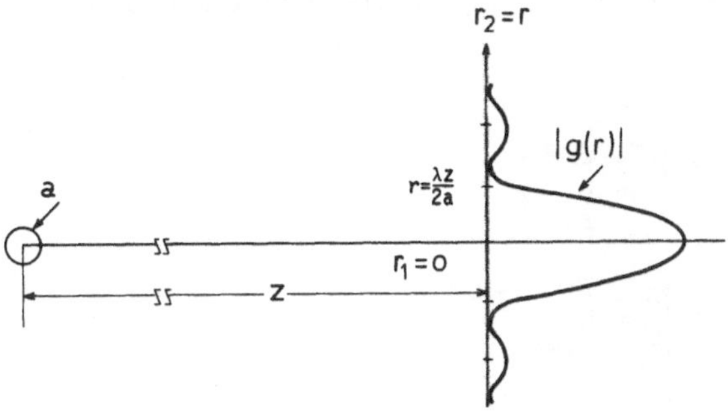

Fig.4.4. Degree of coherence of light due to an incoherent uniform circular source

herence function. The area of coherence of this field is obtained from (4.36)

$$A_c = \lambda^2 z^2/\pi a^2 \quad . \tag{4.52}$$

The mutual coherence function of light radiated by a star should be described by (4.51); it is obvious from Fig.4.4 that the angular diameter of the star can be easily determined from a measurement of the mutual coherence function.

The average intensity at the image plane due to an incoherent object is linearly related to the average intensity distribution of the object by

$$\langle I(\underline{r})\rangle = \int \langle I(\underline{r}')\rangle\, S(\underline{r},\underline{r}')d^2\underline{r}' \quad , \quad S(\underline{r},\underline{r}') = |h(\underline{r},\underline{r}')|^2 \quad , \tag{4.53}$$

where the frequency dependence has been dropped.

Summarizing, we say that the fluctuating fields at the image and object planes of an optical system are related by a linear transformation (4.7).

The average light intensities at the image and object plane are related linearly by (4.53) only if the object radiates incoherently. If the object's illumination is partially coherent, then we encounter problems with non-linearities when average intensities are measured.

Diffraction of Partially Coherent Light from Two Pinholes. Young's Interference Experiment

Another very interesting example of an optical system is Young's interference arrrangement (Fig.4.5). Partially coherent light illuminates a screen with

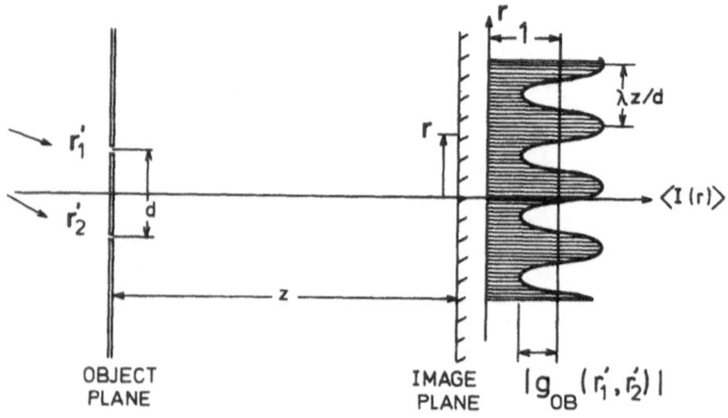

Fig.4.5. Young's interference experiment

two pinholes (whose areas are much smaller than the coherence area of the light at their positions). The intensity of the diffracted light at a far image plane is observed. To find an expression for this intensity distribution, we use the general linear transformation (4.47). Here the double integral reduces to a double sum of the contribution of the two holes; we then get,

$$G_{IM}(\underline{r},\underline{r},\omega) = G_{OB}(\underline{r}_1',\underline{r}_1',\omega)|h(\underline{r},\underline{r}_1',\omega)|^2 + G_{OB}(\underline{r}_2',\underline{r}_2',\omega)|h(\underline{r},\underline{r}_2',\omega)|^2$$

$$+ [G_{OB}(\underline{r}_1',\underline{r}_2',\omega)h^*(\underline{r},\underline{r}_1',\omega)h(\underline{r},\underline{r}_2',\omega) + \text{c.c.}] \quad ,$$

where the first two terms represent the contribution of each hole had the other been closed, and the last two terms represent the effect of their interference. If we use the Fresnel formula for h, (4.8), we get

$$G_{IM}(\underline{r},\underline{r},\omega) \propto G_{OB}(\underline{r}_1',\underline{r}_1',\omega) + G_{OB}(\underline{r}_2',\underline{r}_2',\omega)$$
$$+ 2\,\mathrm{Re}\left|G_{OB}(\underline{r}_1',\underline{r}_2',\omega)\exp(-j\omega\tau_0)\right| \quad ,$$

where

$$\tau_0 = \frac{1}{2cz}\left(|\underline{r}-\underline{r}_2'|^2 - |\underline{r}-\underline{r}_1'|^2\right) \quad .$$

Assuming for simplicity that the power spectra of light at the two pinholes $\underline{r}_1'$ and $\underline{r}_2'$ are identical and taking $r_1' = d/2$ and $r_2' = -d/2$, we get

$$G_{IM}(\underline{r},\underline{r},\omega) \propto 2\,G_{OB}(\underline{r}_i',\underline{r}_i',\omega)\left[1 + \mathrm{Re}\left|g_s(\underline{r}_1',\underline{r}_2',\omega)\exp(-j\omega\tau_0)\right|\right] \quad i = 1,2,$$

where g_s is the degree of spectral coherence and $\tau_0 = (d/cz)r$. The function $G_{IM}(\underline{r},\underline{r},\omega)$ can be observed at the image plane by simply using a filter that passes the frequency ω only. The observed pattern

$$G_{IM}(\underline{r},\underline{r},\omega) \propto 1 + |g_s| \cos\left(\measuredangle g_s - \frac{\omega d}{cz}\,r\right)$$

is a set of fringes with modulation depth or *visibility* (at the frequency ω) equal to the absolute value of the degree of spectral coherence.

It should be noted that the power spectrum at the image point $\underline{r}$ is equal to the power spectrum at the object points $\underline{r}_1',\underline{r}_2'$ only if $g_s\exp(-j\omega\tau_0) =$ constant, i.e., when the field is cross-spectrally pure (cf. (4.32)). Otherwise, propagation and interference modify the spectrum.

Now we examine the average intensity at the image plane.

By transforming to the time domain and putting $\tau = 0$, we get

$$\langle I(\underline{r})\rangle_{IM} \propto \langle I(\underline{r}_1')\rangle_{OB} + \langle I(\underline{r}_2')\rangle_{OB} + 2\,\mathrm{Re}\left|G_{OB}(\underline{r}_1',\underline{r}_2',\tau_0)\right| \quad .$$

Assuming again that the intensities at the holes are equal and taking $r_1' = d/2$ and $r_2' = -d/2$, we get

$$\langle I(\underline{r})\rangle_{IM} \propto 1 + \mathrm{Re}\left|g_{OB}(\underline{r}_1',\underline{r}_2',\tau_0)\right| \quad , \quad \tau_0 = \left(\frac{d}{cz}\right)r \quad .$$

This shows that the intensity distribution of the interference pattern depends directly on the complex degree of coherence of the incident field. In order to see this dependence more clearly, let us take an example of cross-spectrally pure object field with a Lorentzian spectrum

$$g_{OB}(\underline{r}_1',\underline{r}_2',\tau) = g_{OB}(\underline{r}_1',\underline{r}_2')\ \exp(-\Gamma|\tau|)\exp(j\omega_0\tau) \quad .$$

This gives

$$\langle I(r)\rangle_{IM} \propto 1 + \exp\left(-\frac{\Gamma d}{cz}|\underline{r}|\right)|g_{OB}(\underline{r}_1',\underline{r}_2')|\cos\left[\measuredangle g_{OB}(\underline{r}_1',\underline{r}_2') - \frac{\omega_0 d}{cz}\, r\right] \quad .$$

As a function of r, this distribution contains a constant term plus a sinusoidally oscillating function. Its decaying amplitude $|g_{OB}|\exp[-(\Gamma d/cz)|\underline{r}|]$ is proportional to the absolute value of the object's complex degree of coherence and has a decay distance $cz/\Gamma d$. Its phase is determined by the phase of the complex degree of coherence and its frequency is $2\pi cz/d\omega_0 = \lambda_0 z/d$. If the field has a narrow bandwidth ($\Gamma \ll \omega_0$), then the distance between two consecutive fringes is much smaller than the decay distance. Therefore, if we choose our geometry in such a way that details of the fringes are observed, the decay of the fringe amplitude would be very slow.

This provides us with a method of measuring the complex degree of coherence by observing the visibility of these interference fringes and the exact position of the first peak. A review of the methods of measurement of the complex degree of coherence can be found in [146].

Effect of Propagation on the Cross-Spectral Purity

Assuming that light in the object plane of an optical system is cross-spectrally pure, i.e.,

$$G_{OB}(\underline{r}_1',\underline{r}_2',\omega) = G_{OB}(\underline{r}_1',\underline{r}_2')\ g_{OB}(\omega) \quad ,$$

and using (4.47), we immediately see that the light in the image plane is not necessarily cross-spectrally pure,

$$G_{IM}(\underline{r}_1,\underline{r}_2,\omega) = g_{OB}(\omega) \iint G_{OB}(\underline{r}_1',\underline{r}_2')h^*(\underline{r}_1,\underline{r}_1',\omega)\, h(\underline{r}_2,\underline{r}_2',\omega)d\underline{r}_1'd\underline{r}_2' \quad . \tag{4.54}$$

If $h^*(\underline{r}_1,\underline{r}_1',\omega)h(\underline{r}_2,\underline{r}_2',\omega)$ is approximately independent of ω (except for a phase factor $\exp(j\omega\tau_0)$) for all points $\underline{r}'$ that belong to the object and for all points $\underline{r}$ belonging to the region of interest in the image, then the image field is cross-spectrally pure in this region.

To illustrate this, let us take again the example of diffraction. From (4.54) and (4.8),

$$h^*(\underline{r}_1,\underline{r}_1',\omega)h(\underline{r}_2,\underline{r}_2',\omega) \propto \exp\left[j\,\frac{\omega}{2cz}\left(|\underline{r}_1-\underline{r}_1'|^2-|\underline{r}_2-\underline{r}_2'|^2\right)\right] .$$

The frequency ω in this expression can be replaced by the central frequency ω_0, under the condition that

$$\frac{\Delta\nu}{cz}\,a^2 = \frac{a^2}{\ell_c z} \ll 1 \quad , \tag{4.55}$$

where a is the largest dimension of the object and the region of interest in the image, $\ell_c = c\tau_c = c/\Delta\nu$ is the longitudinal coherence length, and z is the distance between the image and the aperture planes. This condition can be satisfied in practice. For a = 1 cm, z = 10 cm, $\Delta\nu = 3 \cdot 10^9$, ℓ_c = 10 cm, $a^2/\ell_c z = 0.01$. The more temporally coherent the field is (i.e., the smaller $\Delta\omega$), the easier it is to satisfy this condition.

In order to give a further illustration of how the temporal and spatial statistical properties of an optical field get mixed together because of the nature of the wave equation, we ask the question: "Is it possible to have an optical field that is spatially completely coherent but temporally partially coherent?" To answer this question let us write the spectral density of the field in the form

$$V(\underline{r},\omega) = \psi(\omega)f(\underline{r}) \quad ,$$

where $f(\underline{r})$ is a deterministic function but $\psi(\omega)$ is a random function. In order for this field to satisfy the wave equation, $f(\underline{r})$ has to satisfy

$$\nabla^2 f + \left(\frac{\omega}{c}\right)^2 f = 0 \quad , \tag{4.56}$$

which means that f must be a function of ω, i.e., our model becomes

$$V(\underline{r},\omega) = \psi(\omega)f(\underline{r},\omega) \quad .$$

The spatial coherence function is then given by

$$G(\underline{r}_1,\underline{r}_2,0) = \int <|\psi(\omega)|^2> f^*(\underline{r}_1,\omega)f(\underline{r}_2,\omega)d\omega \quad . \tag{4.57}$$

It is obvious that the field is not necessarily spatially completely coherent!

Let us take an example. The function

$$f(\underline{r},\omega) = \exp(-j \frac{\omega}{c} z)$$

is a solution of (4.56). It represents a plane wave propagating in the z direction. Substituting in (4.57), we get

$$G(\underline{r}_1,\underline{r}_2,0) = \int <|\psi(\omega)|^2> \exp[-j \frac{\omega}{c} (z_2-z_1)]d\omega \quad .$$

If $<|\psi(\omega)|^2>$ has a Lorentzian shape with bandwidth $\Delta\omega$, we get

$$g(\underline{r}_1,\underline{r}_2,0) = \exp(- \frac{\Delta\nu}{c} |z_2-z_1|), \quad \Delta\nu = \Delta\omega/2\pi \quad . \tag{4.58}$$

This shows that all points in a plane perpendicular to the direction of propagation are completely correlated. But points with different z are not. The field has a *longitudinal coherence distance* $z_c = c/\Delta\nu = c\tau_c$, where τ_c is its coherence time. This is the result to be expected. If A_c is the area of coherence of the beam in the plane of its wavefront then we may define a cylinder parallel to the beam with area A_c and height z_c which has the property that any two points inside it are significantly correlated. The volume of such a cylinder $V_c = c\tau_c A_c$ is called the *volume of coherence.*

From the above equations, it is concluded that a plane wave propagating in the z direction can be represented by the stochastic signal $V(\underline{r},t) = f(t-z/c)$. Let us now take a more realistic solution of (4.56), a Gaussian beam, (4.5).

For simplicity, consider two points in the plane passing through the waist of the beam (z = 0), one in the center of the beam $\underline{r}_1 = (0,0,0)$ and another offset by a distance x, $\underline{r}_2 = (x,0,0)$. Substituting from (4.5) in (4.57), we get

$$G(\underline{r}_1,\underline{r}_2,0) = \int_0^\infty <|\psi(\omega)|^2>\exp\left(- \frac{\omega}{c} \frac{x^2}{2b}\right)d\omega \quad . \tag{4.59}$$

In order to calculate the order of magnitude of the area of coherence A_c, we assume that $<|\psi(\omega)|^2> = \exp(-|\omega|/\Delta\omega)$ and immediately get from (4.59)

$$G(\underline{r}_1,\underline{r}_2,0) \propto \frac{1}{x^2 + 2cb/\Delta\omega} \quad ,$$

which gives for the area of coherence in the transversal plane

$$A_c \simeq 2\pi cb/\Delta\omega = \ell_c b \quad , \qquad (4.60)$$

where $\ell_c = 2\pi c/\Delta\omega = c\tau_c$ is the longitudinal coherence length. This means that if $\ell_c \gg b$, then $A_c \gg b^2$ and the beam is completely coherent all over its cross section. A single-mode laser beam with ℓ_c = several meters and b less than a millimeter is thus definitely transversally coherent [426].

Michelson Interferometer

A plane wave propagating in the direction z and having a power spectrum $g(\omega)$ is represented by the complex analytic signal

$$V(\underline{r},t) = f(t-z/c) \quad ,$$

where f(t) is a complex stochastic signal with correlation function $g(\tau)$ and a power spectrum $g(\omega)$. If the two signals $V_1 = f(t-z_1/c)$ and $V_2 = f(t-z_2/c)$ are mixed together, the average intensity of the resultant field would be

$$<I(t)> = <|V_1+V_2|^2> \propto 1 + \mathrm{Re}\left|g\left(\tau = \frac{d}{c}\right)\right| \quad , \quad d = z_1 - z_2 \quad .$$

By changing the delay d and measuring the resultant intensity, we can scan the real part of the mutual coherence function. If the mutual coherence function is real (symmetric spectrum), then this measurement is sufficient to determine the spectrum by a process of Fourier-transform inversion. (A review of methods of Fourier spectroscopy can be found in [494].) For example, if the field is Lorentzian, $g(\tau) = \exp(-\Gamma|\tau|)\exp(j\omega_0\tau)$, then the variation of the intensity with d takes the form

$$<I(t)> \propto 1 + \exp(-\Gamma|d|/c)\cos(\omega_0 d/c) \quad ,$$

which is an oscillating function of d with a period $2\pi c/\omega_0 = \lambda$ and a modulation depth (visibility) = $\exp(-\Gamma|d|/c)$.

The measurement can be performed by using the optical arrangement of a Michelson interferometer (Fig.4.6) in which z_2 is changed by displacement of the mirrors in one optical path.

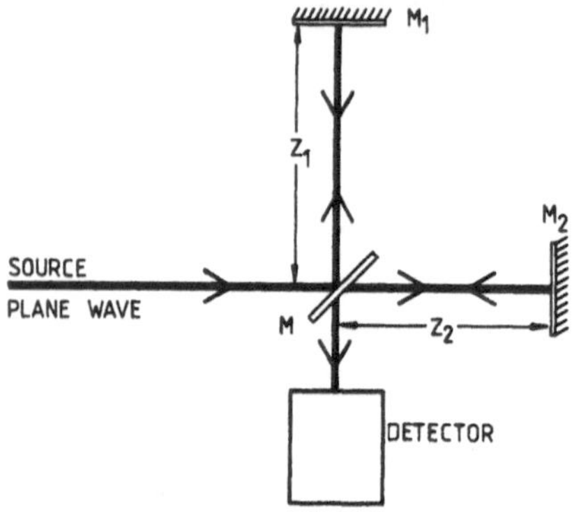

Fig.4.6. Michelson interferometer

4.2 Statistical Properties of some Special Models of Optical Fields

In this section, we introduce several models of optical fields and examine their statistical properties.

4.2.1 Polarized Thermal Light

The first model we consider is polarized light, which is described by an electromagnetic field whose complex amplitude is a stationary complex Gaussian stochastic process. It is also called "thermal" or "chaotic". The light radiated by an incandescent body or a gas-discharge lamp can be very well described by such a model. The Gaussian nature of the field fluctuations can be understood from a thermodynamic analysis of the fluctuations of a radiating source in thermal equilibrium [290]. It can also be visualized as a consequence of the central-limit theorem, because the radiated field is the sum of a very large number of contributions from the independent elements of the radiating source. The model also describes the light scattered when a coherent beam illuminates a rough surface (such as in laser radar [179]) or a thermally fluctuating target (such as the molecules of a liquid or a solution [313]). It describes a free running multimode laser light as well.

In fact, most natural sources of light radiate thermal optical fields.

As is mentioned in Sec. 2.5, the statistical properties of this field are completely described by the second-order correlation function $G(\underline{x}_1,\underline{x}_2)$ (the coherence function of order 1,1). If the field is stationary and cross-spectrally pure, then

$$G(\underline{x}_1,\underline{x}_2) = \langle I(\underline{r}_1)\rangle^{\frac{1}{2}} \langle I(\underline{r}_2)\rangle^{\frac{1}{2}} g(\underline{r}_1,\underline{r}_2)\ g(t_1-t_2) \quad ,$$

and we can say that the field is completely described by its average radiance distribution $\langle I(\underline{r})\rangle$, its normalized spatial coherence function $g(\underline{r}_1,\underline{r}_2)$, and its normalized temporal coherence function $g(t_1-t_2)$.

By definition, the joint probability density of the complex amplitude at several space-time points

$$[V] = \begin{bmatrix} V(\underline{x}_1) \\ V(\underline{x}_2) \\ \vdots \\ V(\underline{x}_N) \end{bmatrix}$$

is given by

$$P([V]) = (2\pi)^{-N}|H|^{-1}\exp(-\tfrac{1}{2}[V]^{\dagger}H^{-1}[V]) \quad , \tag{4.61}$$

where

$$H_{ij} = \frac{1}{2}\, G(\underline{x}_i,\underline{x}_j) \quad . \tag{4.62}$$

The mgf is given by

$$Q_{[V]}([S]) = \exp(\tfrac{1}{2}[S]^{\dagger}H[S]) \tag{4.63}$$

and the higher-order coherence functions are related to the second-order coherence functions by the expansion

$$G^{N,M}(\underline{x}_1,\underline{x}_2,\ \ldots\ \underline{x}_{N+M}) = \delta_{N,M} \sum_{\pi} \prod_{j=1}^{N} G(\underline{x}_j,\underline{x}_{\pi j}) \quad , \tag{4.64}$$

where $\sum_{\pi}$ means summation over the N! permutations of $(\pi_1, \pi_2, \ldots \pi_N)$ over $N+1, N+2, \ldots 2N$. Eqs. (4.61,63,64) are simply (2.50-52) written in different notations.

Intensity Fluctuations

Now let us study the statistical properties of the real process $I(\underline{x}) = |V(\underline{x})|^2$.

At a single space-time point, $I(\underline{x})$ is the sum of two independent squared normal random variables; hence it has a chi_2^2 distribution (chi-square distribution with two degrees of freedom), which is an exponential distribution (Fig.4.7) and is characterized by (Table 2.1)

$$P(I) = \frac{1}{\langle I\rangle} \exp(-I/\langle I\rangle) \tag{4.65}$$

$$Q_I(s) = (1 + s\langle I\rangle)^{-1} \tag{4.66}$$

$$K_I^{(m)} = (m-1)!\langle I\rangle^m \quad , \quad \langle I^m\rangle = m!\,\langle I\rangle^m \quad . \tag{4.67}$$

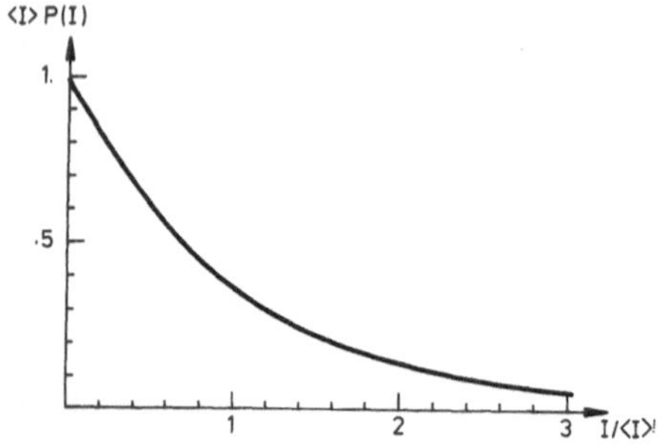

Fig.4.7. PD of the intensity of linearly polarized thermal light

In order to determine the higher-order statistics of $I(\underline{x})$, we consider the set of random variables $[I] = (I_1, I_2, \ldots I_N)$, where $I_j = I(\underline{x}_j)$ $j = 1, \ldots N$.

We start with the moment-generating function

$$Q_{[I]}([s]) = \left\langle \exp\left(-\sum_j s_j I_j\right)\right\rangle = \left\langle \exp\left(-\sum_j s_j |V_j|^2\right)\right\rangle .$$

This expectation value can be evaluated by substituting from (4.61) and performing the 2N-dimensional integral. By use of matrix methods, the result can be written in the form [38]

$$Q_{[I]}([s]) = |\Delta|^{-1} \quad , \tag{4.68}$$

where Δ is the matrix whose elements are

$$\Delta_{ij} = \delta_{ij} + s_i G_{ij} \qquad G_{ij} = G(\underline{x}_i, \underline{x}_j) \tag{4.69}$$

If we put $N = 1$, we reproduce (4.66). For $N = 2$, we get

$$Q_{I_1,I_2}(s_1,s_2) = (1+s_1\langle I_1\rangle)(1+s_2\langle I_2\rangle) - s_1 s_2 |G_{12}|^2 \Big]^{-1} \tag{4.70}$$

and $N = 3$ gives

$$\begin{aligned} Q_{I_1,I_2,I_3}(s_1,s_2,s_3) = \Big\{ &(1+s_1\langle I_1\rangle)(1+s_2\langle I_2\rangle)(1+s_3\langle I_3\rangle) \\ &- s_1 s_2 |G_{12}|^2 - s_2 s_3 |G_{23}|^2 - s_3 s_1 |G_{31}|^2 \\ &- s_1 s_2 s_3 [\langle I_1\rangle |G_{23}|^2 + \langle I_2\rangle |G_{31}|^2 + \langle I_3\rangle |G_{12}|^2 \\ &- 2\,\mathrm{Re}\,\{G_{12} G_{23} G_{31}\}] \Big\}^{-1} \quad . \end{aligned} \tag{4.71}$$

Higher-order cases can be similarly evaluated.

Close examination of (4.66), (4.70) and (4.71) shows that the single-fold statistics depends on the average intensity $\langle I\rangle$ and does not carry information on the spatial or temporal coherence properties of the field. The double-fold statistics carry information on the absolute value of the coherence function and no information on its phase. Thus, if the coherence function is real (corresponding to a symmetric spectrum), complete information on the field is determined from double-fold statistical measurements. The three-fold statistics contain information on the phase and can be used to determine it [167].

The moments of [I] could be determined from the moment-generating function by use of (2.9) and (2.22). It is easier, however, to use directly the expansion (4.64), and get

$$<I_1 I_2 \ldots I_N> = G_I(\underline{x}_1, \underline{x}_2, \ldots \underline{x}_N) = G_V^{N,N}(\underline{x}_1, \ldots \underline{x}_N, \underline{x}_1, \ldots \underline{x}_N)$$

$$= \sum_{\pi} \prod_{j=1}^{N} G_V^{1,1}(\underline{x}_j, \underline{x}_{\pi j}) \quad , \tag{4.72}$$

where $\sum_{\pi}$ is the summation over N! permutations of $\pi_j = 1,2, \ldots N$.

Also the cumulants are given by [79]

$$K_I^N(\underline{x}_1, \ldots , \underline{x}_N) = \sum_{c} \prod_{j=1}^{N} G_V^{1,1}(\underline{x}_j, \underline{x}_{cj}) \quad , \tag{4.73}$$

where $\sum_{c}$ = sum over all cyclic perumtations of the integers (1,2, ... N).

The lower orders of (4.72) and (4.73) are

$$\text{I)} \quad G_I(\underline{x}_1, \underline{x}_2) = <I(\underline{x}_1)><I(\underline{x}_2)> + |G(\underline{x}_1, \underline{x}_2)|^2 \quad , \tag{4.74}$$

which can be written in a more-simple symbolism as

$$<I_1 I_2> = <I_1><I_2> + |G_{12}|^2 \quad , \tag{4.75}$$

$$K_I(\underline{x}_1, \underline{x}_2) = |G(\underline{x}_1, \underline{x}_2)|^2 \quad ,$$

or

$$<\Delta I_1 \Delta I_2> = |G_{12}|^2 \quad . \tag{4.76}$$

$$\text{II)} \quad K_I(\underline{x}_1, \underline{x}_2, \underline{x}_3) = <\Delta I_1 \Delta I_2 \Delta I_3> = 2\,\mathrm{Re}\{G_{12} G_{23} G_{31}\} \quad , \tag{4.77}$$

and so on. The factorization properties of higher-order coherence functions have been experimentally demonstrated [232].

It may seem useless to study higher-order or multifold moments of a thermal field because all information is contained in the single- and double-fold moments (and the triple moment for a field with a nonsymmetric spectrum). Higher-order moments can be used, however, to compare a thermal field with other non-Gaussian fields or to establish that an optical field is indeed thermal.

Moreover, as will be clear in Chaps. 6 and 7, the higher-order moments are needed to calculate the accuracy of measurement of the second-order moments.

Equation (4.74) (or 4.75) is of extreme importance because it shows how intensity-correlation measurement gives direct access to the absolute value of the field-correlation function. In a normalized form, (4.74) can be written in the simple form

$$g_I(\underline{x}_1,\underline{x}_2) = 1 + |g_V(\underline{x}_1,\underline{x}_2)|^2 \quad , \tag{4.78}$$

sometimes known as the Siegert relation [442].

An instrument that measures $g_I(\underline{x}_1,\underline{x}_2)$ or $G_I(\underline{x}_1,\underline{x}_2)$ is known as an "intensity correlator", an "intensity interferometer" or a *"Hanbury Brown and Twiss interferometer"* (HANBURY BROWN and TWISS [193-196]; see also the important work of TITULAER and GLAUBER [481]). If $\underline{x}_1$ and $\underline{x}_2$ correspond to two different positions but equal times, then two photodetectors have to be used and the spatial part of the mutual coherence function can be determined (Fig.4.8a). This, together with the Van Cittert-Zernike theorem (Sec. 4.1.3), which relates the spatial coherence function of light emitted by a far incoherent object to its size, is the principle of stellar interferometry, which aims at measuring the angular diameter of stars [191-196, 493].

If $\underline{x}_1$ and $\underline{x}_2$ correspond to the same position, then two detectors could be used and the time delay could be introduced optically (Fig.4.8b). This is in fact the way HANBURY BROWN and TWISS [193] first demonstrated the effect. Alternatively, one detector could be used, and the time delay could be introduced electrically in the detected electric current (Fig.4.8c). In both arrangements, the temporal part of the coherence function could be obtained, from which the spectrum could be inferred. This subject is discussed in detail in Chap. 7.

It should be realized that, for a thermal field, if $|g(\underline{x}_1,\underline{x}_2)| = 1$, then $g_I(\underline{x}_1,\underline{x}_2) = 2$. But this does *not* correspond to coherent light. For coherent light $|g(\underline{x}_1,\underline{x}_2)| = 1$ but $g_I(\underline{x}_1,\underline{x}_2) = 1$. This demonstrates the importance of higher-order coherence functions in distinguishing between fields of different statistical origins.

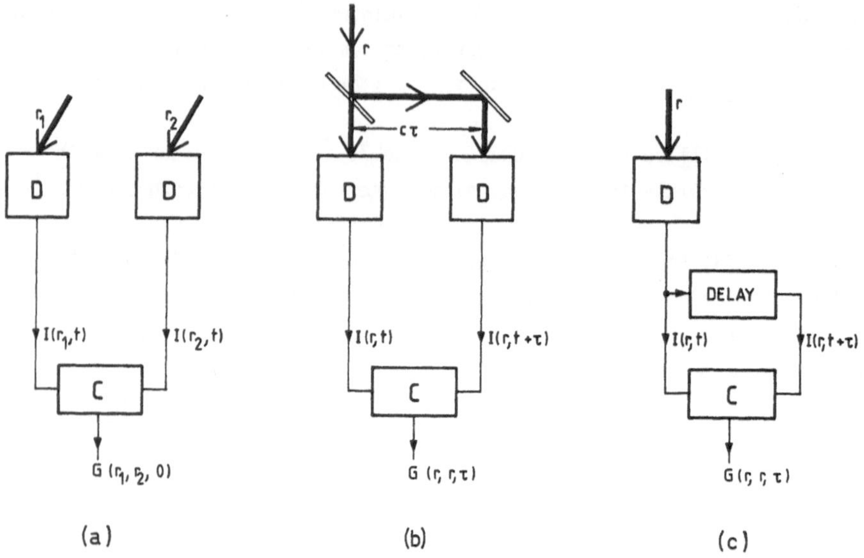

Fig.4.8. Schemes of intensity interferometry; D = detector, C = correlator

The fact that coherent light, which shows the highest visibility or fringe contrast in a Young's interference experiment, shows no intensity correlation (no Hanbury Brown Twiss effect) was rather surprising in the early developments of coherence theory.

4.2.2 Partially Polarized Thermal Light

Consider a beam of light that propagates in the z direction. At a fixed space point in the center of the beam, the electromagnetic field is described by two stochastic processes $V_x(t)$ and $V_y(t)$, which correspond to two perpendicular directions parallel to the arbitrarily chosen coordinates (x,y).

The assumption that the field is thermal means, in this case, that *$V_x(t)$ and $V_y(t)$ are two jointly Gaussian complex stochastic processes*. Therefore their statistical properties are completely defined by assigning the second-order correlations and cross correlations,

$$
\begin{aligned}
&\langle V_x^*(t)V_x(t+\tau)\rangle = \langle I_x\rangle g_{xx}(\tau)\\
&\langle V_y^*(t)V_y(t+\tau)\rangle = \langle I_y\rangle g_{yy}(\tau)\\
&\langle V_x^*(t)V_y(t+\tau)\rangle = (\langle I_x\rangle\langle I_y\rangle)^{\frac{1}{2}} g_{xy}(\tau)
\end{aligned}
\tag{4.79}
$$

It is convenient to put this set of correlation functions in the form of a 2x2 matrix, called the coherency matrix [362, 505, 507, 509]

$$
J(\tau) = \begin{bmatrix} J_{xx}(\tau) & J_{xy}(\tau) \\ J_{yx}(\tau) & J_{yy}(\tau) \end{bmatrix} = \begin{bmatrix} \langle I_x\rangle g_{xx}(\tau) & (\langle I_x\rangle\langle I_y\rangle)^{\frac{1}{2}} g_{xy}(\tau) \\ (\langle I_x\rangle\langle I_y\rangle)^{\frac{1}{2}} g_{yx}(\tau) & \langle I_y\rangle g_{yy}(\tau) \end{bmatrix} . \tag{4.80}
$$

Once this matrix is known, all other higher-order statistics can be determined by use of the known relations that characterize complex Gaussian processes.

A considerable simplification is obtained in the special case when $g_{xx}(\tau)$, $g_{xy}(\tau)$, and $g_{yy}(\tau)$ have the same spectral profile, i.e.,

$$
g_{xx}(\tau) = g_{yy}(\tau) = g_{xy}(\tau)/g_{xy}(0) = g(\tau) \quad ,
$$

because under such a condition (called the cross-spectral purity condition [308] the coherence matrix $J(\tau)$ is the product of a Hermitian matrix $J(0)$ independent of τ and a scalar function $g(\tau)$. We pursue the analysis further in this special case (for a more general treatment, see [25, 357]).

The numerical values of the four elements of $J(0)$ depend on the choice of the axes (x,y) but certain combinations of these elements are invariant to axis rotation, e.g., the determinant $|J(0)|$ and the trace Tr $\{J(0)\}$.

Special choices of the coordinate system (x,y) may simplify the problem. For example, axes (x_0,y_0) can be found such that $V_{x_0}(t)$ and $V_{y_0}(t)$ are statistically uncorrelated. This is accomplished by diagonalizing the matrix $J(0)$. This gives two eigenvalues that determine the average intensities

$$
\langle I_{x_0}\rangle = \langle |V_{x_0}(t)|^2\rangle = \frac{1}{2}\langle I\rangle (1+P)
$$

and

$$<I_{y_0}> = <|V_{y_0}(t)|^2> = \frac{1}{2} <I> (1-P) \quad , \tag{4.81}$$

where

$$<I> = <I_x> + <I_y> = \mathrm{Tr}\{J(0)\} \quad , \tag{4.82}$$

the total intensity, and

$$P = \left\{1 - \frac{4|J(0)|}{[\mathrm{Tr}\{J(0)\}]^2}\right\}^{1/2} \quad , \tag{4.83}$$

are invariant to rotation of the axes.

In this way, our plane beam is now represented by two statistically independent stochastic processes, $V_{x_0}(t)$ and $V_{y_0}(t)$, with variances (1/2)<I> ($1 \pm P$).

I) $P = 1$ means that the x_0-polarized component has the total intensity <I> and the y_0-polarized component is zero. This means that the light is *linearly polarized* in the x_0 direction; its statistics have already been studied in the previous subsection.

II) $P = 0$ means that the x_0-polarized and the y_0-polarized components have equal intensities = (1/2) <I> (degenerate case in the eigenvalue problem). This means that for any orthogonal coordinate system the two components carry equal intensities and the light is termed *unpolarized*. The above considerations justify calling P the *degree of polarization* of the plane beam.

Further analysis of the effect of optical systems (lenses, polarizers, etc.) on a partially polarized thermal beam can be found in several books (e.g., [98, 356, 441, 465]).

Intensity Fluctuations

Each of the components $V_x(t)$ and $V_y(t)$ is a complex Gaussian process, hence their intensity fluctuations can be described as in Sec. 4.2.1. The cross correlation between their intensity fluctuations is given by

$$<I_x I_y> = <|V_x|^2 |V_y|^2> = <I_x><I_y> + J_{xy}(0) J_{yx}(0) \quad , \tag{4.84}$$

where the moment expansion of complex Gaussian variables has been used.

An intensity-coherency matrix $\mathcal{J}$ whose elements are

$$\mathcal{J}_{xx} = \langle\Delta I_x^2\rangle^{1/2} = \langle I_x\rangle = J_{xx}(0)$$

$$\mathcal{J}_{yy} = \langle\Delta I_y^2\rangle^{1/2} = \langle I_y\rangle = J_{yy}(0) \tag{4.85}$$

$$\mathcal{J}_{xy} = \langle\Delta I_x \Delta I_y\rangle^{1/2} = [J_{xy}(0)J_{yx}(0)]^{1/2} = |J_{xy}(0)|$$

can be defined.

It is obvious that J and $\mathcal{J}$ are equal except for the cross-diagonal terms whose absolute values are used in the intensity-coherency matrix. A simple examination shows that both matrices have the same trace and the same determinant. This means that the degree of polarization P which is given by (4.83) is also given by

$$P = \left[1 - \frac{4|\mathcal{J}|}{(\mathrm{Tr}\mathcal{J})^2}\right]^{1/2} \quad . \tag{4.86}$$

Equation (4.86) suggests an interesting and easy method of measuring the degree of polarization of a plane beam, by correlation of intensity fluctuations [84].

Fluctuations of the Total Intensity

The total intensity $I = I_x + I_y = I_{x0} + I_{y0}$ is the sum of two statistically independent random variables; each having a chi_2^2 distribution and with means $\langle I_{x0}\rangle = (1/2)(1 + P)\langle I\rangle$ and $\langle I_{y0}\rangle = (1/2)(1 - P)\langle I\rangle$. Hence, I has a distribution P(I), which is the convolution of $P(I_{x0})$ and $P(I_{y0})$ (see (2.37)). This gives [310]

$$P(I) = (\langle I_{x_0}\rangle - \langle I_{y_0}\rangle)^{-1}\left[\exp(-I/\langle I_{x_0}\rangle) - \exp(-I/\langle I_{y_0}\rangle)\right]$$

$$= \frac{1}{P\langle I\rangle}\left[\exp\left(-\frac{2}{1+P}\frac{I}{\langle I\rangle}\right) - \exp\left(-\frac{2}{1-P}\frac{I}{\langle I\rangle}\right)\right] \quad . \tag{4.87}$$

For $P = 1$ (linearly polarized light), we reproduce the exponential distribution (4.65). For $P = 0$ (unpolarized light), we obtain a chi_4^2 distribution (Table 2.1)

$$P(I) = 4I\langle I\rangle^{-2}\exp(-2I/\langle I\rangle) \quad . \tag{4.88}$$

The PD of I is shown in Fig.4.9. A deviation from the linearly polarized case changes the profile of the distribution and shifts its peak to higher values of I.

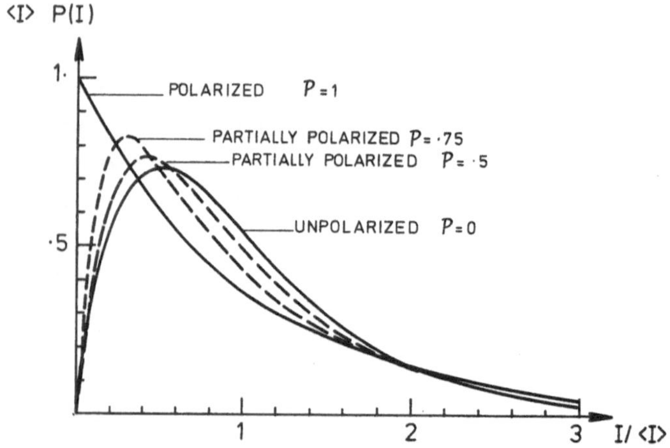

Fig.4.9. Dependence of the PD of the intensity of partially polarized thermal light on the degree of polarization, P

The moment-generating function of I is the product

$$Q_I(s) = Q_{I_{x_0}}(s)Q_{I_{y_0}}(s) = \left[1+\frac{s}{2}(1+P)\langle I\rangle\right]^{-1}\left[1+\frac{s}{2}(1-P)\langle I\rangle\right]^{-1} \quad ; \tag{4.89}$$

its cumulants are therefore given by the sum

$$K_I^{(m)} = K_{I_{x_0}}^{(m)} + K_{I_{y_0}}^{(m)} = (m-1)!\left[\left(\frac{1+P}{2}\right)^m + \left(\frac{1-P}{2}\right)^m\right]\langle I\rangle^m \quad , \tag{4.90}$$

which is plotted in Fig.4.10.

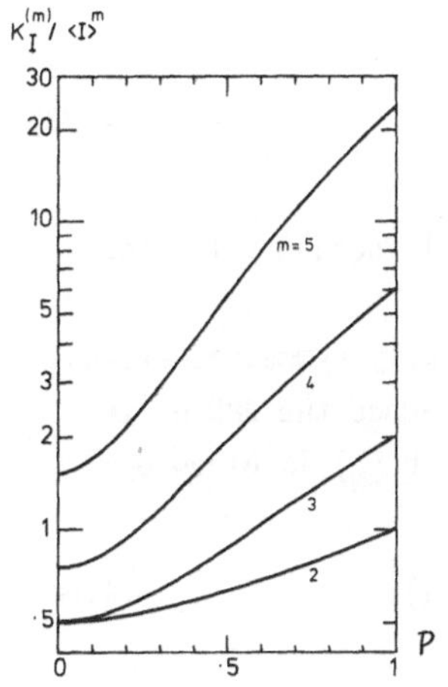

Fig.4.10. Dependence of the normalized cumulants of the intensity of thermal light on its degree of polarization P

For example,

$$\langle \Delta I^2 \rangle = \frac{1}{2}\,(1+P^2)\langle I\rangle^2 \quad , \tag{4.91}$$

which suggests another simple method of measuring P.

4.2.3 Polarized Superposition of Coherent and Thermal Light

The model of a light beam that is composed of a mixture of coherent and thermal light is of extreme practical importance because

I) it describes situations where coherent light is deliberately mixed with thermal light carrying information, in order to increase the signal-to-noise ratio of the measurement (heterodyning);

II) it describes the situation when an amplitude-stabilized laser beam carrying certain information (in an optical communication system) is accidentally mixed with background thermal light (noise);

III) it describes a laser light above threshold [21, 305, 450]; and

IV) it is one of the possible models that describes an amplitude-stabilized laser light that has propagated through a turbulent atmosphere.

Consider a coherent field represented by its analytic signal $A(\underline{x})\exp(j\omega_c t)$ that is mixed with a thermal field $V_{th}(\underline{x}) = \mathcal{V}_{th}(\underline{x})\exp(j\omega_0 t)$ and whose spectrum is centered around ω_0. The resultant field has a complex amplitude

$$\mathcal{V}(\underline{x}) = \mathcal{V}_{th}(\underline{x}) + \mathcal{V}_c(\underline{x}) \quad , \tag{4.92}$$

where

$$V_c(\underline{x}) = A(\underline{x})\exp(j\Delta\omega t) \quad , \quad \Delta\omega = \omega_c - \omega_0 \quad . \tag{4.93}$$

The frequency ω_0 is arbitrarily chosen as the central frequency for the total field.

The amplitude $V_{th}(\underline{x})$ is a complex Gaussian circularly symmetric stochastic process, but $V_c(\underline{x})$ is a deterministic function. Hence the JPD of the amplitudes at different space-time points $V(\underline{x}_1)$, ... $V(\underline{x}_N)$ is given by

$$P([V]) = (2\pi)^{-N}|H|^{-1}\exp\left[-\frac{1}{2}([V]-[V_c])^{\dagger}H^{-1}([V]-[V_c])\right] \quad , \tag{4.94}$$

where

$$H_{ij} = \frac{1}{2}\, G_{th}(\underline{x}_i,\underline{x}_j) \quad ;$$

the subscript "th" denotes the thermal part of the field.

This JPD determines all of the statistical properties of the field. In particular, we are interested in the intensity fluctuations. If the coherent part has a uniformly distributed random phase θ, then the statistics of the intensity are completely unaffected. This can be shown by writing

$$|V_{th}(\underline{x})+V_c(\underline{x})e^{j\theta}| = |V_{th}(\underline{x})e^{-j\theta}+V_c(\underline{x})| \quad ,$$

and noting that the process $V_{th}(\underline{x})\exp(-j\theta)$ is equivalent to the process $V_{th}(\underline{x})$ because $V_{th}(\underline{x})$ itself has a uniformly distributed phase).

Intensity Fluctuations

First Order Statistics

The intensity $I = I(\underline{x})$ is the sum of the squares of two independent Gaussian variables that have means $A(t)\cos(\Delta\omega t)$ and $A(t)\sin(\Delta\omega t)$ and equal variances. By performing the necessary transformations (or by integrating over the JPD (4.94)), we find that I has a Rician-square distribution (Table 2.1) characterized by [172, 326, 368, 400]

$$Q_I(s) = \frac{1}{1+s\langle I_{th}\rangle} \exp\left(-\frac{sI_c}{1+s\langle I_{th}\rangle}\right) \quad (4.95)$$

$$= Q_{I_{th}}(s)\exp\left[-sI_c Q_{I_{th}}(s)\right] \quad (4.96)$$

$$= Q_{I_{th}}(s)\cdot Q_{I_c}(s)\cdot\exp\left(\frac{s^2 I_c\langle I_{th}\rangle}{1+s\langle I_{th}\rangle}\right) ,$$

$$P(I) = \frac{1}{\langle I_{th}\rangle} \exp\left(-\frac{I+I_c}{\langle I_{th}\rangle}\right) I_0\left[2\,\frac{(I_c I)^{1/2}}{\langle I_{th}\rangle}\right] , \quad (4.98)$$

$$\langle I^m\rangle = m!\ \langle I_{th}\rangle^m\, L_m(-I_c/\langle I_{th}\rangle) ,$$

$$\langle \Delta I^2\rangle = \langle I_{th}\rangle^2 + 2\, I_c\langle I_{th}\rangle , \quad (4.99)$$

where I_{th} is the intensity of the thermal part, $I_c = |A(t)|^2$, $I_0[\cdot]$ is the modified Bessel function, and L_m is the Laguerre polynomial.

Figures (4.11) and (4.12) illustrate the dependence of P(I) and $\langle I^m\rangle$ on the ratio $I_c/\langle I_{th}\rangle$.

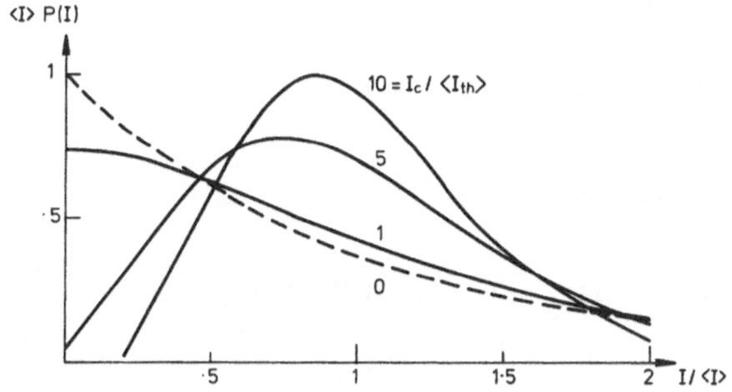

Fig.4.11. PD of the intensity of a mixture of coherent and thermal light for several ratios of their average intensities

Note that these first-order properties are independent of the spectrum of the thermal component and of the frequency difference $\Delta\omega$.

If we examine (4.97), we see that the mgf is the product of 3 terms, the mgf of the thermal part, that of the coherent part, and a third term, which

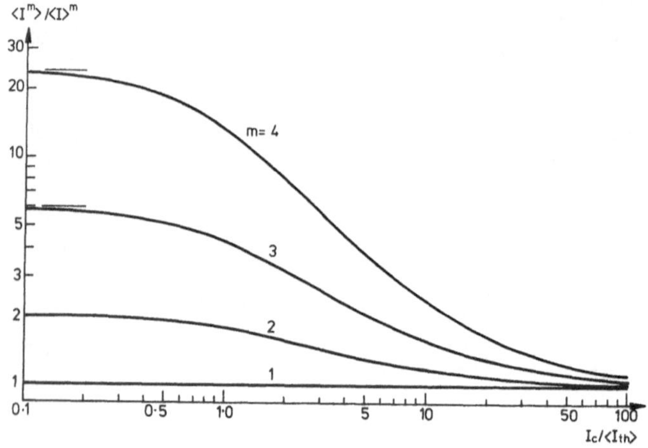

Fig.4.12. Normalized moments of the intensity cf a mixture of coherent and thermal light as a function of the ratio of their average intensities

represents the mixing effect. If the coherent and thermal components were of perpendicular polarizations, then the mgf would have been equal to the product of the first two terms of (4.97) only. In such a case,

$$\langle I^2\rangle = I_c^2 + 2\langle I_{th}\rangle^2 \quad .$$

Higher-Order Statistics

The Moment-Generating Function. Writing

$$Q_{[I]}([s]) = \left\langle e^{-\sum s_j I_j}\right\rangle = \left\langle e^{-\sum s_j |V_j|^2}\right\rangle = \left\langle e^{-[V]^\dagger S[V]}\right\rangle$$

where the elements of the matrix S are $S_{ij} = s_i\delta_{ij}$, and using the expression for the JPD, (4.94), to calculate the expectation value, we can write the result of the multifold integral in the form of a matrix [140]

$$\begin{aligned} Q_{[I]}([s]) &= |\Delta|^{-1}\exp\left[-\tfrac{1}{2}[V_c]^\dagger(\delta-\Delta^{-1})H^{-1}[V_c]\right] \\ &= Q_{[I_{th}]}([s])\exp\left[-\tfrac{1}{2}[V_c]^\dagger(\delta-\Delta^{-1})H^{-1}[V_c]\right] \end{aligned} \quad , \qquad (4.100)$$

where

$$\Delta_{ij} = \delta_{ij} + s_i G_{th}(\underline{x}_i,\underline{x}_j) \quad , \quad H_{ij} = \frac{1}{2} G_{th}(\underline{x}_i,\underline{x}_j) \tag{4.101}$$

and δ is the identity matrix.

As an example, we take the case $N = 2$ and $A(t) = A = I_c^{1/2}$. By simple substitution in (4.100), we get

$$Q_{I_1,I_2}(s_1,s_2) = Q_{I_{th1},I_{th2}}(s_1,s_2)\exp\left[-I_c Q_{I_{th1},I_{th2}}(s_1,s_2)\right.$$
$$\left.\left(s_1+s_2+2s_1s_2\langle I_{th}\rangle\left\{1-\mathrm{Re}\left[g_{th}(\tau)e^{j\Delta\omega\tau}\right]\right\}\right)\right] \quad . \tag{4.102}$$

Here we see that the frequency separation $\Delta\omega$ and the phase of g_{th} contribute to the two-fold statistics. If $g_{th}(\tau)$ is real, (4.102) becomes

$$Q_{I_1,I_2}(s_1,s_2) = Q_{I_{th1},I_{th2}}(s_1,s_2)\exp\left(-I_c Q_{I_{th1},I_{th2}}(s_1,s_2)\right.$$
$$\left.\left\{s_1+s_2+2s_1s_2\langle I_{th}\rangle\left[1-g_{th}(\tau)\cos\Delta\omega\tau\right]\right\}\right) \quad . \tag{4.103}$$

The Moments. The moments of $I(x)$ can be found by relating the moments of the sum $V = V_{th} + V_c$ to those of the components V_{th} and V_c. Using combinatorial analysis and the moment expansion rule of Gaussian variables, CANTRELL [82] found that the cumulants are given by

$$K_I(\underline{x}_1,\underline{x}_2, \dots \underline{x}_N) = \sum_c \prod_{j=1}^{N} G_{th}(\underline{x}_j,\underline{x}_{cj})$$
$$+\sum_{\substack{j,k\\ j\neq k}}^{N} V_c(\underline{x}_j)V_c^*(\underline{x}_k)\sum_{c'(k)}\prod_{\substack{m=1\\ m\neq k}}^{N} G_{th}(\underline{x}_m,\underline{x}_{c'(k)m}) \quad , \tag{4.104}$$

where c is any cyclic permutation of 1,2, ... N; c'(k) is any cyclic permutation of 1,2, ... k-1,k+1, ... N; where j occurs in the permuted set of integers {c'(k)m} it is to be replaced by k.

For example,

$$K_I(\underline{x}_1,\underline{x}_2) = \langle\Delta I_1 \Delta I_2\rangle = |G_{th}(\underline{x}_1,\underline{x}_2)|^2 + 2\,\mathrm{Re}\left\{V_c(\underline{x}_1)V_c^*(\underline{x}_2)G_{th}(\underline{x}_1,\underline{x}_2)\right\} \quad (4.105)$$

or

$$K^{1\ 1}_{I_1,I_2} = |G_{12}|^2 + 2\,\mathrm{Re}\left\{V_{c1}V_{c2}^* G_{12}\right\} \ , \quad (4.106)$$

and

$$K^{1\ 1\ 1}_{I_1,I_2,I_3} = 2\,\mathrm{Re}\left\{G_{12}G_{23}G_{31} + V_{c1}V_{c2}^*G_{12}G_{32} + V_{c1}V_{c3}^*G_{12}G_{23} + V_{c2}V_{c3}^*G_{21}G_{13}\right\} \ . \quad (4.107)$$

If

$$V_c(x) = Ae^{j\Delta\omega t} \ ,$$

then

$$K_I(\underline{x}_1,\underline{x}_2) = |G_{th}(\underline{x}_1,\underline{x}_2)|^2 + 2\,I_c\,\mathrm{Re}\left\{e^{j\Delta\omega\tau}G_{th}(\underline{x}_1,\underline{x}_2)\right\} \ . \quad (4.108)$$

Note that the phase of $G_{th}(\underline{x}_1,\underline{x}_2)$ can be retrieved from $K_I(\underline{x}_1,\underline{x}_2)$ (cf. [335]). If $G_{th}(\underline{x}_1,\underline{x}_2)$ is real (i.e., the spectrum is symmetric)

$$K_I(\underline{x}_1,\underline{x}_2) = |G_{th}(\underline{x}_1,\underline{x}_2)|^2 + 2\,I_c\,G_{th}(\underline{x}_1,\underline{x}_2)\cos(\Delta\omega\tau) \ . \quad (4.109)$$

This corresponds to an intensity-correlation function

$$G_I(\underline{r}_1,\underline{r}_2,\tau) = \langle I_1 I_2\rangle = \langle I\rangle^2 + \langle I_{th}\rangle^2 |g_{th}(\underline{r}_1,\underline{r}_2,\tau)|^2 + 2\,\langle I_{th}\rangle\,I_c\,g_{th}(\underline{r}_1,\underline{r}_2,\tau)\cos(\Delta\omega\tau) \ . \quad (4.110)$$

If $\underline{r}_1 = \underline{r}_2$ and $\tau = 0$, we recover (4.99) as expected.

Further theoretical aspects of mixtures of coherent and thermal light are discussed in [159, 256, 471, 500] and in Sec. 5.2.4 and references therein.

4.2.4 Mixture of Coherent Light and Partially Polarized Thermal Light

Now we consider a mixture of a coherent linearly polarized light and a thermal partially polarized light.

Let us first choose the coordinate system x,y, which makes the components of the thermal part statistically independent, and assume that the coherent field is polarized in a direction that makes an angle θ with the x axis (Fig.4.13). Thus, our total detected intensity is the sum of two independent terms, I_1 and I_2. The component I_1 is the intensity of a mixture of coherent

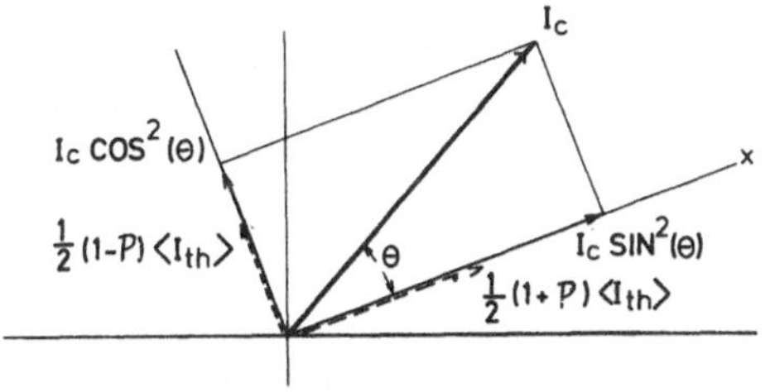

Fig.4.13. Components of a mixture of linearly polarized coherent light and partially polarized thermal light

and thermal contributions with mean intensities $I_{c1} = I_c\cos^2(\theta)$ and $\langle I_{th1}\rangle = (1/2)(1+P)\langle I_{th}\rangle$, respectively. Similarly, I_2 is the intensity of a mixture of coherent and thermal contributions with mean intensities $I_{c2} = I_c \sin^2(\theta)$ and $\langle I_{th2}\rangle = (1/2)(1-P)\langle I_{th}\rangle$. From this, it follows directly that the statistics of the total intensity are given by

$$Q_I(s) = Q_{I_1}(s)Q_{I_2}(s)$$

$$Q_{I_j}(s) = (1+s\langle I_{thj}\rangle)^{-1}\exp\left[-sI_{cj}/(1+s\langle I_{thj}\rangle)\right] , \quad j = 1,2 \qquad (4.111)$$

$$P(I) = P_1(I)\circledast P_2(I)$$

$$P_j(I) = \frac{1}{\langle I_{thj}\rangle}\exp\left(-\frac{I+I_c}{\langle I_{thj}\rangle}\right)I_0\left[2\frac{(I_cI)^{1/2}}{\langle I_{thj}\rangle}\right] , \qquad (4.112)$$

$$K_I^m = K_{I_1}^m + K_{I_2}^m , \quad \langle I^m\rangle = \sum_\ell \binom{m}{\ell}\langle I_1^\ell\rangle\langle I_2^{m-\ell}\rangle$$

$$\langle I_j^m\rangle = m!\,\langle I_{thj}\rangle^m L_m(-I_{cj}/\langle I_{thj}\rangle) . \qquad (4.113)$$

4.2.5 Quasi-Stationary Gaussian Light

In Sec. 4.2.1 it was shown that thermal light (or stationary Gaussian light) has a complex envelope that is a stationary, circularly symmetric, complex Gaussian stochastic process (cf. also Sec. 2.6). This means that its real and imaginary parts are statistically independent and identically distributed. We may inquire if light exists that can be described by a complex envelope whose components are Gaussian but not necessarily uncorrelated or identical. Let us take an interesting example. A coherent light beam is intensity modulated by a square of a stationary Gaussian signal (a chi_1^2 distributed signal). Such a field has a complex envelope that is real. It is therefore not a thermal field.

Formally, let us introduce the following model. *A quasi-stationary Gaussian optical field*, $V(\underline{x}) = \mathcal{V}(\underline{x})\exp(j\omega_0 t)$, *has a complex amplitude* $\mathcal{V}(\underline{x})$ *whose components are stationary and Gaussian* (but not necessarily uncorrelated or identical). The reader may like to review Sec. 2.6.4 in which these concepts are discussed in general terms.

Statistical Properties of the Field

I) The CAS of the field at N points form a set of N complex Gaussian random variables that are not necessarily circularly symmetric (cf. Sec. 2.3.2).

II) The coherence functions of order N = M are time-shift invariant (TSI) but those of order N ≠ M are not TSI and do not necessarily vanish. This can be directly seen by relating the correlation function of V to that of $\mathcal{V}$

$$G_V^{N,M}(\underline{x}_1, \dots \underline{x}_{N+M}) = G_{\mathcal{V}}^{N,M}(\underline{x}_1, \dots \underline{x}_{N+M})\exp\left[-j\omega_0(t_1 + \dots t_N - t_{N+1} \dots t_{N+M})\right]$$

and noting that $\mathcal{V}$ is stationary by definition.

III) The necessary and sufficient condition for the field to be stationary is that $G_{\mathcal{V}}^{2,0} = 0$. This of course implies that $G_{\mathcal{V}}^{N,M} = 0$ for all N ≠ M.

Statistical Properties of the Intensity

The intensity correlation function can be written in the form

$$G_I^{(N)}(\underline{x}_1, \ldots \underline{x}_N) = \langle |V(\underline{x}_1)|^2 \ldots |V(\underline{x}_N)|^2 \rangle$$

$$\langle [V_c^2(\underline{x}_1)+V_s^2(\underline{x}_1)] \ldots [V_c^2(\underline{x}_N)+V_s^2(\underline{x}_N)] \rangle ,$$

where the subscripts c and s denote the real and imaginary parts. By expanding this product and using the fact that the real and imaginary parts of V are jointly Gaussian, we can write this in terms of $G^{1,1}(\underline{x}_i,\underline{x}_j)$ and $G^{2,0}(\underline{x}_i,\underline{x}_j)$. For example [425],

$$G_I^{(2)}(\underline{x}_1,\underline{x}_2) = \langle I(\underline{x}_1)\rangle \langle I(\underline{x}_2)\rangle + |G_V^{1,1}(\underline{x}_1,\underline{x}_2)|^2 + |G_V^{2,0}(\underline{x}_1,\underline{x}_2)|^2 . \quad (4.114)$$

Comparing this with its corresponding expression for a thermal field (4.74), we see that we have here an extra term $|G^{2,0}|^2$.

It is interesting to see that if the field has a *real amplitude*, then $G_V^{2,0} = G_V^{1,1}$ and we get

$$G_I(\underline{x}_1,\underline{x}_2) = \langle I(\underline{x}_1)\rangle \langle I(\underline{x}_2)\rangle + 2\,|G_V(\underline{x}_1,\underline{x}_2)|^2 \quad (4.115)$$

or

$$g_I(\underline{x}_1,\underline{x}_2) = 1 + 2|g(\underline{x}_1,\underline{x}_2)|^2 \quad (4.116)$$

(instead of (4.78) for a thermal field). This function has a value 3 at $\underline{x}_1 = \underline{x}_2$ and decreases to a value 1 for $\underline{x}_1 - \underline{x}_2 = \infty$, whereas, for a thermal light the same function starts from a value 2 and drops to a value 1. For coherent light, the normalized intensity correlation function is constant at a value 1. This indicates that in some sense (and in the second order) the real-amplitude light is more chaotic than the thermal light [377, 379].

Moment-Generating Functions

The correlation functions of higher orders can be determined from the moment-generating function. We shall derive this only for the case of quasi-stationary field with *real amplitude*.

Writing

$$Q_{[I]}([s]) = \left\langle \exp\left(-\sum s_i I_i\right)\right\rangle = \left\langle \exp\left[-\sum_i s_i |V(\underline{x}_i)|^2\right]\right\rangle ,$$

we can find the expectation by integrating over the joint Gaussian distribution of the real variables $\{V(x_i)\}$. The result is [379]

$$Q_{[I]}([s]) = |\Delta|^{-1/2} , \tag{4.117}$$

where

$$\Delta_{ij} = \delta_{ij} + 2\, s_i\, G^{1,1}(\underline{x}_i,\underline{x}_j) , \tag{4.118}$$

e.g.,

$$Q_I(s) = (1 + 2s\, \langle I\rangle)^{-1/2} \tag{4.119}$$

$$Q_{I_1,I_2}(s_1,s_2) = \Big[(1+2s_1\langle I_1\rangle)(1+2s_2\langle I_2\rangle)-4s_1s_2\langle I_1\rangle\langle I_2\rangle \, |g(\underline{x}_1,\underline{x}_2)|^2\Big]^{-1/2} . \tag{4.120}$$

By comparing (4.119) and (4.120) to (4.66) and (4.70), we immediately see that two statistically independent modes of a Gaussian field with real amplitude, each having the same intensity, make a thermal field.

Probability Densities

For a field with *real amplitude*, the intensity is the square of a Gaussian variable, i.e., a chi_1^2 variable with PD

$$P(I) = \frac{1}{(2\pi I\langle I\rangle)^{\frac{1}{2}}} \exp(-I/2\langle I\rangle) . \tag{4.121}$$

This, incidentally, corresponds to the mgf given by (4.119) and has moments

$$\langle I^m\rangle = (2m-1)!!\, \langle I\rangle^m , \tag{4.122}$$

e.g.,

$$\langle I^2 \rangle = 3\langle I \rangle^2 \quad .$$

The chi_1^2 distribution has a normalized variance twice as large as the chi_2^2 distribution (see Fig.4.14). This supports the claim that a Gaussian field with real amplitude is more chaotic than a thermal field.

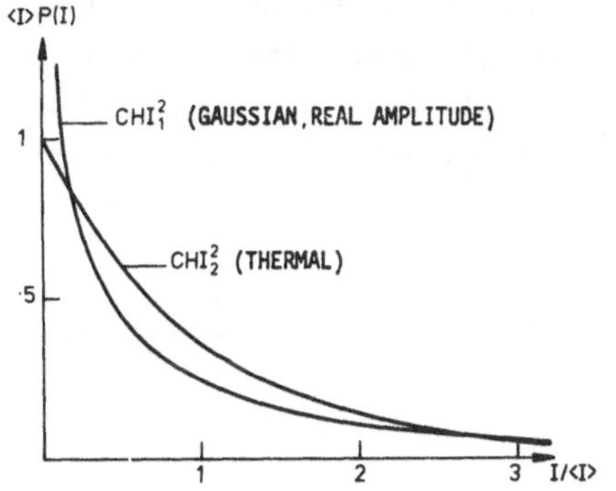

Fig.4.14. PD of the intensity of quasi-stationary Gaussian light compared to that of thermal light

The PD of the intensity of *a general quasi-stationary light* would be something between the chi_1^2 and the chi_2^2 distributions. In this case, I is the sum of the squares of two correlated Gaussian variables. For simplicity, we assume that the two variables have equal variances and a correlation coefficient ρ. By performing the necessary transformation, we get

$$P(I) = \frac{1}{\langle I \rangle (1-\rho^2)^{\frac{1}{2}}} I_0\left[\left(\frac{\rho}{1-\rho}\right) \frac{I}{\langle I \rangle}\right] \exp\left(- \frac{1}{1-\rho^2} \frac{I}{\langle I \rangle}\right)$$

which is a distribution that lies between the chi_1^2 and the chi_2^2 distribution of Fig.4.14.

4.2.6 Transient Thermal Light

Consider an optical field whose complex envelope is described by the stochastic differential equation

$$\frac{dV}{dt} = -\Gamma V(t) + W(t) \quad , \tag{4.123}$$

where $\mathcal{W}(t)$ is stationary white complex Gaussian stochastic process satisfying

$$\langle \mathcal{W}(t)\rangle = 0$$

$$G_{\mathcal{W}}(t_1,t_2) = P_e\Gamma\delta(t_1-t_2) \quad , \quad G_{\mathcal{W}}^{2,0}(t_1,t_2) = 0 \quad .$$

This implies that each of the quadrature components of V satisfies an equation similar to (2.84) and is an Ornstein-Uhlenbeck stochastic process. Because the components of $\mathcal{W}$ are statistically independent, the components of V must also be statistically independent. Using the results of Sec. 2.4.5, it becomes straightforward to show that V is a complex Gaussian process with zero mean and correlation function

$$G_V(t_1,t_2) = P_e\exp(-\Gamma|t_1-t_2|)+(P_i-P_e)\exp[-\Gamma(t_1+t_2)] \; , \quad G_V^{2,0}(t_1,t_2) = 0 \; ,$$

where

$$P_i = G_V(0,0) \quad \text{and} \quad P_e = G_V(\infty,\infty) \quad .$$

At each instant, the random variable $V(t)$ is a circularly symmetric complex Gaussian RV and hence the intensity $I(t) = |V|^2$ has an exponential distribution

$$\begin{aligned} P[I(t)] &= \langle I(t)\rangle^{-1}\exp[-I(t)/\langle I(t)\rangle] \\ \langle I(t)\rangle &= \langle I(\infty)\rangle + [\langle I(0)\rangle - \langle I(\infty)\rangle]\exp(-2\Gamma t) \quad . \end{aligned} \tag{4.124}$$

The mean intensity decays from an initial value $\langle I(0)\rangle = P_i$ to an equilibrium value $\langle I(\infty)\rangle = P_e$ with a time constant $1/2\Gamma$.

As Γt becomes large, the optical field reaches an equilibrium state and becomes thermal light with Lorentzian spectrum and bandwidth Γ (as defined in Sec. 4.2.1).

The intensity correlation of this nonstationary light is also of interest. It can be easily verified that it satisfies the usual relation

$$G_I(t_1,t_2) = \langle I(t_1)\rangle \langle I(t_2)\rangle + |G_V(t_1,t_2)|^2 \quad .$$

The above equations describe a model of nonstationary transient light which approaches thermal light in the steady state. Aside from the interest in this model as our only example of nonstationary light, it has some interesting physical applications.

The model could describe laser light scattered from a linearly unstable molecular system. It also describes the early transient state of a single-mode laser above threshold as it is switched on (cf. the discussion in the next subsection).

4.2.7 Van der Pol's Nonlinear-Oscillator Classical Model of Laser Light

The laser is an oscillator for light. An oscillator is a system that contains an amplification mechanism, a feedback mechanism, and a resonator that selects the desired frequency. The system unavoidably contains a loss mechanism and a source of noise. In a laser, the amplification is obtained by interaction with matter, through a process of stimulated emission; the feedback and the frequency selection are obtained by a cavity formed by two parallel mirrors, which also introduce losses; and the noise is due to spontaneous emission from the atoms of the medium.

A full quantum treatment is necessary to understand the processes of radiation-matter interaction that take place inside a laser. But again, a semiclassical approach (in which the radiation is treated classically, but the matter is treated quantum mechanically) is very successful. Indeed, the quantum correction is extremely small. It should be noted, nevertheless, that the theory of laser-light fluctuations is very involved and we attempt here only to give some simplified results of one possible model. (For further reading, the reader is referred to the book by SARGENT et al. [428], and to the articles [10, 11, 15, 22, 187-189, 294, 403]. We start with the phenomenological model of a Van der Pol oscillator described by the nonlinear stochastic differential equation

$$\ddot{v} - \xi(v)\dot{v} + \omega^2 v = w(t) \quad , \tag{4.125}$$

where ω is the oscillation frequency, $\xi(v)$ is the gain parameter - assumed to be a decreasing function of the amplitude $v(t)$, and $w(t)$ is a noise source that is included to account for the fluctuations of the resultant excitation. The noise starts the oscillation, which is then amplified by the gain mechanism in the system. But as v increases, the gain parameter decreases (because more losses are introduced). This slows down the growth of oscillation; eventually, a steady state is reached. Of course, because of the noise-driving term, even in the steady state, v(t) will be a fluctuating function of time. The simplest function $\xi(v)$ is the quadratic function

$$\xi(v) = \gamma - 2\beta v^2 \quad ,$$

where the higher harmonics due to the product $v^2\dot{v}$ should be neglected (the dot denotes time derivative). Also, we can assume that $w(t)$ represents white noise with correlation

$$\langle w(t_1)w(t_2)\rangle = \frac{1}{2}\, q\delta(t_1-t_2) \quad . \tag{4.126}$$

We are thus faced with the very interesting problem of solving the stochastic differential equation

$$\ddot{v} - (\gamma-2\beta v^2)\dot{v} + \omega^2 v = w(t) \quad . \tag{4.127}$$

Because v(t) is a narrow-bandpass signal centered around the frequency ω, it is advantageous to write (4.127) in terms of the complex envelope V. By substituting $v(t) = V(t)e^{j\omega t} + c\cdot c$ in (4.127) and neglecting terms that contain $\ddot{V}$ and $\dot{V}$ unless they are multiplied by ω, we get

$$\frac{dV}{dt} = f(V) + W(t) \ , \qquad f(V) = \frac{1}{2}\,(\gamma-2\beta|V|^2)V \quad ,$$

where

$$\langle W^*(t_1)W(t_2)\rangle = q\delta(t_1-t_2) \quad .$$

This is a complex stochastic differential equation, which is a generalization of the class studied in Sec. 2.4.5.

The scaling

$$\bar{V} = (\beta/q)^{1/4}\,V \ , \qquad \bar{t} = \sqrt{q\beta}\,t \quad , \tag{4.128}$$

$$a = \gamma/2\sqrt{q\beta} \quad , \tag{4.129}$$

gives

$$\frac{d\bar{V}}{d\bar{t}} = (a - |\bar{V}|^2)\bar{V} + \bar{w}(\bar{t}) \quad , \tag{4.130}$$

where

$$\langle \bar{w}^*(\bar{t}_1)\bar{w}(\bar{t}_2)\rangle = \delta(\bar{t}_1-\bar{t}_2)$$

and simplifies the presentation of the analysis. The parameter a (called the pumping parameter) is proportional to the gain parameter γ which is determined by the net gain in the laser cavity. The pumping parameter is positive or negative depending on whether the laser is above or below its threshold.

Now we turn to the solution of (4.130).

In the early stage of the oscillation buildup, the term $|\bar{V}|^2$ in (4.130) can be neglected. The equation then reduces to the linear equation (4.123) discussed in the previous section (with $\Gamma = -a$). The complex amplitude grows as a transient thermal light, i.e., $\bar{V}$ is a complex Gaussian process whose intensity $\bar{I} = |\bar{V}|^2$ is exponentially distributed with an increasing mean

$$\bar{I}(\bar{t}) = \frac{2}{a}\,(e^{2a\bar{t}} - 1) \quad .$$

The buildup of laser intensity occurs with a time constant

$$(1/2a)/\sqrt{q\beta} = 1/\gamma \quad .$$

As V grows, the nonlinear term in (4.130) can no longer be ignored and the equation has to be solved exactly. The mathematical manipulations used in Sec. 2.4.5, page 36, can be followed with some modification. We start by writing the Fokker-Planck equation that corresponds to the two-dimensional processes $\bar{V} = (\bar{V}_1,\bar{V}_2)$ described by (4.130)

$$\frac{\partial P}{\partial \bar{t}} = -\sum_{i=1}^{2} \frac{\partial}{\partial \bar{V}_i} J_i \quad , \quad J_i = -(a-|\bar{V}|^2)\bar{V}_i P + \frac{\partial}{\partial \bar{V}_i} P \tag{4.131}$$

where $\bar{V}_1$ and $\bar{V}_2$ are the real and imaginary components of V.

Here, P stands for $P(\bar{V}) = P(\bar{V}_1,\bar{V}_2)$ or for the transition probability

$$P[\bar{V}(\bar{t}_2)|\bar{V}(\bar{t}_1)] = P(\bar{V}_1,\bar{V}_2\ ;\ \bar{V}_1',\bar{V}_2').$$

Because of the Markovian nature of the process,

$$P[\bar{V}(\bar{t}_1),\bar{V}(\bar{t}_2), \ldots \bar{V}(\bar{t}_N)] = \left\{\prod_{i=2}^{N} P[\bar{V}(\bar{t}_i)|\bar{V}(\bar{t}_{i-1})]\right\} P[\bar{V}(\bar{t}_1)] \; . \tag{4.132}$$

The solution is greatly facilitated by a transformation to polar coordinates, $\bar{V} = |\bar{V}|\exp(j\phi)$.
By separation of variables, RISKEN [401-403] has found the stationary solutions of (4.131)

$$P(|\bar{V}|) \propto |\bar{V}|\exp[-\tfrac{1}{4}(|\bar{V}|^2-a)] \; , \qquad |\bar{V}| \geq 0 \tag{4.133}$$

and

$$P(\bar{V}|\bar{V}') = \frac{1}{2\pi|\bar{V}|}\;\frac{\psi_{00}(|\bar{V}|)}{\psi_{00}(|\bar{V}'|)}\sum_{m,n=0}^{\infty}\psi_{nm}(|\bar{V}|)\psi_{nm}(|\bar{V}'|)e^{jn(\phi-\phi')}e^{-\lambda_{nm}t} \; , \tag{4.134}$$

where λ_{nm} and ψ_{nm} are the eigenvalues and the orthonormal eigenfunctions of the Schrödinger equation

$$\frac{\partial^2}{\partial r^2}\psi_{nm} - V_n(r)\psi_{nm} = \lambda_{nm}\psi_{nm} \quad , \tag{4.135}$$

where

$$V_n(r) = \frac{n^2}{r^2} + \frac{\psi_{00}''(r)}{\psi_{00}(r)} \tag{4.136}$$

and

$$\psi_{00}(r) \propto \sqrt{r}\,\exp\left(-\frac{1}{8}r^4 + \frac{1}{4}ar^2\right) \quad . \tag{4.137}$$

With these equations, we are (at least formally) in a position to determine the statistical properties of the radiated field. In particular, we are interested in the PD of the intensity and the correlation functions of the field and the intensity.

Intensity Fluctuations

Equation (4.133) for the PD of $|\tilde{V}|$ gives for the intensity, $I = |V|^2$, a truncated Gaussian distribution

$$P(I) = \frac{N}{\sqrt{2\pi}\sigma} \exp[-(I-\mu)^2/2\sigma^2] , \quad I > 0 , \tag{4.138}$$

where

$$\mu = \left(\frac{q}{\beta}\right)^{1/2} a = \frac{\gamma}{\beta} \tag{4.139}$$

$$\sigma^2 = 2q/\beta \tag{4.140}$$

and N is a normalization constant.

Note that the pumping parameter

$$a = \gamma/2\sqrt{q\beta} = \sqrt{2}\mu/\sigma \tag{4.141}$$

is proportional to the ratio between the mean and the standard deviation of the intensity distribution. As Fig.4.15 shows, for a large negative a (laser

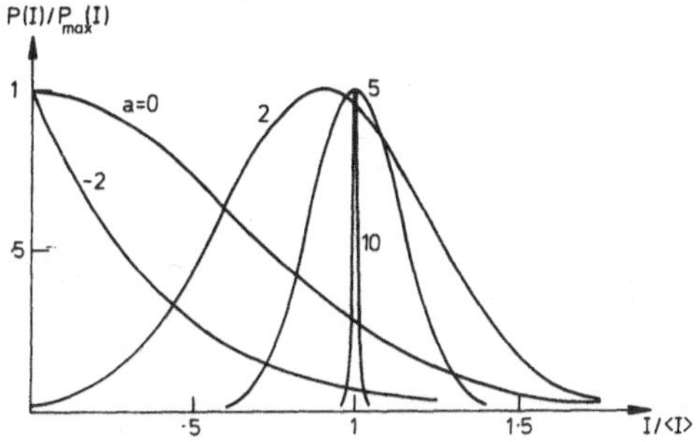

Fig.4.15. Normalized PD of the intensity of laser light for several values of the pumping parameter, a

below threshold), $P(I)$ is the tail of the Gaussian distribution, which looks like an exponential distribution (characterizing a thermal field). On the other hand, for large positive a (laser above threshold), $P(I)$ is a narrow Gaussian distribution (charaterizing the sum of coherent and thermal light). At threshold ($a = 0$), we have

$$P(I) = \frac{2}{\sqrt{2\pi}\sigma} \exp(-I^2/2\sigma^2) , \tag{4.142}$$

which has a mean

$$I_0 = \sqrt{\frac{2}{\pi}}\,\sigma \quad . \tag{4.143}$$

The truncated Gaussian distribution (4.138) is characterized by an mgf

$$Q_I(s) = R_0(a - \sqrt{\pi} I_0 s)/R_0(a) \quad , \tag{4.144}$$

where

$$R_0(a) = \sqrt{\pi}\, e^{a^2/4}\left[1 + \Phi\left(\frac{a}{2}\right)\right] \quad , \qquad \Phi(x) = \frac{2}{\sqrt{\pi}} \int_0^x e^{-y^2} dy \quad . \tag{4.145}$$

Its moments are obtained from

$$\langle I^m \rangle = \left(\frac{\sqrt{\pi}}{2} I_0\right)^m R_m(a)/R_0(a) \quad , \tag{4.146}$$

where $R_m(a)$ is obtained from the recurrence relation

$$\begin{aligned} R_m(a) &= 2(m-1)R_{m-2}(a) + a\, R_{m-1}(a) \quad , \quad m > 1 \\ R_1(a) &= 2 + a\, R_0(a) \quad . \end{aligned} \tag{4.147}$$

In particular,

$$\langle I \rangle = I_0 \left\{ \frac{\sqrt{\pi}}{2} a + e^{-(a/4)^2} [1 + \Phi(\tfrac{a}{2})] \right\} \quad . \tag{4.148}$$

The cumulants can be obtained by the rule

$$K_I^{(m+1)}(a) = \sqrt{\pi}\, I_0 \frac{d}{da} K_I^{(m)}(a) \quad , \tag{4.149}$$

which can be proved by determining the cgf from the logarithm of (4.144) and taking its derivatives. A plot of some low-order cumulants is given in Fig.4.16. It shows how the mean increases continuously with increase of a, whereas the variance increases and then saturates at a constant level.

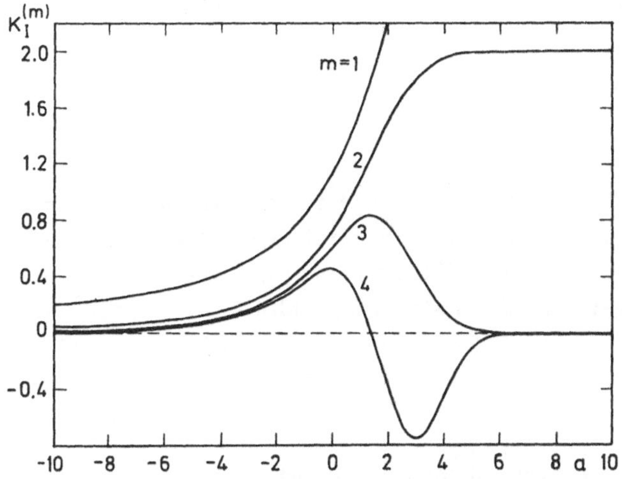

Fig.4.16. Variation of the cumulants of the intensity of laser light with the pumping parameter a (after RISKEN [403])

The normalized moments $\langle I^m\rangle/\langle I\rangle^m$ for coherent light, thermal light, and laser at threshold (a=0) are compared in Table 4.1. It shows that the statistical properties of light from a laser at threshold is intermediate between coherent and thermal light.

Table 4.1 Comparison between the normalized moments of intensity, $\langle I^m\rangle/\langle I\rangle^m$, for coherent, thermal, and laser light at threshold

m	1	2	3	4
Coherent	1	1	1	1
Laser at threshold	1	$\pi/2$	π	$3\pi^2/4$
Thermal	1	2	6	24

Field Correlation

The correlation function $G_V(\tau)$ can be obtained by integrating $V^*(t)V(t+\tau)$ over the PD given by (4.132-134). This gives, after scaling and normalization,

$$g(\tau) = \sum_{m=0}^{\infty} c_m \exp(-\lambda_{1m}\sqrt{q\beta}|\tau|) \quad , \tag{4.150}$$

where

$$c_m = \left[\int_0^\infty r\psi_{00}(r)\psi_{1m}(r)dr\right]^2 \quad ;$$

i.e., the correlation function takes the form of a sum of exponentials. However, numerical calculations show that only the first exponential is of importance for the whole range of values of the pumping parameter, a. Therefore, the spectrum is approximately Lorentzian with a bandwidth

$$\Delta\nu = \sqrt{\beta q}\,\lambda_{10} \quad .$$

The eigenvalue λ_{10} has a value about 10 for a = -10 and drops rapidly with increase of a to a value much less than one for a = 10. This shows how the bandwidth decreases as the laser is pumped. This has been experimentally confirmed.

Intensity Correlation Function

The intensity correlation function

$$G_I(\tau) = \langle|V(t)|^2|V(t+\tau)|^2\rangle = \langle|V(t)|^2|V(t+\tau)|^2\rangle$$

can also be obtained by averaging over the PD given by (4.132-134). Again, this gives a sum of exponentials

$$g_I(\tau) = 1 + \sum_m c'_m \exp(-\lambda_{0m}\sqrt{q\beta}|\tau|) \quad , \tag{4.151}$$

where

$$c'_m = \left[\int_0^\infty r^2\psi_{00}(r)\psi_{0m}(r)dr\right]^2 \quad .$$

Approximately 4 terms of this series have to be considered for a pump parameter slightly above threshold. For large pump parameters, only one exponential prevails. For large negative values of a (laser below threshold) the asymptotic values of λ_{0m}, λ_{1m}, c_m and c'_m have been shown to lead to

$$g_I(\tau) = 1 + |g_V(\tau)|^2 \quad ,$$

which is the same relation that is satisfied by a thermal field (4.78).

4.2.8 The Sum of a Small Number of Independent Modes of Light

The model that we discuss in this section is an optical field that is the sum of N statistically independent contributions, i.e.,

$$V(\underline{x}) = \sum_{j=1}^{N} V_j(\underline{x}) \quad , \tag{4.152}$$

each assumed to be stationary and circularly symmetric. The total field itself is of course stationary and circularly symmetric.

As we know from the central-limit theorem, in the limit of very large N, this field must be Gaussian, irrespective of the statistics of the components $V_j(\underline{x})$. Hence, in this limit, $V(\underline{x})$ must represent thermal light.

It would definitely be useful to study the convergence of the distribution of V to the complex Gaussian distribution (or the convergence of the distribution of the intensity I to the exponential distribution) as N increases and to find the influence of the statistics of $\{V_j\}$ on this convergence.

Aside from the general mathematical interest of this problem, it has some practical importance. For example, the light scattered when a coherent light beam illuminates a small number of independent particles in a liquid can be described by this model, and the amount of deviation from Gaussian statistics should carry information on the number of scattering particles [247-249, 432-434].

The model could also describe the statistics of multimode laser light (see, e.g., [217]). The case N=2 corresponds to the interference of two independent optical fields [485].

At a single space-time point $V_j(\underline{x}) = V_j$ is described by a circularly symmetric complex random variable. Hence its mgf is given by

$$Q_{V_j}(S) = h_j(|S|) \tag{4.153}$$

(cf. Sec. 2.3.2), where h_j is determined by the actual statistical distribution of V_j. Since $\{V_j\}$ are statistically independent, we have

$$Q_V(S) = \prod_{j=1}^{N} h_j(|S|) \quad . \tag{4.154}$$

Because V itself is circularly symmetric, the PD of its absolute value $|V|$ is given by the Hankel transform, (2.42),

$$P(|V|) = |V| \int_0^\infty du\, u\, J_0(u|V|)h(u) \quad , \quad h(u) = \prod_{j=1}^{N} h_j(u) \quad . \tag{4.155}$$

The PD of the intensity $I = |V|^2$ is obtained from the transformation rule $P(I)dI = P(|V|)d|V|$, which gives

$$P(I) = \frac{1}{2} \int_0^\infty du\, u\, J_0(u\sqrt{I})h(u) \quad . \tag{4.156}$$

To be more specific about V_j, assume that

$$V_j = A_j \exp(j\theta_j) \quad , \tag{4.157}$$

where A_j are real random variables and $\{\theta_j\}$ are uniformly distributed random variables.

This gives

$$h_j(u) = \langle J_0(A_j u)\rangle \quad , \tag{4.158}$$

where $\langle\cdot\rangle$ is an expectation value over the distribution of A_j, hence

$$(P(I) = \frac{1}{2} \int_0^\infty du\, u\, J_0(u\sqrt{I})h(u) \quad ,$$
$$h(u) = \prod_{j=1}^{N} \langle J_0(A_j u)\rangle \quad . \tag{4.159}$$

This distribution corresponds to a mgf [247]

$$Q_I(s) = \langle \Psi_2(1;1,1, \ldots 1; -A_1^2 s, -A_2^2 s, \ldots A_N^2 s)\rangle \quad , \tag{4.160}$$

where Ψ_2 is the confluent hypergeometric function of N variables, and to moments given by

$$\langle I^m\rangle = (m!)^2 \underbrace{\sum_{m_1=0}^{\infty} \sum_{m_2=0}^{\infty} \cdots \sum_{m_N=0}^{\infty}}_{\sum_i m_i = m} \frac{\prod_i \langle A_i^{2m_i}\rangle}{\left(\prod_i m_i!\right)^2} , \tag{4.161}$$

$$\langle I\rangle = \sum_{j=1}^{N} \langle A_j^2\rangle \quad . \tag{4.162}$$

If the components are assumed to be identically distributed, then (4.159) gives

$$P_N(I) = \frac{1}{2}\int_0^{\infty} du\, u\, J_0(u\sqrt{I})h(u) \quad ,$$

$$h(u) = \langle J_0(Au)\rangle^N \quad , \tag{4.163}$$

and (4.161) gives

$$\langle I\rangle = N\,\langle A^2\rangle \tag{4.164}$$

$$\langle I^2\rangle = \langle I\rangle^2 \left[2! + \frac{1}{N}\left(\frac{\langle A^4\rangle}{\langle A^2\rangle^2} - 2\right)\right] \tag{4.165}$$

$$\langle I^3\rangle = \langle I\rangle^3 \left\{3! + \frac{9}{N}\left(\frac{\langle A^4\rangle}{\langle A^2\rangle^2} - 2\right) + \frac{1}{N^2}\left[\left(\frac{\langle A^6\rangle}{\langle A^2\rangle^3} - 6\right) - 9\left(\frac{\langle A^4\rangle}{\langle A^2\rangle^2} - 2\right)\right]\right\} \quad . \tag{4.166}$$

A series expansion of (4.163) gives

$$P_N(I) = \frac{1}{\langle I\rangle} e^{-I/\langle I\rangle}\left[1 + \frac{1}{4N}\left(\frac{\langle A^4\rangle}{\langle A^2\rangle^2} - 2\right)\left(2 - \frac{4I}{\langle I\rangle} + \frac{I^2}{\langle I\rangle^2}\right) + \frac{1}{N^2}(\ \ldots\ ) + \ldots\right] \quad . \tag{4.167}$$

It should be noted that if A takes only values less than a finite value α, then the maximum value that I could take is $N^2\alpha^2$ (corresponding to the case when all vectors in the sum (4.152) take their maximum value and are parallel). This means that

$$P_N(I) = 0 \quad , \qquad I > N^2\alpha^2 \quad .$$

Because $P_N(I)$ is the Fourier-Bessel transform of h(u), h(u) is therefore band-limited and a sampling series for $P_N(I)$ is possible. BARAKAT [27] showed that $P_N(I)$ can be written as the Fourier-Bessel series

$$P_N(I) = \sum_{n=1}^{\infty} c_n(N,\alpha,\langle A\rangle)J_0(\gamma_n\sqrt{I}/N\alpha) \quad , \tag{4.168}$$

$$c_n(N,\alpha,\langle A\rangle) = [J_0(\gamma_n\langle A\rangle/N\alpha)]^N/[N\alpha J_1(\gamma_n)]^2 \quad ,$$

where $\gamma_1, \gamma_2, \ldots$ are the positive roots of J_0.

Examining (4.164-168), we immediately see that as $N \to \infty$, P(I) becomes exponential, having moments $\langle I^m\rangle = m!\langle I\rangle^m$, and corresponding to thermal light. This holds irrespective of the statistics of A. If *A has a Rayleigh distribution*, having

$$\langle A^{2m}\rangle = m! \ \langle A^2\rangle^m$$

(Table 2.1), then terms other than the first in (4.167) cancel and P(I) also becomes exponential. This is to be expected because as we know, the sum of a set of complex Gaussian variables is also complex Gaussian. Now let us take other distributions for A. For example, assume that *A is deterministic*, i.e.,

$$\langle A^{2m}\rangle = \langle A^2\rangle^m \quad . \tag{4.169}$$

This gives a distribution of I that has moments

$$\langle I^2\rangle = \langle I\rangle^2\left(2 - \frac{1}{N}\right) \quad ,$$

$$\langle I^3\rangle = \langle I\rangle^3\left(6 - \frac{9}{N} + \frac{4}{N^2}\right) \quad . \tag{4.170}$$

A general expression of $\langle I^m\rangle$ for arbitrary m can be found in [88].

Let us now take another interesting distribution for A, *the log-normal distribution*. From Table 2.1, we obtain

$$\langle A^{2m}\rangle = e^{2(m^2-m)\sigma^2}\langle A^2\rangle^m \quad .$$

By substitution in (4.165), we get

$$\langle I^2\rangle = \langle I\rangle^2\left[2 + \frac{1}{N}\left(e^{4\sigma^2}-2\right)\right] \quad .$$

This means that for $\sigma = 0.8$ and $N = 10$, the normalized second moment is 50 % greater than that for thermal light. Indeed, it can be shown that the convergence in this case is very slow. This is related to the so-called "permanence" property of the log-normal distribution [28, 349]. BARAKAT [28] showed that this property is a consequence of the fact that the log-normal distribution has a nonzero coefficient of skewness (cf. Sec. 2.1.2 page 9).

When A is constant, the PD of I, (4.163), becomes

$$P_N(I) = \frac{1}{2}\int_0^\infty du\, u\, J_0(u\sqrt{I})[J_0(Au)]^N \quad , \tag{4.171}$$

which is the expression obtained by KLUYVER [269], PEARSON [367], and RAYLEIGH [393].
For $N = 1$, this gives $P_1(I) = \delta(I-\langle I\rangle)$ as we expect for a constant intensity.
For $N = 2$, it gives

$$P_2(I) = \frac{1}{\pi}\,[I(2\langle I\rangle - I)]^{-1/2} \quad , \qquad I \leq 2\langle I\rangle \quad . \tag{4.172}$$

This shows that when two independent beams with constant amplitudes and uniformly distributed phases interfere, the intensity I can take values between zero and $2\langle I\rangle$ with the probability of taking any of these extreme values much higher than for the intermediate values (see Fig.4.17a,b). This means that the probability of complete destructive or complete constructive interference is much greater than the gray state in between.

Fig.4.17b shows how $P_N(I)$ approaches the exponential distribution as N exceeds 5. For $N > 7$, the infinities and discontinuities disappear and we can use the expansion (4.167) [392].

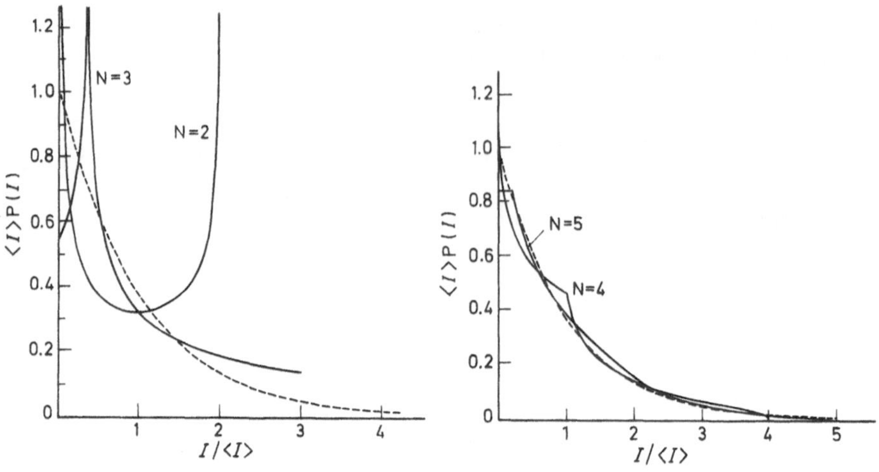

Fig.4.17a and b. PD of the intensity of light that is composed of N independent components each of constant amplitude and uniformly random phase (after PUSEY et al. [392]

Joint Statistics

Here we limit ourselves to the calculation of the intensity correlation function $G_I(\underline{x}_1,\underline{x}_2) = \langle I(\underline{x}_1)I(\underline{x}_2)\rangle$. Other statistics are much more complicated and will not be considered here (see [88]). By substitution from (4.152), we get

$$G_I(\underline{x}_1,\underline{x}_2) = \sum_{i,j}\sum_{\ell,m} \langle V_i^*(\underline{x}_1)V_j(\underline{x}_1)V_\ell^*(\underline{x}_2)V_m(\underline{x}_2)\rangle \quad .$$

Since the components V_i are circularly symmetric and statistically independent, it becomes immediately clear that the only nonzero contributions come from the terms with $i = j$, $\ell = m$ and $i = \ell \neq j = m$. We thus have

$$\begin{aligned} G_I(\underline{x}_1,\underline{x}_2) = &\sum_{i\neq\ell} \langle V_i^*(\underline{x}_1)V_i(\underline{x}_1)\rangle\, \langle V_\ell^*(\underline{x}_2)V_\ell(\underline{x}_2)\rangle \\ &+ \sum_i \langle V_i^*(\underline{x}_1)V_i(\underline{x}_1)V_i^*(\underline{x}_2)V_i(\underline{x}_2)\rangle \\ &+ \sum_{i\neq\ell} \langle V_i^*(\underline{x}_1)V_i(\underline{x}_2)\rangle\, \langle V_\ell^*(\underline{x}_2)V_\ell(\underline{x}_1)\rangle \end{aligned}$$

$$= N(N-1)\langle I_i(\underline{x}_1)\rangle \langle I_i(\underline{x}_2)\rangle + N\langle I_i(\underline{x}_1)I_i(\underline{x}_2)\rangle$$

$$+ N(N-1)|G_{V_i}(\underline{x}_1,\underline{x}_2)|^2 ,$$

i.e.,

$$G_I(\underline{x}_1,\underline{x}_2) = \langle I(\underline{x}_1)\rangle \langle I(\underline{x}_2)\rangle + |G_V(\underline{x}_1,\underline{x}_2)|^2 \\ + N[\langle \Delta I_i(\underline{x}_1)\Delta I_i(\underline{x}_2)\rangle - |G_{V_i}(\underline{x}_1,\underline{x}_2)|^2] \tag{4.173}$$

This is an interesting result, because if the components V_i are Gaussian, $\langle \Delta I_i(\underline{x}_1)\Delta I_i(\underline{x}_2)\rangle = |G_{V_i}(\underline{x}_1,\underline{x}_2)|^2$ and we get for the total field

$$G_I(\underline{x}_1,\underline{x}_2) = \langle I(\underline{x}_1)\rangle\langle I(\underline{x}_2)\rangle + |G_V(\underline{x}_1,\underline{x}_2)|^2 , \tag{4.174}$$

which is characteristic of a thermal field (4.74). On the other hand, if the components V_i have constant amplitudes, the cumulant $\langle \Delta I_i(\underline{x}_1)\Delta I_i(\underline{x}_2)\rangle$ vanishes and we get

$$G_I(\underline{x}_1,\underline{x}_2) = \langle I(\underline{x}_1)\rangle \langle I(\underline{x}_2)\rangle + (1-\tfrac{1}{N})|G_V(\underline{x}_1,\underline{x}_2)|^2 . \tag{4.175}$$

Again, in the limit of large N, (4.174) is reproduced. In the limit N = 1, the light intensity is constant and the second term of (4.175) vanishes.

Number Fluctuations

Sometimes the number of components N in the sum (4.152) is itself a random variable. This occurs, for example, when light is scattered from a small number of particles in a dilute solution. When the scattering volume, which is determined by the illuminating beam, is smaller than the size of the cell, then as the particles freely enter and leave the scattering volume, the number N fluctuates randomly.

One should therefore average all of the previous distributions and moments over the probability distribution of N. Several authors have studied the effect of number fluctuations, using a Poisson distribution for N [29, 392, 431, 433] theoretically and experimentally. The averaging of $P_N(I)$ gives

$$P(I) = \frac{1}{2}\int_0^\infty du\, u\, J_0(u\sqrt{I})\exp\left\{\langle N\rangle[J_0(u\sqrt{\langle I\rangle/\langle N\rangle})-1]\right\} \quad . \tag{4.176}$$

Averaging of the moments in (4.170) gives

$$\langle I^2\rangle = \langle I\rangle^2(2 + 1/\langle N\rangle) \quad , \tag{4.177}$$

$$\langle I^3\rangle = \langle I\rangle^3(6 + 9/\langle N\rangle + 1/\langle N\rangle^2) \quad , \tag{4.178}$$

$$\ldots ,$$

and so on.

The averaging over fluctuations in N obviously smooths the probability distribution of I and broadens it. The distinctions between the distributions that correspond to different values of $\langle N\rangle$ remain clear, provided that $\langle N\rangle$ is small.

The fluctuations of I due to interference of the individual waves and those due to number fluctuations could also be distinguished by their time scales. In experiments on light scattering from macromolecules in dilute solutions, number fluctuations are usually much slower.

In conclusion, if N is small, it can be measured by a statistical analysis of the intensity of the non-Gaussian light, If N is large, the light is Gaussian and N cannot be estimated.

4.2.9 Phase-Fluctuating (or Diffused) Light

Consider an optical field described by the model

$$V(\underline{x}) = V\exp[-j\phi(\underline{x})] \quad , \tag{4.179}$$

where V is a constant and $\phi(\underline{x})$ is a stochastic function of space and time.

This model is often encountered in laser-radar applications, in which the phase fluctuations are cause by range fluctuations or steady but slow target motion. Also, coherent light that passes through a random scattering medium (phase screen) suffers phase fluctuations and can be described by this model.

The field fluctuations are governed by the statistics of $\phi(\underline{x})$. A most common situation is that in which $\phi(\underline{x})$ is a zero-mean Gaussian process. The process is characterized completely by the correlation function

$$G_\phi(\underline{x}_1,\underline{x}_2) = \sigma^2 g_\phi(\underline{x}_1,\underline{x}_2) \quad . \tag{4.180}$$

The field itself is obviously non-Gaussian and is not circularly symmetric. In order to determine the coherence functions of the field V, we need to determine the expectation value

$$\left\langle \exp\left[-j\sum_\ell \phi(\underline{x}_\ell)\right]\right\rangle .$$

This quantity is recognized as the multifold mgf of the process $\phi(\underline{x})$ at $s_\ell = j$, $\ell = 1,2, \ldots$. From (2.30), we can write

$$\begin{aligned} \left\langle \exp\left[-j\sum_\ell \phi(\underline{x}_\ell)\right]\right\rangle &= \exp\left[-\frac{1}{2}\sum_\ell\sum_k \langle\phi(\underline{x}_\ell)\phi(\underline{x}_k)\rangle\right] \\ &= \exp\left[-\frac{1}{2}\sum_\ell\sum_k G_\phi(\underline{x}_\ell,\underline{x}_k)\right] \quad . \end{aligned} \tag{4.181}$$

This relation can immediately be used to show that

$$g_V(\underline{x}_1,\underline{x}_2) = \exp\left\{-\sigma^2[1-g_\phi(\underline{x}_1,\underline{x}_2)]\right\} \quad . \tag{4.182}$$

This function decays from a value of 1 when $\underline{x}_1 = \underline{x}_2$ to a value of $\exp(-\sigma^2)$ when $\underline{x}_1$ and $\underline{x}_2$ are widely separated. It does not decay to zero, as in all the models of light that were previously described. This is because

$$\langle V(\underline{x})\rangle = \mathcal{V}\exp(-\tfrac{1}{2}\sigma^2)$$

is not zero, as in all circularly symmetric fields. Another peculiar aspect of g_V is that it does not factorize into spatial and temporal functions, even if g_ϕ does. This makes it difficult to attain cross-spectral purity. Higher-order coherence functions can be similarly obtained by using (4.181), e.g.,

$$G_V^{2,2}(\underline{x}_1,\underline{x}_2,\underline{x}_3,\underline{x}_4) = \exp\Big\{-\sigma^2\Big[2 + g_\phi(\underline{x}_1,\underline{x}_2) + g_\phi(\underline{x}_3,\underline{x}_4) - g_\phi(\underline{x}_1,\underline{x}_3) - g_\phi(\underline{x}_1,\underline{x}_4) - g_\phi(\underline{x}_2,\underline{x}_3) - g_\phi(\underline{x}_2,\underline{x}_4)\Big]\Big\} \quad . \tag{4.183}$$

We have assumed above that $\phi(\underline{x})$ has a Gaussian distribution with zero mean and variance σ^2. If $\sigma >> 2\pi$, then we can assume that $\phi(\underline{x})$ is practically uniformly distributed over the interval $[0,2\pi]$. This means that the field is circularly symmetric, from which $<V(\underline{x})> \simeq 0$. Also, under this assumption, $g(\underline{x}_1,\underline{x}_2)$ varies from a value 1 when $g_\phi = 1$ to a value zero when $g_\phi = 0$.

The light intensity that corresponds to (4.179) is obviously constant. Direct detection of such a field would therefore not reveal the phase-fluctuation statistics.

Two situations are considered, in which phase fluctuations are manifest by intensity measurements:
1) When the light is mixed with a coherent beam, as in heterodyne detection of laser radar signals; 2) when the light propagates through a nonideal system (e.g., when it is diffracted).

Phase-Fluctuating Field Mixed with a Coherent Field

When a coherent field of analytic signal V_0 and intensity $I_0 = |V_0|^2$ is mixed with a phase-fluctuating field $V_s\exp[-j\phi(\underline{x})]$ of intensity I_s, the resultant field

$$V(\underline{x}) = V_0 + V_s\exp[-j\phi(\underline{x})]$$

has an instantaneous intensity

$$I(\underline{x}) = I_0 + I_s + 2\sqrt{I_0 I_s}\,\cos\phi(\underline{x}) \quad . \tag{4.184}$$

The moments of I are given by

$$<I^m> = \sum_{r=0}^{m} \binom{m}{r} (I_0+I_s)^{m-r}(2\sqrt{I_0 I_s})^r <\cos^r(\phi)> \quad . \tag{4.185}$$

By using (4.181), we obtain

$$<\cos^r(\phi)> = \sum_{\ell=0}^{r} \binom{r}{\ell} \exp\left[-\sigma^2(r-2\ell)^2/2\right] \quad .$$

This gives, for example,

$$\langle I\rangle = I_0+I_s+\sqrt{I_0I_s}\,\exp(-\sigma^2/2) \quad ,$$

$$\langle I^2\rangle = (I_0+I_s)^2+\sqrt{I_0I_s}(I_0+I_s)\exp(-\sigma^2/2)+2I_0I_s[1+\exp(-2\sigma^2)] \quad , \tag{4.186}$$

and so on.

Note that if $\sigma >> 2\pi$, the uniform distribution of ϕ gives

$$\langle\cos^r(\phi)\rangle = (r-1)!!/r! \quad r \text{ even}$$

$$= 0 \quad r \text{ odd} \quad ,$$

from which

$$\begin{aligned}\langle I\rangle &= I_0 + I_s\\ \langle I^2\rangle &= \langle I\rangle^2 + 2I_sI_0\\ \langle I^3\rangle &= \langle I\rangle^3 + 6I_sI_0^2 + 6I_s^2I_0\\ \langle I^4\rangle &= \langle I\rangle^4 + 12I_sI_0^2 + 18I_s^2I_0 + 12I_s^3I_0\\ &\ldots .\end{aligned} \tag{4.187}$$

In the case of a uniformly distributed ϕ, we can also compute the mgf that corresponds to (4.184) and obtain

$$Q_I(s) = \exp(-s\langle I\rangle)I_0(m\langle I\rangle s) \quad , \quad m = \frac{4\sqrt{I_0I_s}}{\langle I\rangle} \quad , \tag{4.188}$$

where $I_0(\cdot)$ is the modified Bessel function. The moments in (4.187) can be obtained from (4.188) by using the usual differentiation rule.

Now we turn to the mulifold statistics. The intensity correlation function is given by

$$\begin{aligned}\langle I\rangle^2 g_I(\underline{x}_1,\underline{x}_2) = \langle I(\underline{x}_1)I(\underline{x}_2)\rangle &= (I_0+I_s)^2+4I_0I_s\langle\cos\phi(\underline{x}_1)\cos\phi(\underline{x}_2)\rangle\\ &+ 4(I_0+I_s)\sqrt{I_0I_s}\langle\cos\phi(\underline{x}_1)\rangle \quad .\end{aligned} \tag{4.189}$$

Using (4.181), we obtain

$$\langle I\rangle^2 g_I(\underline{x}_1,\underline{x}_2) = (I_0+I_s)^2+4(I_0+I_s)\sqrt{I_0I_s}\exp(-\sigma^2/2)$$
$$+2I_0I_s\left(\exp\left|-\sigma^2[1+g_\phi(\underline{x}_1,\underline{x}_2)]\right|+\exp\left|-\sigma^2[1-g_\phi(\underline{x}_1,\underline{x}_2)]\right|\right). \quad (4.190)$$

If $\sigma^2 >> 1$, we can write

$$g_I(\underline{x}_1,\underline{x}_2) \simeq 1 + \frac{2I_0I_s}{\langle I\rangle^2}\exp\left|-\sigma^2[1-g_\phi(\underline{x}_1,\underline{x}_2)]\right|$$
$$= 1 + \frac{2I_0I_s}{\langle I\rangle^2} g_V(\underline{x}_1,\underline{x}_2) \quad (4.191)$$

Compare this relation with the corresponding ones in the case of thermal field, (4.78), and the case of a mixture of coherent and thermal fields, (4.110).

Phase-Fluctuating Light after Propagation Through an Optical System. The Phenomenon of Speckle

A phase-fluctuating field $V_s(\underline{r}) = \mathcal{V}_s(\underline{r})\exp[j\phi(\underline{r})]$ has a nonfluctuating intensity $I_s(\underline{r}) = |\mathcal{V}_s(\underline{r})|^2$. When it propagates through an optical system that has a point spread function $h(\underline{r},\underline{r}')$, the resultant field (cf. Sec. 4.1.1).

$$V(\underline{r}) = \int h(\underline{r},\underline{r}')V_s(\underline{r}')d\underline{r}' \quad (4.192)$$

has an instantaneous intensity

$$I(\underline{r}) = \iint h^*(\underline{r},\underline{r}')h(\underline{r},\underline{r}'')V_s^*(\underline{r}')V_s(\underline{r}'')d\underline{r}'d\underline{r}'' \quad (4.193)$$

that, in general, fluctuates with $\underline{r}$.

For an ideal system, i.e., $h(\underline{r},\underline{r}') \sim \delta(\underline{r}-\underline{r}')$, the image is an exact replica of the object, it suffers no intensity fluctuations. A nonideal system has a finite resolution (a finite width of the function h). Every point on the object contributes to the field at many points in the image; interference between the propagated waves results in a random intensity distribution, or a random interference pattern known as speckle. The speckle phenomenon takes place whenever coherent light undergoes random phase fluctuations (e.g., due to scattering from a rough surface) and then propagates through a nonideal system. The speckle phenomenon was observed a long time ago but has only recently become the subject of extensive research and wide applications (for review articles, cf. [111]).

We concern ourselves here with the statistics of a speckle pattern $I(\underline{r})$ in the case when $\phi(\underline{r})$ is Gaussian. By use of (4.192,193,182,183), we can relate the moments of $V(\underline{r})$ and $I(\underline{r})$ to those of $\phi(\underline{r})$,

$$G_V(\underline{r}_1,\underline{r}_2) = \iint h^*(\underline{r}_1,\underline{u}_1)h(\underline{r}_2,\underline{u}_2)V_s^*(\underline{u}_1)V_s(\underline{u}_2) \exp\left\{-\sigma^2[1-g_\phi(\underline{u}_1,\underline{u}_2)]\right\} d\underline{u}_1 d\underline{u}_2 \tag{4.194}$$

$$\langle I(\underline{r})\rangle = G_V(\underline{r},\underline{r}) \tag{4.195}$$

$$G_I(\underline{r}_1,\underline{r}_2) = \iiiint h^*(\underline{r}_1,\underline{u}_1)h(\underline{r}_1,\underline{u}_2)h^*(\underline{r}_2,\underline{u}_3)h(\underline{r}_2,\underline{u}_4) V_s^*(\underline{u}_1)V_s(\underline{u}_2)V_s^*(\underline{u}_3)V_s(\underline{u}_4) \exp\left[-\sigma^2(2+g_{12}+g_{34}-g_{13}-g_{14}-g_{23}-g_{24})\right] d\underline{u}_1 d\underline{u}_2 d\underline{u}_3 d\underline{u}_4 \quad , \tag{4.196}$$

$$g_{ij} = g_\phi(\underline{u}_i,\underline{u}_j)$$

$$\langle I^2(\underline{r})\rangle = G_I(\underline{r},\underline{r}) \quad . \tag{4.197}$$

Higher-order moments can similarly be related by lengthier expressions.

It remains to evaluate these integrals for a given source distribution V_s, phase correlation function g_ϕ, and optical system h.

We first note that if the phase correlation length ξ (width of g_ϕ) is small compared to the resolution of the system, then light at any point in the image plane is the sum of many contributions of independent components from the object and must be Gaussian. If h represents simple diffraction, then every point in the object contributes to all points in the image. The size of the object a then plays the role of the resolution length. If $\xi \ll a$, the diffracted light must be Gaussian. To illustrate possible non-Gaussian behaviour of speckled light, we discuss briefly an example.

A coherent beam of light, $V_s(\underline{r}) = A\exp(-r^2/a^2)$, that has a Gaussian intensity profile of width a propagates through a phase screen that introduces Gaussian random phase fluctuations $\phi(\underline{r})$. The far-field diffracted light is observed. The optical system has a psf, h, given by (4.9) (cf. the geometry in Fig.4.1). Assuming that $\sigma^2 \gg 1$ (deep phase screen) and approximating g_ϕ by $g_\phi(\underline{r}) = 1 - r^2/\xi^2$, JAKEMAN and PUSEY [247, 249] evaluated the integrals (4.194-197) and obtained

$$\langle I\rangle \propto \left(\frac{a\xi}{\sigma}\right)^2 \exp\left[\left(\frac{\omega\xi}{2c\sigma}\right)^2 \sin^2(\theta)\right] \quad (4.198)$$

$$\langle I^2\rangle/\langle I\rangle^2 \simeq 2(1-\xi^2/a^2)+\left(\frac{\xi\sigma}{2a}\right)^2 \exp\left[\left(\frac{\omega\xi}{2c\sigma}\right)^2 \sin^2(\theta)\right] \quad (4.199)$$

$$g_V(\underline{r}_1,\underline{r}_2) \simeq \exp\left[-\frac{1}{8}\left(\frac{\omega a}{c}\right)^2 u_-^2\right] \quad (4.200)$$

$$\begin{aligned} g_I(\underline{r}_1,\underline{r}_2) \simeq\ & (1-\xi^2/a^2)[1+|g_V(\underline{r}_1,\underline{r}_2)|^2] \\ & +\left(\frac{\xi\sigma}{2a}\right)^2 J\left(\frac{\omega\xi}{2c}\,u_-\right)\exp\left[\left(\frac{\omega\xi}{4c\sigma}\right)^2 (u_+^2+2u_-^2)\right], \end{aligned} \quad (4.201)$$

where

$$J(x) = 2J_1(x)/x \quad \text{and} \quad u_\pm = |\sin(\theta_1)\underline{r}_1/r_1 \pm \sin(\theta_2)\underline{r}_2/r_2|$$

(cf. Fig.4.1).

The studies of GALLAGHER and LIU [166], MENZEL and STOFFREGEN [345] and BERTOLOTTI et al. [55] are also relevant to this problem.

An examination of these equations leads to the following conclusions.

The mean intensity exhibits a broad angular spread and an almost total absence of the direct beam. This is typical of radiation scattered by a deep random phase screen or a very rough surface. The mean intensity distribution is determined by the ratio σ/ξ, the root-mean-square slope of the wavefront of the diffused light.

Departure from the Gaussian statistics is evident from (4.199) and (4.201). In the limit of very large a/ξ, Gaussian statistics are obtained. In (4.199), the deviation from the Gaussian value of 2 is proportional to $(\xi\sigma/a)^2$; therefore even for large a/ξ, the deviation may be considerable.

The spatial coherence function g_V has an angular width $(c/\omega a)$. This characterizes the far field of an incoherent source of size a. The second term of g_I (the non-Gaussian term) has the decay length that is expected from a source of radius ξ; it will fall more slowly if $\xi < a$.

Experimental verification of these effects was done by JAKEMAN and PUSEY [248], PUSEY and JAKEMAN [391] and SCUDIERI et al. [436]. JAKEMAN and MCWHIRTER [230] also studied the effect in the Fresnel region (see also [466, 504]. The effect of temporal fluctuations in ϕ was studied by JAKEMAN and PUSEY [249].

The effect of time fluctuations in the primary source V_s (speckle in polychromatic light) was also studied by several authors (e.g., [244, 329, 363]).

The statistics of a coherent addition of a uniform beam to a speckle pattern was examined by DAINTY [110]. BARAKAT [26] investigated the statistics of mixtures of speckle patterns.

Interest in the speckle phenomenon and its applications is growing.

5. Photoelectron Events: A Doubly Stochastic Poisson Process or Theory of Photoelectron Statistics

In the previous chapter, we studied the classical statistical description of optical fields without any reference to the question of their measurement. In this chapter we study the statistical properties of photoelectrons that are generated when light illuminates a photodetector. If the light level is sufficiently low, then the photoelectrons can be treated as individual events, the statistics of whose occurrences have to be described by the techniques of point processes (Chap. 3).

When an optical field illuminates the cathode of a photomultiplier tube, it excites its electrons from their ground state to an excited state where they are emitted and collected by the anode. Such a transition can then be electronically recorded, counted or processed. Although light can be detected by many other means (the human eye, the photographic film, or photoconductors), the photomultiplier tube plays an important role in measuring the statistical properties of light because of its high sensitivity and ability to record the individual occurrences of photoelectrons.

It is now well established that photoelectron emissions follow a doubly stochastic Poisson point process (DSP.PP) (Sec. 3.5) whose rate density is proportional to the instantaneous rate of electromagnetic energy collected by the detector (the instantaneous intensity, as defined in Sec. 4.1.2 integrated over the detector's area).

Taking this fact for granted, we are in a position to determine the statistics of photoelectrons generated by all optical fields of the various statistical models of Sec. 4.2, simply by using the properties of a DSP.PP (Sec. 3.5). This is exactly what we shall do in Sec. 5.2. But before we embark on this task, let us discuss some of the background behind the doubly stochastic Poisson nature of photoelectron events.

Classical arguments [306, 307, 309], a semiclassical analysis [317], and a fully quantum analysis [171, 260, 275, 298], have shown that *the probability of transition of an electron in a photodetector's surface from its ground state to an unbounded state* (where it can be emitted) *during a time interval* Δt (much longer than a period of electromagnetic oscillation but much shorter than any other time of experimental interest) *is proportional to the time interval* Δt *and to the instantaneous intensity* $I(\underline{r},t)$.

For a single realization of the instantaneous intensity $I(\underline{r},t)$ (i.e., if $I(\underline{r},t)$ is assumed deterministic) and *if the photoemissions are statistically independent of each other*, then it follows (cf. Sec. 3.4) that the photoelectron events form a Poisson point process (P.PP) with rate density $\lambda(t) \propto \int_A I(\underline{r},t)d\underline{r}$, where A is the area of the detector. By allowing $I(\underline{r},t)$ (and hence $\lambda(t)$) to be stochastic, we conclude that photoelectron occurrence is governed by a DSP.PP (cf. Sec. 3.5).

An extensive collection of experimental evidence supports the DSP.PP nature of photoelectron events by confirming predictions based on it.

In Sec. 5.1, the semiclassical derivation of the linear relation between $\lambda(t)$ and $I(t)$ of MANDEL et al. [317] is outlined. Then, in Secs. 5.2 and 5.3, the statistical properties of photoelectron events that result from the detection of light whose statistical properties follow the various models of Sec. 4.2 are derived. Some effects that lead to deviations from the ideal DSP.PP description are discussed in Sec. 5.4.

5.1 The Photoelectric Detection of Light

5.1.1 Semiclassical Derivation of the Photodetection Equation

Consider the interaction of an electromagnetic field with an electron in an atom of the photodetector material. The *electromagnetic field*, which is treated classically, is described by its vector potential $v(t)$ at the location of the atom ($v(t)$ is a real function). The *unperturbed electron* is described quantum mechanically by a Hamiltonian H_0 that has a set of eigenstates $|\psi_n\rangle$ that correspond to energy levels E_n. We assume that there exist a bound state (ground state) $|\psi_b\rangle$ at an energy level E_b and a quasi-continuum of excited states $|\psi_k\rangle$ with energy levels $E_k = E_b + \hbar\omega_k$. In units of frequency $\omega = E/\hbar$, the excited states have a density $\rho(\omega_k)$.

We study the process of excitation from the ground state (ionization or photoemission) and assume that no intermediate levels participate in the photoemission process.

Let the *interaction* between the field and the electron start at time $t = 0$ and assume that the radiation creates a time-dependent perturbation Hamiltonian $H_1(t)$. In the *dipole approximation* $H_1(t) \sim v(t).P$, where P is the momentum of the electron. According to the *first-order perturbation theory* (see any textbook on quantum mechanics) the probability of the electron's transition from the state $|\psi_b\rangle$ to a state $|\psi_k\rangle$ within a time Δt equals $|C_k(\Delta t)|^2$, where

$$C_k(t) = \frac{1}{j\hbar}\int_0^{\Delta t} \langle\psi_k|H_1(t')|\psi_b\rangle \exp(j\omega_k t')dt'$$

$$\propto \int_0^{\Delta t} P_{kb}\; v(t') \exp(j\omega_k t')dt' \quad ,$$

where $P_{kb} = \langle\psi_k|P|\psi_b\rangle$. If we assume that P_{kb} is approximately independent of k, we have

$$C_k(t) \propto \int_0^{\Delta t} v(t') \exp(j\omega_k t')dt' \quad . \tag{5.1}$$

The probability that a transition to any of the unbound states occurs in a time Δt is then given by an integration over the density of unbound states

$$\lambda(t)\Delta t = \int_0^{\infty} |C_k(\Delta t)|^2 \rho(\omega_k)d\omega_k \quad , \tag{5.2}$$

where $\lambda(t)$ is the probability density of electron excitation.

We shall prove that, under certain conditions, $\lambda(t)$ is proportional to the instantaneous intensity of the radiation $I(t) \propto |V(t)|^2$ (where $V(t)$ is the complex analytic signal that corresponds to $v(t)$). For simplicity, let us start by assuming that the radiation is monochromatic with frequency ω_0,

$$v(t) = A_0 \exp(-j\omega_0 t) + A_0^* \exp(j\omega_0 t) \quad .$$

Then by substituting in (5.1), we get

$$C_k(t) \propto A_0 \exp[-j(\omega_0-\omega_k)\Delta t/2]\frac{\sin[(\omega_k-\omega_0)\Delta t/2]}{(\omega_k-\omega_0)}$$

$$+ A_0^* \exp[j(\omega_0+\omega_k)\Delta t/2]\;\frac{\sin[(\omega_k+\omega_0)\Delta t/2]}{(\omega_k+\omega_0)}$$

If we assume that $\Delta t \gg \omega_0^{-1}$, i.e., *$\Delta t$ is much larger than the period of oscillation of the radiation,* then we immediately conclude that the second term in $C_k(t)$ is negligible. Using (5.2), we can then write

$$\lambda(t)\Delta t \propto |A_0|^2 \int_0^\infty \rho(\omega_k) \frac{\sin^2[(\omega_k-\omega_0)\Delta t/2]}{(\omega_k-\omega_0)^2} d\omega_k \quad .$$

If, moreover, we assume that Δt^{-1} is much smaller than the bandwidth of the density of unbound states $\Delta\Omega$, i.e., $\Delta t\Delta\Omega \gg 1$, we can put

$$\frac{\sin^2[(\omega_k-\omega_0)\Delta t/2]}{(\omega_k-\omega_0)^2} \to \frac{\pi}{2}\Delta t\delta(\omega_k-\omega_0)$$

and get

$$\lambda(t) \propto |V|^2 \rho(\omega_0) \propto I(t) \quad , \quad I = |V|^2 = |A|^2 \quad .$$

This shows that the probability rate of excitation is proportional to the radiation intensity.

In order to repeat the derivation for a general (not necessarily monochromatic) radiation, we write

$$v(t) = \int_{-\infty}^{\infty} A(\omega)\, e^{j\omega t} d\omega \quad .$$

Substituting in (5.1), we see that when Δt is much longer than the radiation period of oscillation, the negative-frequency part of $A(\omega)$ does not contribute, and we get

$$C_k(\Delta t) \propto \int_0^\infty A(\omega)\, \exp[j(\omega-\omega_k)\Delta t/2] \frac{\sin[(\omega_k-\omega)\Delta t/2]}{(\omega_k-\omega)} \quad .$$

By substituting in (5.2) and assuming that

I) the radiation has a bandwidth $\Delta\omega$ much smaller than the bandwidth of unbound states $\Delta\Omega \gg \Delta\omega$ (see Fig.5.1)

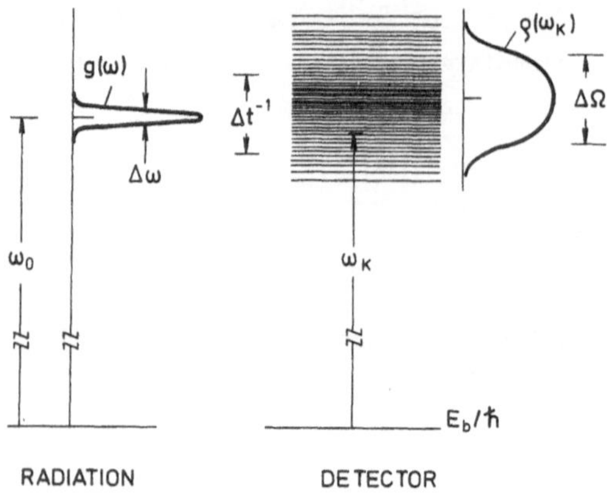

Fig.5.1. Energy levels of a photodetector

II) the time Δt *is much shorter than the inverse of the radiation bandwidth*, i.e., shorter than the field's coherence time, $\Delta t^{-1} \gg \Delta\omega$, we get

$$\lambda(t) \propto \left| \int_0^\infty A(\omega)e^{j\omega t}d\omega \right|^2 = |V(t)|^2 = I(t) \quad ,$$

where $I(t) = |V(t)|^2$ is the instantaneous radiation intensity. Of course, a photodetector surface is composed of many atoms. If we assume that *the atoms are independent*, then the probability rate of transition of an electron in the whole detector surface is given by

$$\lambda(t) = \alpha \int_A I(\underline{r},t)d\underline{r} \quad , \tag{5.3}$$

where α is a constant, the quantum efficiency of the detector, and A is the area of the detector.

The above derivation is oversimplified. Rigorous semiclassical and fully quantum analyses can be found in the references mentioned in the introduction to this chapter. More recently, ROCCA [404] and ARNEDO and ROCCA [23] studied the effect of interaction between the atoms of the detectors and showed that, in practical experiments, these effects can be negligible. ROUSSEAU [412] studied the case in which $\Delta\omega$ is not necessarily much smaller than $\Delta\Omega$ (see Fig.5.1) and concluded that the emission of photoelectrons is still a DSP.PP with a rate density different from that given by (5.3).

5.1.2 Photoelectrons versus Photons

In the semiclassical treatment of photoelectron detection, the electromagnetic field is treated classically and the concept of the photon is not invoked [289, 437]. Atoms on the surface of a photodetector absorb energy from the incident electromagnetic radiation in units of quanta of energy, $h\nu$. Each absorbed quantum instantaneously excites an electron, called a photoelectron. The process of energy absorption and consequent photoelectron emission is probabilistic in nature. The number of photoelectrons emitted due to coherent electromagnetic radiation, whose intensity is deterministic, follows a random Poisson process. Statistical fluctuations in the radiation intensity modify the photoelectron statistics as will be shown in the rest of this book.

According to the fully quantum theory, the electromagnetic radiation is described by its Hamiltonian, eigenvalues of which have the discrete values $E_n = n\hbar\omega$, $n = 0,1,2, \ldots$, where $\hbar\omega$ is the quantum of energy called the photon. The values E_n are the only values that could possibly be obtained when the energy is measured. The state of an electromagnetic field is described by a vector in a Hilbert space, whose projections on the eigenvectors of the Hamiltonian operator determine the probability law of the outcome of a measurement of energy. A given state corresponds to a certain probability distribution of n, the number of photons. For example, an electromagnetic field in a coherent state [168, 169] corresponds to a Poisson distributed n. One can therefore conceive the electromagnetic radiation as a collection of photons of different frequencies, polarizations, and directions of propagation, each with an assigned probability distribution.

When the combined radiation-atoms system is analyzed quantum theoretically, it turns out that the number of emitted photoelectrons is proportional to the number of photons. There exists a one-to-one correspondence between photoelectrons and photons.

We have already mentioned that correct photoelectron distribution could be obtained from a semiclassical analysis. Due to the correspondence between photons and photoelectrons, these distributions also describe the photons.

The photoelectron statistics presented in this chapter and in the entire book could also be regarded as photon statistics.

5.2 Single-Fold Photoelectron Statistics of some Special Optical Fields

In this section, the statistical properties of photoelectron events due to optical fields described by several statistical models are derived. These properties, which involve the statistics of the number of counts in specified time intervals and the statistics of times of occurrence of the events, as defined in Chap. 3, can be easily determined once the statistics of the integrated rate of events $W = \int_0^T \lambda(t)dt$ in an interval [0,T] are found. W is proportional to the total energy collected by the detector, $\int_A \int_0^T I(\underline{r},t) d\underline{r}dt$, and therefore depends on the statistics of the light intensity I. To economize on mathematical expressions (but without any loss of generality), we assume that the quantum efficiency is one, i.e., $\lambda(t) = \int_A I(\underline{r},t)d\underline{r}$. This can be regarded as if the light intensity were measured in units of photoelectron counts per second per unit area. Because of its importance, we begin the analysis in each case by determining the moment-generating function of the integrated rate of events W. Using the rules (3.55-57, 68, 81, 93-96), we directly determine the statistics of counts and times.

Besides the nature of the statistical model that describes the light fluctuations, a number of parameters play an important role in determining the statistics of the corresponding photoelectrons.
These are the three characteristic times:

I) The coherence time of the field τ_c. This determines the "memory" scale of the exciting field.

II) The mean time between two consecutive photoelectrons, $T_s = 1/\langle\lambda\rangle$, where $\langle\lambda\rangle$ is the mean number of photoelectrons per second.
The ratio between τ_c and T_s, $\bar{n}_d = \tau_c/T_s = \langle\lambda\rangle\tau_c$, is the average number of photoelectrons per coherence time. It represents the "population" of events in one "memory" time, or one "generation".

III) The counting time T. This is the time interval chosen by the experimeter to observe the events. It represents the resolution of the measuring instrument. The parameter $\gamma = T/\tau_c$ determines how many independent samples of the field are lumped together when the measurement is taken. The mean number of counts in [0,T], $\langle n\rangle$, plays an important role in determining the counting statistics. In experiments in which time intervals between events are measured, the counting time T is of no significance and the instrument is assumed to have ideal resolution.

Moreover, two characteristic areas should be considered:

I) The coherence area of the optical field A_c. This represents the spatial extent of points, fluctuations at which have a significant degree of "correlatedness". The mean number of counts per coherence time per coherence area is also a parameter of fundamental importance. This is sometimes called the degeneracy parameter and is denoted r.

II) The area of the detector A, which determines the spatial resolution of the instrument. The ratio A/A_c is the number of independent samples of the field covered by the detector. It determines how much the phenomenon under integration is "washed out" by spatial integration.

5.2.1 Coherent Light

Because coherent light is assumed to have a nonfluctuating (deterministic) intensity, the point process of detected photoelectron events must be Poisson. This is true for all values of T, A, or λ. The statistical properties of a P.PP have already been fully discussed in Sec. 3.4 and will not be mentioned here again.

5.2.2 Linearly Polarized Thermal Light

As we have seen in Sec. 4.2.1, thermal light has a randomly fluctuating intensity. The energy collected by a photodetector has therefore random fluctuations and consequently the photoelectron process is no longer Poisson. The statistics of thermal light are completely described by its mutual coherence function. The photoelectron statistics are therefore completely described by the complex degree of coherence $g(\underline{r}_1,\underline{r}_2,\tau)$, the average count rate $\langle\lambda\rangle$, and the detector parameters A and T. Because of the complexity of the analysis, we present it in three steps. First, we assume that the counting time is short and the detector area is small. The statistics then depend on the average number of counts $\langle\lambda\rangle$ only. Then we allow T to be arbitrary and study the effect of the field's power spectrum. Finally, we allow A to take an arbitrary value and study the role played by spatial coherence in determining photoelectron statistics.

Short Sampling Time, Small Detector $T \ll \tau_c$, $A \ll A_c$

If the counting time interval T is short compared to the field's coherence time τ_c and the detector area A is small compared to the coherence area A_c, then the received energy W is approximately proportional to the intensity I.

Statistics of W

Because I is exponentially distributed (Sec. 4.2.1), W also has an exponential distribution characterized by (Table 2.1)

$$P(W) = \frac{1}{\langle W \rangle} e^{-W/\langle W \rangle} \tag{5.4}$$

$$Q_W(s) = (1 + s \langle W \rangle)^{-1} \tag{5.5}$$

$$K_W^{(m)} = (m-1)! \langle W \rangle^m \quad , \quad \langle W^m \rangle = m! \langle W \rangle^m \quad . \tag{5.6}$$

Counting Statistics. Statistics of n

P(n) can be found by calculating the Poisson transform of P(W) (Sec. 3.5.2) or by finding the nth derivative of $Q_W(s)$ (3.81). The result is the geometric distribution (better known among physicists as the Bose-Einstein distribution),

$$P(n) = \frac{\langle n \rangle^n}{(1 + \langle n \rangle)^{n+1}} \tag{5.7}$$

This distribution can be generated by the recurrence relation

$$P(n+1) = \frac{\langle n \rangle}{1 + \langle n \rangle} P(n) \quad , \qquad P(0) = \frac{1}{1 + \langle n \rangle} \tag{5.8}$$

and has the factorial moments

$$F_n^{(m)} = m! \langle n \rangle^m \quad . \tag{5.9}$$

For m = 2, this gives

$$\langle \Delta n^2 \rangle = \langle n \rangle^2 + \langle n \rangle \quad . \tag{5.10}$$

Comparing this with $\langle \Delta W^2 \rangle = \langle W \rangle^2$, we immediately see that the detection process has introduced an extra uncertainty (the term $\langle n \rangle$, the so-called quantum noise or shot noise). The first term ($\langle n \rangle^2$) is the uncertainty that results from the field fluctuations (sometimes called "excess photon noise" or "photon-bunching noise").

The distributions P(n) for thermal and coherent light are compared in Fig.5.2.

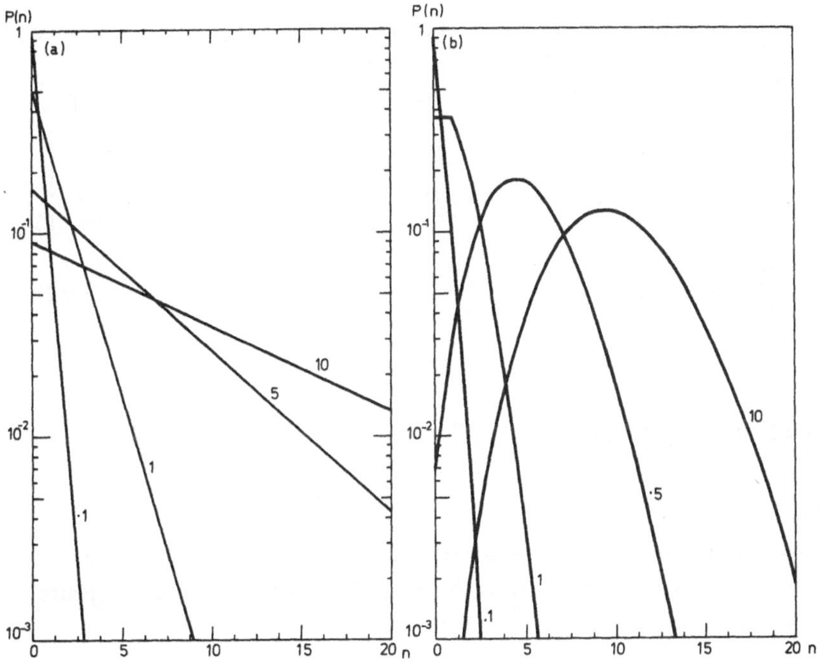

Fig.5.2a and b. PD of photoelectron counts of: a) thermal light, b) coherent light, for the indicated average number of counts <n>, in the limit of short sampling time

Note that if $\langle n \rangle \gg 1$, the uncertainty (or noise) due to field fluctuations dominates. On the other hand, if $\langle n \rangle \ll 1$, the uncertainty due to the detection process dominates and the Poisson distribution is recovered. The Bose-Einstein distribution of the number of photoelectrons/photons can be physically interpreted in terms of fluctuations of an assembly of bosons in one cell of the phase space [306, 307, 310, 318]. Such interesting physical interpretations are, however, outside the scope of this book.

Statistics of Times

In an experiment in which the times of occurrence of events are observed, the meaning of the condition $T \ll \tau_c$ has to be reexamined, because, in such an experiment, there is no prior-fixed time interval during which the events are counted.

Instead, we assume that $T_s \ll \tau_c$, where $T_s = 1/\langle\lambda\rangle$ is the mean time interval between two consecutive events. This is equivalent to $\langle\lambda\rangle\tau_c = \bar{n}_d \gg 1$ which implies that the degeneracy parameter $r = \bar{n}_d A_c/A$ is large. Consequently, the integrated light intensity W over a time not much longer than T_s is proportional to the intensity itself and therefore has the statistics described by (5.4-6). Using (5.4-6) in (3.93-96), we get the PD's for the time of arrival of the first photoelectron, the inter-event time, and the time of arrival of the qth photoelectron,

$$P_f(t) = T_s^{-1}(1 + t/T_s)^{-2} \quad , \tag{5.11}$$

$$P_i(\tau) = 2T_s^{-1}(1 + \tau/T_s)^{-3} \quad , \tag{5.12}$$

and

$$P(t_{q0}) = qT_s^{-1}(t_{q0}/T_s)^{q-1}(1 + t_{q0}/T_s)^{q+1} \quad , \tag{5.13}$$

respectively.

These probability distributions are plotted in Fig.5.3, where they are compared with their corresponding distributions for a Poisson process (photoelectron event process due to coherent light).

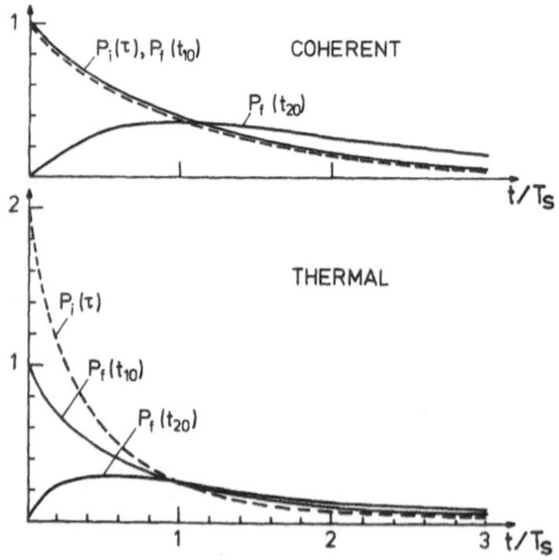

Fig. 5.3. PD of the inter-event time and the time of arrival of the first and second photoelectron due to coherent light, and thermal light of large degeneracy parameters. Ordinate is scaled by T_s

The most striking difference is manifested in the probability distribution of inter-event times. For thermal light, photoelectrons have a higher probability of being closer together. There is a greater tendency for photoelectrons to arrive in "bunches". The photoelectron bunching phenomena may at first seem rather puzzling. But it can be simply viewed as follows. The fluctuating intensity of thermal light results in additional fluctuations in the inter-event time. The higher the intensity, the shorter the inter-event time. Because the exponential distribution of the intensity is highly skewed to the right-hand side (unsymmetric around its mean, with coefficient of skewness=2), then the probability distribution of the inter-event time skews toward the same direction, which causes the bunching effect.

Arbitrary Sampling Time, Small Detector $A \ll A_c$

In this section, the detector is assumed to be small (a point detector) and the effect of time integration is studied.

I) General Spectrum

Statistics of W. The integrated intensity W is given by

$$W = \int_0^T I(t)dt \quad , \qquad I(t) = |V(t)|^2 \quad , \tag{5.14}$$

where $V(t)$ is the complex amplitude of the field at the detector. To find the statistics of an integral over a stochastic process is, in general, a formidable task. The fact that $V(t)$ is Gaussian simplifies the problem. Actually, a similar problem arose and was solved in the field of radio engineering [255, 397-400, 442, 443, 447]. The solution [36, 171, 240, 380] is based on expanding $V(t)$ in a basis orthonormal over the time interval [0,T], i.e.,

$$V(t) = \sum_k \alpha_k \Phi_k(t), \quad \alpha_k = \int_0^T V(t)\Phi_k^*(t)dt \quad . \tag{5.15}$$

Because $V(t)$ is a complex Gaussian process, then $\{\alpha_k\}$ are joint complex Gaussian with a covariance matrix Λ whose elements are given by

$$\Lambda_{\ell k} = \langle\alpha_\ell^*\alpha_k\rangle \quad \int_0^T\int_0^T dt_1dt_2G(t_1,t_2)\Phi_\ell^*(t_1)\Phi_k(t_2), \; G(t,t')=\langle V^*(t)V(t')\rangle . \tag{5.16}$$

By substituting from (5.15) in (5.14) and using the orthonormal property of the basis,

$$\int_0^T \Phi_\ell^*(t)\, \Phi_k(t)dt = \delta_{\ell k} \quad ,$$

we can write the integrated intensity as

$$W = \sum_k W_k \quad , \quad W_k = |\alpha_k|^2 \quad . \tag{5.17}$$

This expansion enables us to find the statistics of W.

The Moment-Generating Function of W.

$$Q_W(s) = \left\langle e^{-s\sum\limits_k W_k} \right\rangle = \int d^2\alpha_1 \ldots \int d^2\alpha_\ell \ldots e^{-s\sum\limits_k |\alpha_k|^2} P(\{\alpha_\ell\}) \quad .$$

By use of the expression of $P(\{\alpha_\ell\}$ for Gaussian statistics (2.50), we get

$$Q_W(s) = |\delta + sT\Lambda|^{-1} \quad , \tag{5.18}$$

where δ is the unit matrix. Eq. (5.18) can, in principle, be used to determine $Q_W(s)$, e.g., by using numerical techniques. The method is: given $G(t_1,t_2)$, an orthonormal basis is chosen. Then the matrix Λ is determined by using (5.16); then (5.18) is used to determine $Q_W(s)$.

A simpler approach is to choose the basis $\phi_k(t)$ in such a way that $\{\alpha_k\}$ are statistically independent, or so that the matrix Λ is diagonal. This means using a Karhunen-Loève expansion with $\phi_k(t)$ satisfying the eigenvalue equation (cf. Sec. 2.5.2)

$$\int_0^T G(t,t')\Phi_j(t)dt = m_j\Phi_j(t') \quad . \tag{5.19}$$

Thus, we can write W as a sum of a set of statistically independent variables, which means that the mgf is the product of the mgf's of these variables,

$$Q_W(s) = \langle e^{sW} \rangle = \langle e^{s\sum W_k} \rangle = \prod_k Q_{W_k}(s) \quad . \tag{5.20}$$

Now, it suffices to determine the statistics of each mode W_k separately. Here we use the fact that the field is thermal, from which it is concluded that the coefficients of its linear expansion are circularly symmetric complex Gaussian with variances m_k, i.e.,

$$P(\alpha_k) = (\pi m_k)^{-1} \exp(-|\alpha_k|^2/m_k)$$

or

$$P(|\alpha_k|) = 2|\alpha_k| m_k^{-1} \exp(-|\alpha_k|^2/m_k) \quad .$$

Therefore, $W_k = |\alpha_k|^2$ are exponentially distributed,

$$P(W_k) = m_k^{-1} \exp(-W_k/m_k) \quad ,$$

with averages m_k. We know (cf. Table 2.1) that the mgf of an exponential distribution is

$$Q_{W_k}(s) = (1 + sm_k)^{-1} \quad .$$

From (5.20), we finally obtain the important equation,

$$Q_W(s) = \prod_k (1 + sm_k)^{-1} \quad , \tag{5.21}$$

from which the cumulant-generating function is

$$Q_W^c(s) = -\sum_k \ln (1 + sm_k) \quad . \tag{5.22}$$

Two problems remain to be solved: finding the eigenvalues m_k and computing the infinite product in (5.21) or the sum in (5.22).

Because the field is stationary, it is more convenient to write the eigenvalue equation (5.19) in the form

$$\int_0^T g(t-t')\Phi_k(t')dt' = \lambda_k\Phi_k(t) \quad , \tag{5.23}$$

where $g(t-t') = G(t-t')/G(0) = TG(t-t')/\langle W\rangle$
and $\lambda_k = Tm_k/\langle W\rangle$,
which results in

$$Q_W(s) = \prod_k (1 + s\langle W\rangle\lambda_k/T)^{-1} \quad , \tag{5.24}$$

and

$$Q_W^c(s) = -\sum_k \ln(1+s\langle W\rangle\lambda_k/T) = \sum_{j=1}^{\infty} \frac{(-1)^j}{j}\left[\sum_k \left(\frac{\langle W\rangle\lambda_k}{T}\right)^j\right] s^j \quad . \tag{5.25}$$

Several attempts have been made to find a general procedure for obtaining a closed-form expression for $Q_W(s)$.

For example, SIEGERT [442] showed that

$$Q_W(s) = \exp\left[-\int_0^{s\langle W\rangle/T} du \int_0^T h(t,t;u)dt\right] \quad , \tag{5.26}$$

where $h(t',t;u)$ is the solution of the integral equation

$$h(t',t;u) + u\int_0^T G(t'-s)h(s,t;u)ds = G(t'-t), \quad 0 < t,t' < T \quad . \tag{5.27}$$

When the power spectrum (the Fourier transform of $G(t)$) is a rational function, this equation can be solved by transform techniques. SRINIVASAN and SUKAVANAM [458, 459] and SRINIVASAN [456] developed a general mathematical technique for solving this problem.

The Moments of W. The relations between the moments of $W = \int_0^T I(t)dt$ and the moments of $I(t)$, $t\in[0,T]$ have been stated in Sec. 3.5.2. Of course, these relations are valid irrespective of the statistical properties of the field. But, as discussed in Sec. 4.2.1, for a Gaussian field, the higher-order moments of $I(t)$ are related to the second-order moments. By using (4.72), we obtain

$$K_W^{(m)} = (m - 1)!\ B_m \langle W\rangle^m \quad , \tag{5.28}$$

where

$$B_1 = 1$$

$$B_m = \frac{1}{T^m}\int_0^T dt_1 \ldots \int_0^T dt_m\, g(t_1-t_2)g(t_2-t_3) \ldots g(t_m-t_1) \qquad m > 1 \ . \tag{5.29}$$

The cumulants of W can also be obtained directly from the cumulant-generating function given by (5.25) and the rule (2.12). This gives

$$K_W^{(m)} = (m - 1)!\ \langle W\rangle^m \frac{1}{T^m}\sum_k \lambda_k^m \quad , \tag{5.30}$$

from which (5.28) and (5.29) can be derived [225].

If the eigenvalues λ_k are known, then it is easier to compute $K_W^{(m)}$ from (5.30) rather than to evaluate the multiple integrals of (5.29).

The Probability Distribution of W. Because W is the sum of many independent exponential variables its PD can be written as a multiple convolution,

$$\begin{aligned} P(W) &= \circledast_j P_j(W) \\ &= \circledast_j m_j^{-1} \exp(-W/m_j) \quad . \end{aligned} \tag{5.31}$$

Alternatively, P(W) can be obtained by finding the inverse Laplace transform of (5.21). Thus,

$$P(W) = \sum_j m_j^{-1} \exp(-W/m_j) \prod_{k\neq j} m_j(m_j-m_k)^{-1} \quad , \tag{5.32}$$

where m_j are assumed distinct.

Equations (5.31) and (5.32) can be useful only when numerical values for m_j are available. In this case, computer techniques can be used to find P(W). Now we consider some limiting cases.

Limiting Case $T \ll \tau_c$. It has been shown in Sec. 2.4.4 that, in this limit, the integral equation (5.19) has only one eigenvalue $m_1 = \langle W \rangle$ (or $\lambda_1 = T$). Hence, the sum in (5.17) is reduced to one component with exponential distribution (5.4), and the product in (5.20) reduces to one component, reproducing (5.5). Also, in this limit, (5.29) gives $B_m = 1$, which reproduces (5.6).

Limiting Case $T \gg \tau_c$. This limit has also been discussed in Sec. 2.4.5. A very large number $N = T/\tau_c$ of approximately equal eigenvalues $m = \langle W \rangle / N$ (or $\lambda = T/N$) exists. This leads to the approximate expression,

$$Q_W(s) = (1 + s \langle W \rangle / N)^{-N} , \quad N = T/\tau_c \gg 1 \quad . \tag{5.33}$$

By taking the mth derivative of $Q_W(s)$ at $s = 0$ and using (2.9), we get

$$\langle W^m \rangle = \frac{\Gamma(N+m)}{\Gamma(N)} \frac{1}{N^m} \langle W \rangle^m \quad . \tag{5.34}$$

We can also put $\lambda_k = T/N$, $k = 1, \ldots N$ and $\lambda_k = 0$, $k > N$ in (5.30) and get

$$K_W^{(m)} = \frac{(m-1)!}{N^{m-1}} \langle W \rangle^m \quad , \tag{5.35}$$

$$\langle \Delta W^2 \rangle = \langle W \rangle^2 / N \quad .$$

Of course, (5.34) and (5.35) confirm the relation between the cumulants and the ordinary moments (2.5).

The probability density $P(W)$ corresponding to (5.33) can be obtained by finding the inverse Laplace transform of $Q_W(s)$. But it is easier to argue that W is the sum of N statistically independent variables; each having a chi-square distribution of order 2 (an exponential distribution). Therefore, it has a chi-square distribution of order $2N$

$$P(W) = \frac{N^N}{\Gamma(N)} \langle W \rangle^{-N} W^{N-1} \exp(-NW/\langle W \rangle) \tag{5.36}$$

[310, 325, 484]. Note that if N becomes very large, we should expect W to have a Gaussian distribution (according to the central-limit theorem). This can also be seen by examining (5.35). If $\langle W \rangle / N$ is small, then the cumulants

$K_W^{(m)}$ decrease rapidly with m. If N is sufficiently large, then all higher moments except the first two, can be neglected, i.e., W is approximately Gaussian with mean $\langle W\rangle$ and variance $\langle W\rangle^2/N$.

In the limit $N \to \infty$, (5.33) becomes

$$Q_W(s) = e^{-s\langle W\rangle} ,$$

corresponding to a deterministic W (having zero variance). This takes place because the long integration time washes out the statistical fluctuations in W.

Figure 5.4 is a plot of P(W) against $W/\langle W\rangle$ for several values of N. It illustrates how the PD changes from an exponential distribution for $N = 1$ to a Gaussian distribution for larger N and ultimately to a delta function centered around $\langle W\rangle$ for $N \to \infty$.

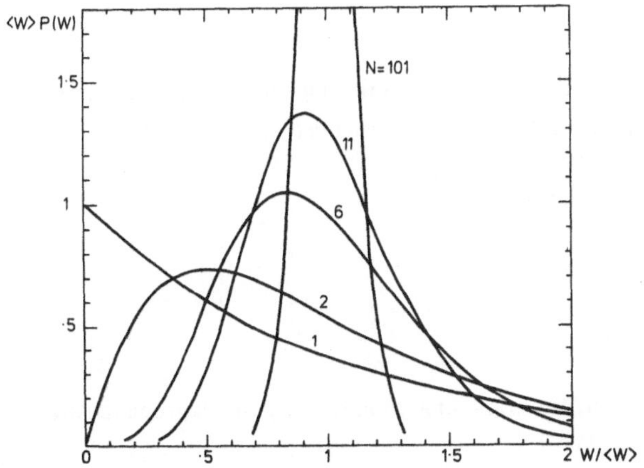

Fig.5.4. Dependence of the PD of the integrated intensity of thermal light on the number of collected modes N

Counting Statistics. The probability distribution P(n) can be obtained by finding the Poisson transform of P(W), or by finding the nth derivative of $Q_W(s)$ at s = 1 and using (3.81). This leads to

$$P(n) = P(0) \sum_{(k)} \prod_{j=1}^{n} m_{k_j}(1-m_{k_j})^{-1} \qquad n > 0$$

$$P(0) = Q_W(1) = \prod_k (1+m_k)^{-1} \quad , \tag{5.37}$$

where (k) represents all integers ($0 < k_1 < k_2 \ldots < \infty$). Another more useful expression is the recurrence relation

$$(n+1)P(n+1) = \sum_{\ell=0}^{n} P(n-\ell) \sum_k \left(\frac{m_k}{1+m_k}\right)^{\ell+1} , \tag{5.38}$$

obtained by taking the logarithm of $Q_W(s)$ and then the nth derivative.

A third way of obtaining P(n) would be to find the multiple convolution of $P_i(n)$, the distribution resulting from each component W_i in the sum (5.17). Because $\{W_i\}$ are exponential, their Poisson transform is geometric and we get

$$P(n) = \underset{k}{\circledast}\, m_k^n/1+m_k)^{1+n} \quad . \tag{5.39}$$

The factorial moments $F_n^{(m)}$ can be obtained by finding the nth derivative of $Q_W(s)$ at s = 0. In this way, we can obtain the recurrence relation

$$F_n^{(m+1)} = \sum_{\ell=0}^{m} \frac{m!}{(m-\ell)!} \quad \sum_k (m_k)^{\ell+1} F_n^{(m-\ell)}$$

$$F_n^{(1)} = \langle n \rangle \quad . \tag{5.40}$$

<u>Limiting Case $T \ll \tau_c$</u>. Using (5.39) with one eigenvalue, we reproduce the Bose-Einstein distribution (5.7).

<u>Limiting Case $T \gg \tau_c$</u>. From the approximate expression (5.33) of $Q_W(s)$, we can take the nth derivative at s = 1 and obtain the negative binomial distribution

$$P(n) = \binom{n+N-1}{n} \left(1+ \frac{\langle n\rangle}{N}\right)^{-N} \left(1+\frac{N}{\langle n\rangle}\right)^{-n} \quad . \tag{5.41}$$

This is recognized as the Poisson transform of the chi-square distribution with 2*N* degrees of freedom (Table 3.1). This can also be obtained by computing the *N*-fold convolution of *N* identical Bose-Einstein distributions. The factorial moments are obtained by rewriting (5.34)

$$F_n^{(m)} = \frac{\Gamma(N+m)}{N^m\Gamma(N)} <n>^m \tag{5.42}$$

(cf. (3.69)).

From the approximate normal distribution of W, which is valid for large N, we take the Poisson transform and obtain [330]

$$P(N) \approx \frac{<n>^n}{n!} e^{-<n>} \left\{1+ \frac{1}{2N} [(n-<n>)^2-n]\right\} \quad . \tag{5.43}$$

As $T \to \infty$, $N \to \infty$ and both (5.41) and (5.43) yield a Poisson distribution, as expected. Fig.5.5 illustrates the change of P(n) as N increases. Time integration washes out the field fluctuations but not the uncertainties introduced by the process of detection.

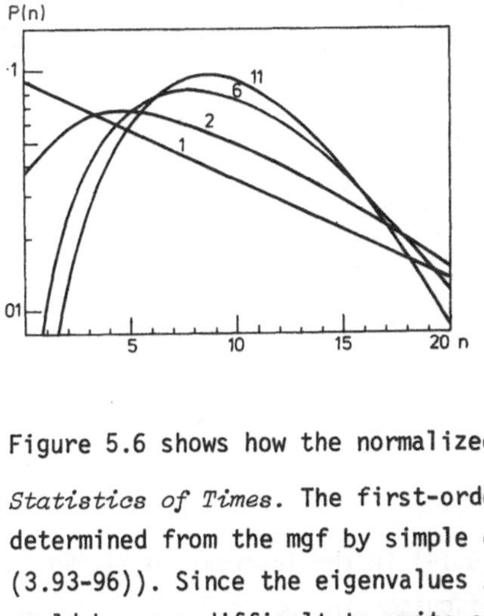

Fig.5.5. Photoelectron counting PD of thermal light for the indicated values of N, the number of collected modes

Figure 5.6 shows how the normalized factorial moments $F_n^{(m)}/<n>^m$ vary with N.

Statistics of Times. The first-order time statistics can, in general, be determined from the mgf by simple differentiation with respect to T (cf. (3.93-96)). Since the eigenvalues λ_k themselves are functions of T, it would be very difficult to write explicit formulae for the time probability densities in the most general situation. We can however return to the original definitions (3.88-90) and use the expansion (5.15) as we did with the moment-generating function. This gives [409]

$$P_f(t) = T_s^{-1}Q(1)\eta(0,0) \tag{5.44}$$

and

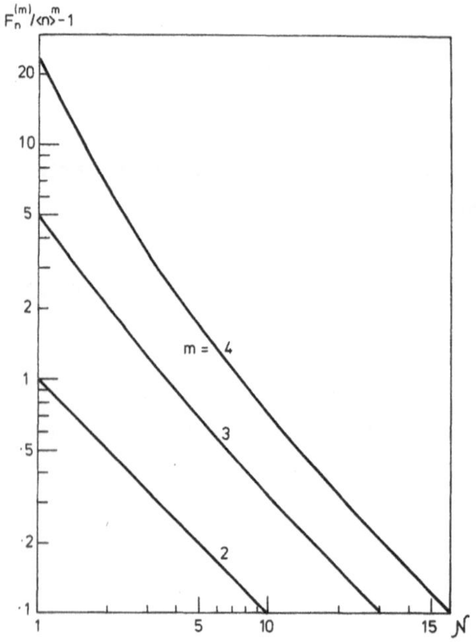

Fig.5.6. Dependence of the photoelectron counting normalized factorial moments on the number of modes of thermal light N

$$P_i(t) = T_s^{-1}Q(1)\left\{[\eta(0,0)]^2 + |\eta(0,t)|^2\right\} \quad , \tag{5.45}$$

where

$$\eta(t',t'') = T_s \sum_k m_k(1+m_k)^{-1}\Phi_k^*(t')\Phi_k(t'')$$
$$Q(1) = \prod_k (1+m_k)^{-1} \quad . \tag{5.46}$$

Here, m_k and $\Phi_k(t')$ are the solutions of (5.19) in the interval [0,t]. Some properties of the function $\eta(t',t'')$ are discussed in Sec. 5.3.1.

Limit $T_s \gg \tau_c$. This limit means very few counts per coherence time, $\bar{n}_d \ll 1$. Because we assume that $A \ll A_c$, this limit also means a very small degeneracy parameter r. We can use (5.33) and apply the differentiation rule to get

$$P_f(t) = T_s^{-1}\,\frac{\ln(1+\bar{n}_d)}{\bar{n}_d}\left[(1+\bar{n}_d)^{1/\bar{n}_d}\right]^{-t/T_s} \quad , \quad \bar{n}_d = \langle\lambda\rangle\tau_c \quad . \tag{5.47}$$

As $\bar{n}_d$ becomes very small, (5.47) gives an exponential distribution that corresponds to a Poisson process, as we should expect.

Because $P_i(t)$ is the derivative of $P_f(t)$, we conclude that it also tends to an exponential function in the limit of very small degeneracy parameter. Because most thermal radiations have very low $\bar{n}_d$, the phenomenon of bunching described in Sec. 5.2.2 cannot be easily observed. Thermal radiations with large degeneracy parameter have been obtained by scattering laser light from a rotating rough surface [324] and by using a laser below its threshold (cf. Sec. 4.3.7).

Now let us discuss some optical fields that have specific spectra (correlation functions). We begin with the Lorentzian spectrum because it is the easiest to analyze mathematically, and because of its frequent occurrence in physics.

II) Lorentzian Spectrum

Statistics of W

Moment-Generating Function. For a Lorentzian spectrum, or exponential correlation function, we have

$$g(\tau) = \exp(-\Gamma|\tau|) \quad , \qquad \tau_c = \Gamma^{-1} \quad , \tag{5.48}$$

and (5.23) becomes

$$\int_0^T \exp(-\Gamma|t-t'|)\Phi_k(t')dt' = \lambda_k\Phi_k(t) \quad . \tag{5.49}$$

The solution of this eigenvalue problem has been presented in Sec. 2.4.4,

$$\lambda_k/T = 2\gamma/(\gamma^2 + \omega_k^2T^2) \quad , \tag{5.50}$$

where ω_k are determined from the transcedental equation (2.74)

$$(\omega_k^2-\Gamma^2)\tan(\omega_kT) = 2\omega_k\Gamma \tag{5.51}$$

and $\gamma = \Gamma T$.

We now face the problem of determining the product

$$Q_W(s) = \prod_k (1+s \langle W\rangle\lambda_k/T)^{-1}$$

given (5.50) and (5.51). We start by changing the product into a sum by taking the logarithm and differentiating,

$$\frac{Q'_W(s)}{Q_W(s)} = -\sum_k \langle W\rangle \frac{\lambda_k}{T}\left(1+s\langle W\rangle\frac{\lambda_k}{T}\right)^{-1} = -2\gamma\langle W\rangle \sum_k \left(z^2+\omega_k^2T^2\right)^{-1} , \tag{5.52}$$

where

$$z^2 = \gamma^2 + 2\gamma\langle W\rangle s . \tag{5.53}$$

The techniques of contour integration can be used to determine the sum of the series (5.52). We write (5.51) in the form

$$\cos\omega T + \frac{1}{2}\left(\frac{\gamma}{\omega T} - \frac{\omega T}{\gamma}\right)\sin\omega T = 0 ,$$

and define the function of the complex variable ξ,

$$F(\xi) = \frac{1}{\xi^2+z^2}\frac{d}{d\xi}\ln\left[\cos\xi + \frac{1}{2}\left(\frac{\gamma}{\xi} - \frac{\xi}{\gamma}\right)\sin\xi\right] ;$$

we then consider the integral over a closed contour

$$\oint_c F(\xi)d\xi .$$

The only finite singularities of $F(\xi)$ are simple poles at $\xi = \pm iz$ and at $\xi = \pm\omega_kT$. If c is properly chosen, it can be shown that the integral tends to zero as the number of poles inside c tends to infinity. Hence the sum of the residues at all poles vanishes and we can write

$$2\sum_k \frac{1}{z^2+\omega_k^2T^2} = \frac{1}{z}\frac{d}{dz}\ln\left[\cosh z + \frac{1}{2}\left(\frac{\gamma}{z} + \frac{z}{\gamma}\right)\sinh z\right] .$$

Substituting in (5.52) and integrating (using the initial condition $Q_W(0) = 1$), we get

$$Q_W(s) = e^{\gamma}\left[\cosh z + \sinh z \left(\frac{\gamma}{2z} + \frac{z}{2\gamma}\right)\right]^{-1}$$

$$z^2 = \gamma^2 + 2\gamma\langle W\rangle s \quad , \quad \gamma = \Gamma T \quad . \tag{5.54}$$

This confirms the solution of SIEGERT [442] based on (5.26) and (5.27).

The limits $\gamma \ll 1$ and $\gamma \gg 1$ recover (5.5) and (5.33), respectively.

Moments. The moments of W can be obtained either from $Q_W(s)$ or from (5.28) by finding the values of B_m (5.29). The results are [225, 346]

$$K_W^{(m)} = (m-1)!\, B_m\langle W\rangle^m$$

$$B_1 = 1$$

$$B_2 = (e^{-2\gamma} + 2\gamma - 1)/(2\gamma^2) \tag{5.55}$$

$$B_3 = 3[(1+\gamma)e^{-2\gamma}+\gamma-1]/(2\gamma^3)$$

$$B_4 = [e^{-4\gamma}+4(7+10\gamma+4\gamma^2)e^{-2\gamma}+20\gamma-29]/(8\gamma^4) \quad .$$

Probability Distribution. The probability distribution P(W) can be obtained from the mgf $Q_W(s)$, given by (5.54), by finding the inverse Laplace transform

$$P(W) = \frac{1}{2\pi j}\int_{c-j\infty}^{c+j\infty} Q_W(s)e^{Ws}ds \quad . \tag{5.56}$$

Noting that $Q_W(s)$ has simple poles at $s_\ell = -T/\langle W\rangle\lambda_k$ (cf. (5.24)), where λ_k are given by (5.50,51), we then obtain

$$P(W) = \sum_\ell R(s_\ell)$$

where

$$R(s_\ell) = e^{Ws_\ell}\left(\frac{dQ^{-1}}{ds}\right)^{-1}_{s=s_\ell}$$

are the residues of the integrand in (5.56). By substituting from (5.54), we obtain [240, 380]

$$P(W) = \frac{e^{\gamma}}{\langle W\rangle}\sum_{\ell=0}^{\infty} (-1)^{\ell}\left(\frac{2-\lambda_{\ell}\gamma/T}{1+\lambda_{\ell}/T}\right)\exp\left(-\frac{T}{\lambda_{\ell}}\frac{W}{\langle W\rangle}\right) , \tag{5.57}$$

where λ_{ℓ} are given by (5.50, 51).

In (5.57) it is assumed that λ_{ℓ} are arranged in increasing order. JAKEMAN and PIKE [240], who first derived (5.57), noted that for reasonable values of its parameters, the series is rapidly convergent. For example, when $\gamma \approx 1$ and $W > 0.4 \langle W\rangle$, $P(W)$ is very well approximated by the first two terms.

In the limit $\gamma \ll 1$ ($T \ll \tau_c$), (5.57) reduces to

$$P(W) \approx \left[1 + \frac{\gamma}{4}\left(3-\frac{2W}{\langle W\rangle}\right)\right]\frac{1}{\langle W\rangle}\, e^{-W/\langle W\rangle} , \tag{5.58}$$

which becomes the expected exponential distribution in the asymptotic limit $\gamma \to 0$.

On the other hand, in the limit $\gamma \gg 1$ (5.57) gives [240]

$$P(W) \approx \left(\frac{\gamma}{2\pi W\langle W\rangle}\right)^{1/2} \exp\left[-\frac{\gamma}{2W\langle W\rangle}(W-\langle W\rangle)^2\right] . \tag{5.59}$$

In the limit $\gamma \to \infty$, this gives a delta function. Fig.5.7 shows the variation of the distribution with $\gamma = \Gamma T = T/\tau_c$. Compare this with Fig.5.4 and note that the curve for $\gamma \to 0$ corresponds to that for $N = 1$; the distribution for $\gamma \gg 1$ corresponds to $N \simeq \gamma$.

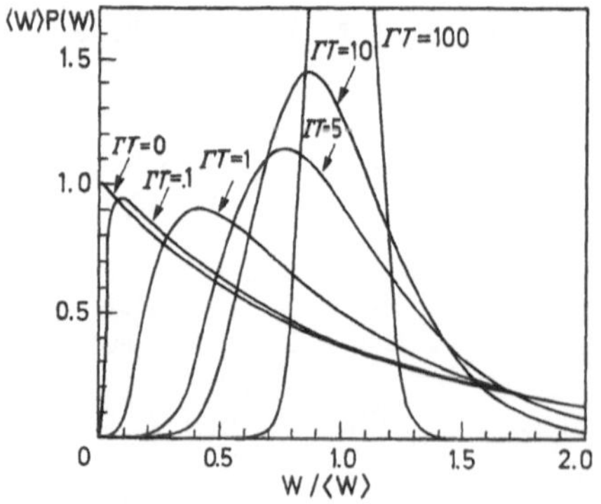

Fig.5.7. Dependence of the PD of the integrated intensity of thermal light that has a Lorentzian spectrum on the time-bandwidth product, $\gamma=\Gamma T=T/\tau_c$ (after JAISWAL and MEHTA [225])

Counting Distribution. Applying the Leibnitz differentiation rule to $Q_W(s)$ of (5.54) and using (3.81) and (3.68), we obtain the recurrence relations for P(n) and $F_n^{(m)}$ [36]

$$P(n) = \sum_{r=0}^{n-1} \frac{(-1)^{n+r+1}}{(n-r)!} D_{n-r}(1)P(r) \quad , \tag{5.60}$$

$$P(0) = Q_W(1) \quad ,$$

$$F_n^{(m)} = \sum_{r=0}^{m-1} (-1)^{n+r+1}\binom{m}{r} D_{m-r}(0)F_n^{(r)}$$

$$F_n^{(0)} = 1 \quad , \quad F_n^{(1)} = \langle n\rangle \quad , \tag{5.61}$$

where

$$D_\ell(s) = Q(s)e^{-\gamma}\left(\frac{\gamma\langle n\rangle}{z}\right)^\ell\left[\left(\frac{\gamma}{2} + \frac{z^2}{2\gamma}\right) i_\ell(z)+z\left(1+ \frac{\ell}{\gamma}\right)i_{\ell-1}(z)\right] \quad , \tag{5.62}$$

$$z^2 = \gamma^2 + 2\gamma\langle n\rangle s \quad ,$$

in which i_ℓ are the modified spherical Bessel functions of the first kind and order ℓ. These can be generated from the recurrence relation

$$i_{\ell+1}(z) = - \frac{2\ell+1}{z} i_\ell(z) + i_{\ell-1}(z) \tag{5.63}$$

and the initial conditions

$$i_0(z) = \sinh(z)/z$$

$$i_{-1}(z) = \cosh(z)/z \quad . \tag{5.64}$$

The factorial moments can be obtained from (5.55) by

$$F_n^{(m)} = (m - 1)!\ B_m \langle n\rangle^m \quad , \tag{5.65}$$

where the coefficients B_m are listed in (5.55). In Fig.5.8 P(n) are plotted for several values of γ. Similarity with Fig.5.5 is noted, where $\gamma \to 0$ cor-

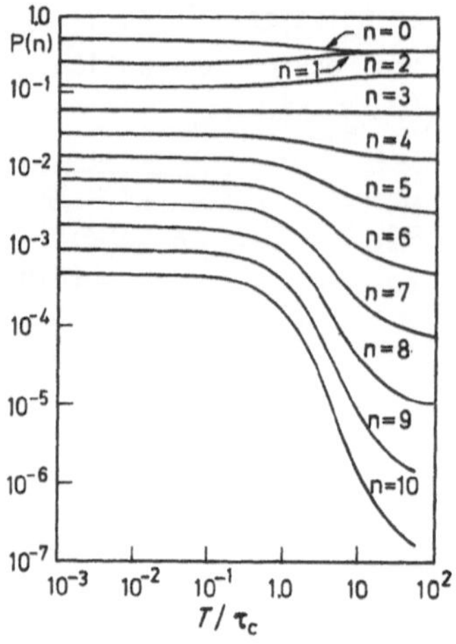

Fig.5.8. Dependence of the photoelectron counting PD of thermal light that has a Lorentzian spectrum on the time-bandwidth product, $\langle n\rangle=1$ (after BEDARD et al. [42])

responds to $N = 1$ and γ corresponds to N for large γ and N. Fig.5.9 illustrates the variation of $F_n^{(m)}$ with γ and should be compared to Fig.5.6.

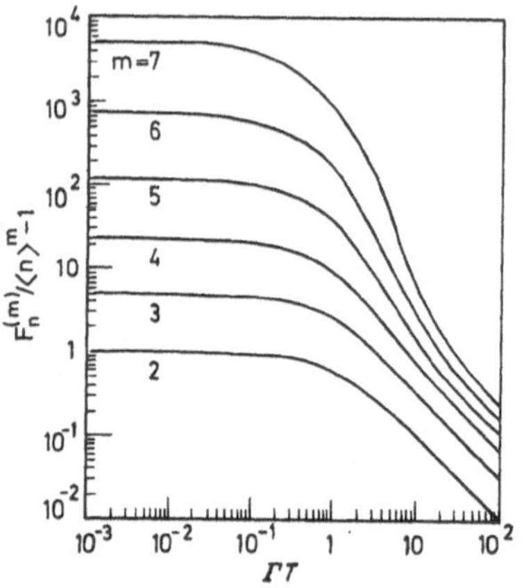

Fig.5.9. Dependence of the photoelectron counting normalized factorial moments of thermal light that has a Lorentzian spectrum on the time-bandwidth product (after BEDARD [36])

Statistics of Times. Now that we have an explicit expression for $Q_W(s)$, (5.54), we can apply the differentiation rules (3.93) and (3.94) and obtain the time of arrival and the inter-event probability densities [409, 415],

$$P_f(t) = 2\,T_s^{-1}\gamma\,\frac{A(t)}{B(t)}\,Q(t) \tag{5.66}$$

$$P_i(\tau) = 4T_s^{-1}\gamma^2\,\frac{A^2(\tau)+4\alpha^2}{B^2(\tau)}\,Q(\tau)\quad , \tag{5.67}$$

where

$$Q(t) = e^{\gamma t/T_s}\left[\cosh(\alpha t/T_s)+\frac{1}{2}(\gamma/\alpha+\alpha/\gamma)\sinh(\alpha t/T_s)\right]^{-1}$$

$$A(t) = (\gamma+\alpha)e^{\alpha t/T_s} - (\gamma-\alpha)e^{-\alpha t/T_s}$$

$$B(t) = (\gamma+\alpha)^2 e^{\alpha t/T_s} - (\gamma-\alpha)^2 e^{-\alpha t/T_s} \qquad (5.68)$$

$$\alpha^2 = \gamma^2 + 2\gamma$$

$$\gamma = \Gamma T_s = T_s/\tau_c = 1/\bar{n}_d \quad .$$

These expressions can also be obtained from the general expressions (5.44-46), [409]. The probabilities are plotted in Fig.5.10 in which the cases for $\gamma \ll 1$ and $\gamma = 1$ are compared. As γ increases, the distribution of P_f bends

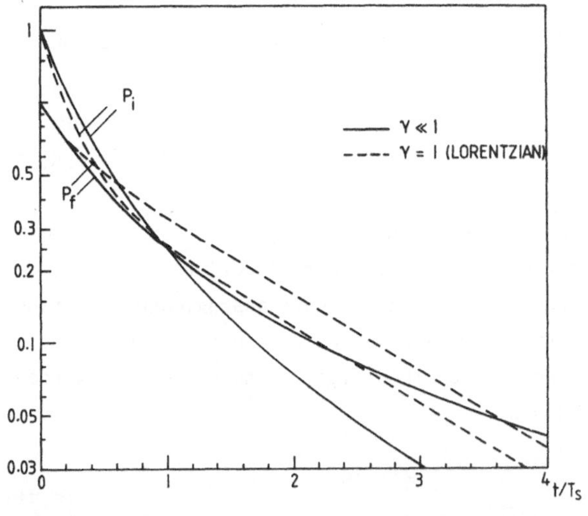

Fig.5.10. PD of the inter-event time and the time of arrival of the first photoelectron due to thermal light that has a Lorentzian spectrum for the indicated values of the average counts per coherence time $\gamma = \Gamma T_s$

downwards, but that of P_i bends upwards, corresponding to a lower degree of bunching of the photoelectron events. In the limit $\gamma \gg 1$ (or small degeneracy parameter), $P_f(t)$ becomes approximately constant in the time interval $[0,\tau_c]$, whereas $P_i(\tau)$ takes the form

$$P_i(\tau) = T_s^{-1}[1 + \exp(-2\Gamma\tau)] \quad ,$$

i.e., is proportional to the intensity correlation function. This is the result obtained by neglecting the exponential terms in (3.88) and (3.90). This proportionality suggests a method of estimating the parameter Γ from measurement of $P_i(\tau)$ (cf. Sec. 7.2.2).

III) Rectangular Spectrum

Light with a rectangular power-spectrum profile,

$$\begin{aligned} g(\omega) &= g(0) \qquad |\omega| \leq \pi\Gamma \\ &= 0 \qquad |\omega| > \pi\Gamma \end{aligned} \tag{5.69}$$

(sometimes called band-limited white noise) has a normalized correlation function,

$$g(\tau) = \mathrm{sinc}(\Gamma\tau) \tag{5.70}$$

where Γ is the bandwidth of the spectrum (in cycles/s).
As has already been mentioned (Sec. 2.4.4), the eigenfunctions of (5.23) are the spheroidal functions and the eigenvalues have approximately rectangular shapes (cf. Fig.2.2b). The first N eigenvalues are approximately equal; subsequent values are nearly negligible. The effective number of modes (degrees of freedom) can be defined by

$$N = \left(\sum\lambda_k\right)^2 / \sum\lambda_k^2 \quad , \tag{5.71}$$

and can be approximated by

$$N \approx \Gamma T + 1 \quad . \tag{5.72}$$

If we adopt this approximation, we get

$$Q_W(s) = (1 + s\langle W\rangle/N)^{-N} \quad , \tag{5.73}$$

which is the approximate mgf for the general case $T \gg \tau_c$, corresponding to a chi^2_{2N} distribution of W, and a negative binomial photocounting distribution (Figs.5.4-6). MEHTA and MEHTA [340] computed $Q_W(s)$ and P(n) for a rectangular spectrum by using tabulated values of λ_k and (5.24) and (5.37). They also compared the exact $Q_W(s)$ to approximate values calculated from (5.73) and (5.71). They found that the approximation is excellent.

Arbitrary Sampling Time and Arbitrary Detector Area. Cross-Spectrally Pure Light

Here we consider the most general case of arbitrary T/T_c and A/A_c. In this situation, both the spectral and the spatial-coherence properties of the field play important roles in determining the photoelectron statistics. Again, we use a Karhunen-Loève expansion of the field $V(\underline{r},t)$, but this time in both time and space,

$$V(\underline{r},t) = \sum_{j,\ell} \alpha_{j\ell}\Phi_{j\ell}(\underline{r},t) \tag{5.74}$$

where $\phi_{j\ell}(\underline{r},t)$ is a basis orthonormal over [0,T] and A, i.e.,

$$\iint_{A\,0}^{\;\;T} \Phi^*_{j\ell}(\underline{r},t)\Phi_{j'\ell'}(\underline{r},t)dt\,d\underline{r} = \delta_{jj'}\delta_{\ell\ell'} \quad . \tag{5.75}$$

The coefficients $\alpha_{j\ell}$ are circularly symmetric complex Gaussian random variables if $V(\underline{r},t)$ is a thermal field. They can be made statistically independent by choosing $\{\phi_{j\ell}\}$ that satisfy the integral equation,

$$\iint_{A\,0}^{\;\;T} G(\underline{r},t;\underline{r}',t')\Phi_{j\ell}(\underline{r},t)d\underline{r}\,dt = m_{j\ell}\Phi_{j\ell}(\underline{r}',t') \quad . \tag{5.76}$$

The coefficients $\alpha_{j\ell}$ have variances equal to the eigenvalues $m_{j\ell}$.

The energy W can now be written as a sum

$$W = \sum_{j,\ell} |\alpha_{j\ell}|^2$$

and its moment-generating function as

$$Q_W(s) = \prod_{j,\ell} (1 + s\, m_{j\ell})^{-1} \quad . \tag{5.77}$$

The above analysis is merely a straightforward generalization of the calculations on page 171. The main problem now is to solve the integral equation (5.76). The problem is greatly simplified in the not very restictive case of cross-spectrally pure light (cf. Sec. 4.1.2), i.e., when

$$G(\underline{r}_1, \underline{r}_2; t_2 - t_1) = g(\underline{r}_1, \underline{r}_2) G(t_2 - t_1) \quad .$$

This permits breaking the eigenfunctions and the eigenvalues into spatial and temporal parts,

$$\begin{aligned} \Phi_{j\ell}(\underline{r}, t) &= \xi_j(\underline{r}) \chi_\ell(t) \quad , \\ m_{j\ell} &= \mu_j \lambda_\ell \frac{\langle W \rangle}{T} \quad , \end{aligned} \tag{5.78}$$

where $\chi_\ell(t)$ and λ_ℓ are the eigenfunctions and eigenvalues of

$$\int_0^T g(t-t') \chi_\ell(t) dt = \lambda_\ell \chi_\ell(t') \quad , \tag{5.79}$$

and $\xi_j(\underline{r})$ and μ_j are the eigenfunctions and eigenvalues of

$$\int_A g(\underline{r}, \underline{r}') \xi_j(\underline{r}) d\underline{r} = \mu_j \xi_j(\underline{r}') \quad . \tag{5.80}$$

We have the freedom to choose the condition

$$\sum_j \mu_j = 1 \quad . \tag{5.81}$$

The significance of such a model expansion was discussed by HELSTROM [205].

Now our mgf becomes

$$Q_W(s) = \prod_j \prod_\ell \left(1 + s\lambda_\ell \mu_j \frac{\langle W\rangle}{T}\right)^{-1} \quad . \tag{5.82}$$

Let us discuss some special and limiting cases.

I) $\underline{A \ll A_c}$: In this case, only one spatial mode $\mu_j = 1$ is significant and we reproduce the results on pages 171-189.

II) On the other hand, if $\underline{T \ll \tau_c}$, only one temporal mode exists $\lambda_\ell = 1$ and we have

$$Q_W(s) = \prod_j \left(1 + s\mu_j \frac{\langle W\rangle}{T}\right)^{-1} \quad . \tag{5.83}$$

III) In the limiting case $\underline{A \gg A_c}$ and $\underline{T \gg \tau_c}$, $N_t = T/\tau_c$ temporal modes of approximately equal weight exist together with $N_s = A/A_c$ spatial modes of equal weight. This means that altogether we have $N = N_s N_t$ modes, and the mgf takes the familiar form,

$$Q_W(s) = \left(1 + s\frac{\langle W\rangle}{N}\right)^{-N} \quad , \quad N = \frac{T}{\tau_c} \cdot \frac{A}{A_c} \quad . \tag{5.84}$$

IV) In the case $A \gg A_c$ but $T \ll \tau_c$, we have $N = A/A_c$ spatial modes and one temporal mode, i.e.,

$$Q_W(s) = \left(1 + s\frac{\langle W\rangle}{N}\right)^{-N} , \quad N = \frac{A}{A_c} \quad . \tag{5.85}$$

Similarly if $A \ll A_c$ but $T \gg \tau_c$, we have $N = T/\tau_c$ temporal modes.

V) If $A/A_c \gg 1$ but T/τ_c is arbitrary, we can write $\mu_j = 1/N_s$, $j \le N_s$ and from (5.82) we get

$$Q_W(s) = \left[Q_{W/N_s}(s)\right]^{N_s} \quad , \tag{5.86}$$

where $Q_{W/N_s}(s)$ is given by the formula of a point detector. For example, for a Lorentzian light,

$$Q_W(s) = e^{N_s\gamma}\left[\cosh z + (\gamma/2z+z/2\gamma)\sinh z\right]^{-N_s} , \tag{5.87}$$

where $N_s = A/A_c$
and

$$z = \gamma^2 + 2s\gamma\langle W\rangle/N_s .$$

The probability density P(n) that corresponds to (5.86) (or (5.87)) can best be determined by computing the discrete self-convolution of P(n), due to one spatial mode, N_s times.

VI) If the spatial coherence function $g(\underline{r}_1-\underline{r}_2)$ is the two-dimensional Fourier transform of a function that is uniform over a finite aperture and zero otherwise (this is the case of light in the far field of a uniform incoherent object), as was pointed out by several authors [158, 482] the eigenvalues μ_j have approximately rectangular shapes. In particular, if the aperture is rectangular, then $g(\underline{r}) = \mathrm{sinc}(x/r_c)\mathrm{sinc}(y/r_c)$; we have already seen that this yields eigenvalues with rectangular distribution (cf. Sec. 2.4.4). If the aperture is circular and the detector is also circular, then

$$g(\underline{r}) = 2\, J_1(2r/r_c)/(2r/r_c) ;$$

the eigenvalues show the same effect [262]. This permits us to use an effective number of equal spatial modes

$$N_s = \left(\sum \mu_j\right)^2 / \sum \mu_j^2 \tag{5.88}$$

and use (5.86) or (5.87).

The Moments

If the field is cross-spectrally pure, then exact expressions for the moments of W (or the factorial moments of n) can be determined by integrating over the multifold moment expansion of $I(\underline{r},t)$, or by generalizing the approach that led to (5.28-29). The result is

$$K_W^{(m)} = (m-1)!\, B_m D_m \langle W\rangle^m$$

$$B_m = \frac{1}{T^m} \int_0^T dt_1 \ldots \int_0^T dt_m\, g(t_1-t_2)g(t_2-t_3) \ldots g(t_m-t_1) \tag{5.89}$$

$$D_m = \frac{1}{A^m} \int_A d\underline{r}_1 \ldots \int_A d\underline{r}_m \quad g(\underline{r}_1,\underline{r}_2)g(\underline{r}_2,\underline{r}_3) \ldots g(\underline{r}_m,\underline{r}_1) \quad .$$

For $A \ll A_c$, $g(\underline{r}_i,\underline{r}_j) = 1$, and $D_m \approx 1$, (5.89) reproduces (5.28). On the other hand, if $A \gg A_c$, we see that

$$D_m = 1/N_s^{m-1} \quad , \quad N_s = A/A_c \gg 1 \tag{5.90}$$

(because the integral vanishes unless all variables $\underline{r}_1, \ldots \underline{r}_m$ belong to the same coherence area). This makes the normalized cumulants decrease rapidly with m; if A/A_c is large enough, we find that all cumulants other than the first are negligible. This corresponds to the limit of a deterministic W and a Poisson photoelectron process.

Approximate Statistics for Light with an Arbitrary Spectrum. Arbitrary T/τ_c and A/A_c

So far, we have presented usable expressions for photoelectron statistics in the limits ($T \ll \tau_c$, $A \ll A_c$) and ($T \gg \tau_c$, $A \gg A_c$) for light with an arbitrary spectrum and spatial coherence function. The intermediate range of T/τ_c and A/A_c is still rather difficult to handle. This makes the following approximate formula attractive and useful.

We have found that, for $T \gg \tau_c$ and $A \gg A_c$, the integrated intensity is the sum of N independent exponentially distributed modes, thus having a chi-square distribution with $2N$ degrees of freedom, where $N = (T/\tau_c)(A/A_c)$. This results in the statistics (repeated here for convenience),

$$Q_W(s) = (1 + s\,\langle W\rangle/N)^{-N} \tag{5.91}$$

$$P(W) = \frac{N^N}{\Gamma(N)} \frac{W^{N-1}}{\langle W\rangle^N} \exp\left(-N\frac{W}{\langle W\rangle}\right)$$

$$K_W^{(m)} = \frac{(m-1)!}{N^{m-1}} \langle W\rangle^m \tag{5.92}$$

$$P(n) = \binom{n+N-1}{n}(1 + \langle n\rangle/N)^{-N}(1 + N/\langle n\rangle)^{-n}$$

$$F_n^{(m)} = \frac{\Gamma(N+m)}{\Gamma(N)N^m} \langle n\rangle^m \quad . \tag{5.93}$$

In the other limit, $T \ll \tau_c$, the photoelectron statistics are also given by exactly the same set of equations (5.91-95) but with $N = 1$. Also, in the special case of light with rectangular temporal spectrum and cylindrical spatial spectrum, (5.91-93) apply. This leads us [307, 397, 399, 400] to expect that the same set approximately describes the intermediate range of T/τ_c and A/A_c if an effective value of N is properly chosen. The most logical choice for N is the value that generates the exact second moment $\langle W^2\rangle$. If (5.92) is used, then

$$N = \langle W\rangle^2/\langle\Delta W^2\rangle \quad . \tag{5.94}$$

The exact value of $\langle\Delta W^2\rangle$ can be determined from

$$\langle\Delta W^2\rangle = K_W^{(2)} = \langle W\rangle^2 \frac{1}{T^2}\int_0^T dt_1 \int_0^T dt_2 \frac{1}{A^2}\int_A d\underline{r}_1 \int_A d\underline{r}_2 \; |g(\underline{r}_1,\underline{r}_2,t_2-t_1)|^2 \quad . \tag{5.95}$$

Hence N is given by

$$N^{-1} = \frac{1}{T^2}\frac{1}{A^2}\iint_{00}^{TT} dt_1 dt_2 \iint_{AA} d\underline{r}_1 d\underline{r}_2 |g(\underline{r}_1,\underline{r}_2,t_2-t_1)|^2 \quad . \tag{5.96}$$

For cross-spectrally pure light, (5.96) factorizes into

$$N = N_t N_s \quad ,$$

where

$$N_t^{-1} = \frac{2}{T^2}\int_0^T (T-\tau)|g(\tau)|^2 d\tau \tag{5.97}$$

and

$$N_s^{-1} = \frac{1}{A^2} \iint_{AA} d\underline{r}_1 d\underline{r}_2 |g(\underline{r}_1, \underline{r}_2)|^2 \quad . \tag{5.98}$$

In the limit $T \ll \tau_c$ and $A \ll A_c$, $N_t = N_s = N = 1$. In the limit $T \gg \tau_c$, $N_t = T/\tau_c$ and in the limit $A \gg A_c$, $N_s = A/A_c$. For Lorentzian light, (5.97)gives

$$N_t = 2\gamma^2/(e^{-2\gamma} + 2\gamma - 1) \ , \quad \gamma = \Gamma T \quad . \tag{5.99}$$

This relation is plotted in Fig.5.11, which shows how N_t increases from 1 at $\gamma = 0$ to $N_t = \gamma$ for large γ. The N_t-γ relation depends on the spectrum only in the region $\gamma \sim 1$.

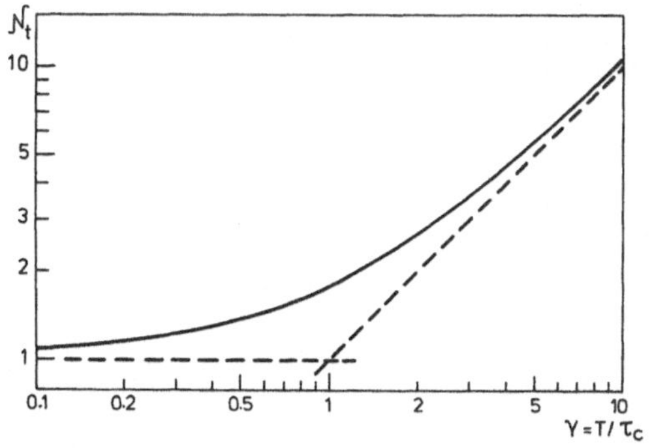

Fig.5.11. Effective number of temporal modes in thermal light, whose spectrum is Lorentzian, as a function of the time-bandwidth product γ

The number of spatial modes N_s has been computed for different geometries and spectral profiles [74, 233, 429, 430]. As an example, we take a circular detector with radius a and a homogeneous optical field with spatial coherence function

$$g(\underline{r}_1, \underline{r}_2) = J(2|\underline{r}_1 - \underline{r}_2|/r_c) \exp\left[j \frac{\pi}{\lambda z} (r_1^2 - r_2^2)\right] , \quad J(x) = 2J_1(x)/x \ . \tag{5.100}$$

This is recognized as the degree of coherence of light radiated by a circular incoherent source of uniform intensity where from (4.51) and (4.52),

$$r_c = \lambda z/\pi a \quad ,$$

a being the radius of the source and z the distance between the source and detector. Substitution in (5.98) gives

$$N_s^{-1} = \sum_{r=0}^{\infty} \frac{[(2r+2)!]^2}{[(r+2)!]^2[(r+1)!]^4} \left(-\frac{A}{A_c}\right)^r \tag{5.101}$$

where A is the area of the detector and $A_c = \pi r_c^2$ is the area of coherence.

In Fig.5.12, the number of spatial modes N_s is plotted against A/A_c. It follows a pattern similar to N_t. It changes from the asymptotic value 1 for $A \ll A_c$ to the asymptotic value A/A_c for $A \gg A_c$.

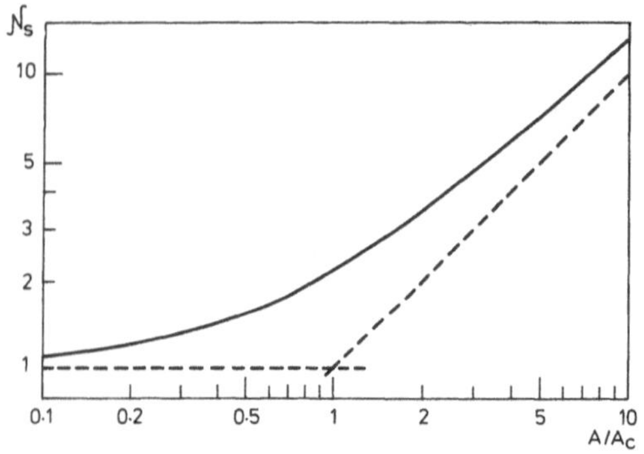

Fig.5.12. Effective number of spatial modes collected from thermal light, whose coherence function is given by (5.100), as a function of the ratio between the area of the detector and the coherence area of the field

An idea of the accuracy of the approximation considered in this section can be obtained only by comparing its predictions with exact results for cases where they are known. This was done by BEDARD et al. [42], LACHS [279] (see comments of MANDEL [314]), and verified experimentally by PEARL and TROUP [366].

Numerical Techniques

So far, we have obtained an exact closed-form expression for $Q_W(s)$ in only one case, the Lorentzian spectrum. Some authors have treated other spectral profiles (the sum of two Lorentzian centered around the same frequency [483] and the sum of two Lorentzians centered around different frequencies [337]). The fact that these mathematical expressions are becoming more and more involved led to the idea of using computer techniques.

Once the eigenvalues m_ℓ are determined, the calculations of P(n) and $F_n^{(m)}$ become a straightforward use of the discrete convolution (5.39) or the recurrence relations in (5.38) and (5.40)

The eigenvalues m_ℓ can be computed if the integral equation (5.23) is transformed into a matrix equation. This can be done [279] by expanding $\phi_k(t)$ and g(t) in Fourier series and substituting in the integral equation. This gives

$$T\sum_{k=-\infty}^{\infty} A_{nk}R_{km} = \lambda_n A_{nm} \quad ,$$

where

$$A_{nk} = \frac{1}{T}\int_0^T e^{jk\Omega t}\phi_n(t)dt \quad ,$$

and

$$R_{km} = \frac{1}{T^2}\int_0^T dt e^{j(k-m)\Omega t}\int_t^{t+T} dt' e^{-jk\Omega t} g(t') \quad .$$

To determine the eigenvalues λ_n, the matrix R_{km} can be truncated and diagonalized. Further details can be found in [279, 285]. By use of this technique, the photoelectron statistics of light with an arbitrary spectrum can be obtained. LACHS and RUGGIERI [285] computed the photoelectron counting distributions of light with a truncated spectrum.

Experimental verification of the photoelectron statistics of coherent and thermal light was done in many laboratories, e.g., [7, 8, 12, 14, 17, 19, 48, 72, 231, 252, 263, 301, 325, 381, 382, 438].

In these experiments, thermal light was obtained either by scattering laser light from a rotating rough disk or from particles in Brownian motion. Thermal light from a tungsten lamp is so weak (degeneracy parameter is very small) that the photoelectron statistics are always Poissonian.

5.2.3 Partially Polarized Thermal Light

It has been shown in Sec. 4.2.2 that under a condition of cross-spectral purity, the two orthogonal components of a partially polarized light field can be chosen such that they are statistically uncorrelated. Because these components are Gaussian, they must also be statistically independent. Thus, the total detected intensity can be written as the sum of two statistically independent components,

$$I = I_1 + I_2 \quad ,$$

$$\langle I_1 \rangle = \frac{1}{2}(1+P)\langle I \rangle \quad ,$$

$$\langle I_2 \rangle = \frac{1}{2}(1-P)\langle I \rangle \quad .$$

Similarly, the integrated intensity can be written as

$$W = W_1 + W_2 \quad ,$$

$$\langle W_1 \rangle = \frac{1}{2}(1+P)\langle W \rangle \quad , \qquad (5.102)$$

$$\langle W_2 \rangle = \frac{1}{2}(1-P)\langle W \rangle \quad ,$$

where W_1 and W_2 are statistically independent. The number of photoelectrons is also an independent sum of the contributions of W_1 and W_2,

$$n = n_1 + n_2 \quad . \qquad (5.103)$$

This enables us to write the statistical properties of W and n simply in terms of the statistics of their components.
From Sec. 2.2 the mgf is the product

$$Q_W(s) = Q_{W_1}(s) \cdot Q_{W_2}(s) \quad ; \qquad (5.104)$$

consequently the cgf is the sum

$$Q_W^c(s) = Q_{W_1}^c(s) + Q_{W_2}^c(s) \quad . \qquad (5.105)$$

Also, the PD is the convolution

$$P_W(W) = P_{W_1}(W) \circledast P_{W_2}(W) = \int_0^W P_{W_1}(W')P_{W_2}(W-W')dW' \quad . \tag{5.106}$$

From the cgf, the cumulants can be determined as the sum

$$K_W^{(m)} = K_{W_1}^{(m)} + K_{W_2}^{(m)} \quad . \tag{5.107}$$

The moments are also related by

$$\langle W^m \rangle = \sum_{\ell=0}^{m} \binom{m}{\ell} \langle W_1 \rangle^{\ell} \langle W_2 \rangle^{m-\ell} \quad .$$

Knowing that the Poisson transform of the convolution of two functions is the discrete convolution of the Poisson transform of the functions (3.106), we conclude that

$$P_n(n) = P_{n_1}(n) \circledast P_{n_2}(n) = \sum_{\ell=0}^{n} P_{n_1}(\ell)P_{n_2}(n-\ell) \quad . \tag{5.108}$$

Also from (3.69), it is shown that

$$F_n^{(m)} = \sum_{\ell=0}^{m} \binom{m}{\ell} F_{n_1}^{(\ell)} \cdot F_{n_2}^{(m-\ell)} \quad . \tag{5.109}$$

Now, we come to the time statistics of the photoelectron point process. By substituting from (5.104) in (3.93), we obtain the PD of the forward recurrence time

$$\begin{aligned} P_f(t) &= Q_{W_1}(1)P_{f_1}(t) + Q_{W_2}(1)P_{f_2}(t) \\ &= P_{n_1}(0)P_{f_1}(t) + P_{n_2}(0)P_{f_2}(t) \end{aligned} \quad . \tag{5.110}$$

Then, by using (3.94), we obtain the PD of the time interval between two events,

$$P_i(\tau) = \frac{\langle\lambda_2\rangle}{\langle\lambda\rangle} P_{n_1}(0)P_{i_2}(\tau) + \frac{\langle\lambda_1\rangle}{\langle\lambda\rangle} P_{n_2}(0)P_{i_1}(\tau) + \frac{2}{\langle\lambda\rangle} P_{f_1}(\tau)P_{f_2}(\tau) \quad . \tag{5.111}$$

With the above relations, and the statistics of the components W_1 and W_2 and n_1 and n_2 already determined from the results of the previous section, we can consider that the problem of the photoelectron statistics of partially polarized light is solved. Below, we write the final results for some of the special cases of Sec. 5.2.2.

The Limit $T \ll \tau_c$

The integrated intensity is now proportional to the intensity itself and its statistics are given by (4.87-91), rewritten now in the form

$$\begin{aligned} P(W) &= \frac{1}{P\langle W\rangle}\left[\exp\left(-\frac{2}{1+P}\frac{W}{\langle W\rangle}\right) - \exp\left(-\frac{2}{1-P}\frac{W}{\langle W\rangle}\right)\right] \\ &= 4W\langle W\rangle^{-2}\exp(-2W/\langle W\rangle) \quad , \qquad P = 0 \\ &= \langle W\rangle^{-1}\exp(-W/\langle W\rangle) \quad , \qquad P = 1 \end{aligned} \tag{5.112}$$

$$K_W^{(m)} = (m-1)!\left[\left(\frac{1+P}{2}\right)^m + \left(\frac{1-P}{2}\right)^m\right]\langle W\rangle^m \tag{5.113}$$

$$\langle \Delta W^2\rangle = \frac{1}{2}(1 + P^2)\langle W\rangle^2 \quad .$$

Comparing this to (5.35), we see that $\langle\Delta W^2\rangle = \langle W\rangle^2/N_{eff}$, where $N_{eff} = 2/(1+P^2)$ represents an effective number of modes ("degrees of freedom"). For polarized light $N_{eff} = 1$ and for unpolarized light $N_{eff} = 2$.

For the photoelectron-counting distribution P(n), the discrete convolution of two Bose-Einstein distributions with means

$$\langle n_1\rangle = \frac{1}{2}(1 + P)\langle n\rangle \quad \text{and} \quad \langle n_2\rangle = \frac{1}{2}(1 - P)\langle n\rangle$$

gives

$$P(n) = \frac{1}{\langle n_1\rangle - \langle n_2\rangle} \left[\left(\frac{\langle n_1\rangle}{1+\langle n_1\rangle}\right)^{n+1} - \left(\frac{\langle n_2\rangle}{1+\langle n_2\rangle}\right)^{n+1} \right] , \qquad (5.114)$$

which can also be obtained by taking the Poisson transform of (5.112).

This correspond to

$$F_n^{(m)} = \frac{m!}{P} \left[\left(\frac{1+P}{2}\right)^{m+1} - \left(\frac{1-P}{2}\right)^{m+1} \right] \langle n\rangle^m \qquad (5.115)$$

$$\langle \Delta n^2\rangle = \langle n\rangle + \frac{1}{2}(1+P^2)\langle n\rangle^2 = \langle n\rangle + \langle n\rangle^2/N_{eff} . \qquad (5.116)$$

Notice that the higher the degrees of freedom N_{eff} is, the lower the contribution of the field fluctuation to the uncertainty of n. The contribution of the detection process is unaffected.

Limit $T \gg \tau_c$

In this limit, W_1 and W_2 have approximately chi-square distributions with $2N$ degrees of freedom, $N = T/\tau_c$. The convolution of these two distributions gives

$$P(W) = \frac{N^{2N}}{\Gamma(N)} \frac{W^{2N-1} e^{-N \frac{W}{\langle W_2\rangle}}}{(x\langle W_1\rangle\langle W_2\rangle)^N} \sum_{r=0}^{N-1} \frac{(-1)^r}{r!} \frac{1}{x^r} \left[1 - e^{-x} \sum_{m=0}^{N+r-1} \frac{x^m}{m!}\right] \qquad (5.117)$$

$$x = N\left(\frac{W}{\langle W_1\rangle} - \frac{W}{\langle W_2\rangle}\right)$$

and the cumulants are given by

$$K_W^{(m)} = \frac{(m-1)!}{N^{m-1}} \left[\left(\frac{1+P}{2}\right)^m + \left(\frac{1-P}{2}\right)^m \right] \langle n\rangle^m \qquad (5.118)$$

$$\langle \Delta W^2\rangle = \langle W\rangle^2/N_{eff} , \quad N_{eff} = 2N/(1+P^2) . \qquad (5.119)$$

Similarly, in this limit, n_1 and n_2 have negative-binomial distributions whose discrete convolution gives [198]

$$P(n) = \left(1 + \frac{\langle n_1 \rangle}{N}\right)^{-N} \left(1 + \frac{\langle n_2 \rangle}{N}\right)^{-N} \sum_{\ell=0}^{n} \binom{N+\ell-1}{N-1} \binom{n+N+\ell-1}{N-1}$$

$$\left(1 + \frac{N}{\langle n_1 \rangle}\right)^{\ell} \left(1 + \frac{N}{\langle n_2 \rangle}\right)^{n-\ell} \quad , \tag{5.120}$$

$$\langle n_1 \rangle = \frac{1}{2}(1 + P)\langle n \rangle \quad ,$$

$$\langle n_2 \rangle = \frac{1}{2}(1 - P)\langle n \rangle \quad ,$$

which has the factorial moments

$$F_n^{(m)} = \frac{m!\ \langle n \rangle^m}{(2N)^m} \sum_{\ell=0}^{m} \binom{N+\ell-1}{N-1} \binom{N+m-\ell-1}{N-1} (1+P)^{\ell}(1-P)^{m-\ell} \quad . \tag{5.121}$$

In the limit $T/\tau_c \to \infty$, $P(n)$ becomes a Poisson distribution independent of P.

As we have seen in Sec. 5.2.2, if N is properly adjusted, (5.117-121) provide good approximations for the whole range of T/τ_c. If the effect of the detector area is also included, we obtain [521]

$$N = N_t N_s\ 2/(1 + P^2) \quad , \tag{5.122}$$

where N_t and N_s are given by (5.97) and (5.98).

Partially Polarized Thermal Light with Lorentzian Spectrum

Using the exact expression for $Q_W(s)$, $K_W^{(m)}$, $P(n)$, $F_n^{(m)}$ for polarized thermal light with Lorentzian spectrum (5.54, 55), we can immediately apply the relations (5.104-107) and obtain [225]

$$P(W) = \frac{2e^{\gamma}}{\langle W \rangle} \sum_{\ell=0}^{\infty} (-1)^{\ell} \left(\frac{2-\gamma\lambda_{\ell}/T}{1+\lambda_{\ell}/T}\right) \left\{ \frac{Q(y_{\ell})}{1+P} \exp\left[\frac{-2T}{(1+P)\lambda_{\ell}} \frac{W}{\langle W \rangle}\right] + \frac{Q(y_{\ell}')}{1-P} \right.$$

$$\left. \exp\left[\frac{-2T}{(1-P)\lambda_{\ell}} \frac{W}{\langle W \rangle}\right] \right\} \tag{5.123}$$

where

$$Q(y) = e^{\gamma}\left[\cosh y + \tfrac{1}{2}(y/\gamma + \gamma/y)\sinh y\right]^{-1} \quad , \tag{5.124}$$

$$y_\ell^2 = \gamma^2 - \frac{1-P}{1+P}\frac{2\gamma T}{\lambda_\ell} \quad , \quad y_\ell'^2 = \gamma^2 - \frac{1+P}{1-P}\frac{2\gamma T}{\lambda_\ell} \quad , \tag{5.125}$$

and λ_ℓ are the eigenvalues given by (5.50, 51).
This distribution is plotted in Fig.5.13 for $P = 0$. Also,

$$K_W^{(m)} = (m-1)!\left[\left(\frac{1+P}{2}\right)^m + \left(\frac{1-P}{2}\right)^m\right]B_m \langle W\rangle^m \quad , \tag{5.126}$$

where B_m are given by (5.55).

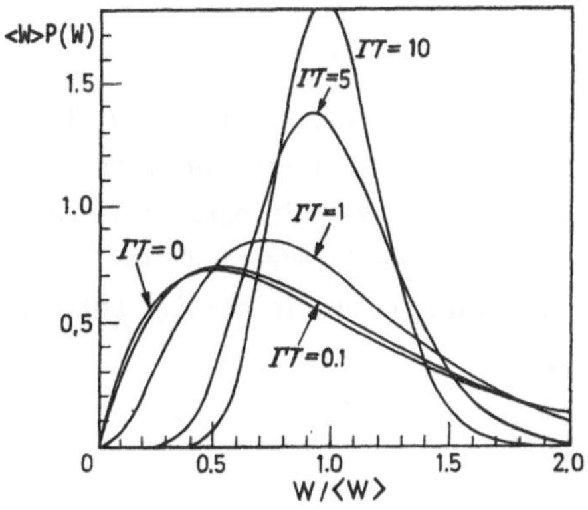

Fig.5.13. PD of the integrated intensity of unpolarized thermal light that has a Lorentzian spectrum, for several values of time-bandwidth product (after JAISWAL and MEHTA [225])

The moments $\langle W^m\rangle$ can be obtained from $K_W^{(m)}$ by using (2.5). The photoelectron-counting distribution $P(n)$ is given by the discrete convolution of $P_{n_1}(n)$ and $P_{n_2}(n)$ determined from the recurrence relation (5.60) and (5.62) with average counts $(1/2)(1+P)\langle n\rangle$ and $(1/2)(1-P)\langle n\rangle$. The factorial moments are determined from $F_n^{(m)} = \langle W^m\rangle$. These are plotted in Fig.5.14 as functions of the degree of polarization P for several values of γ and for m=2-5.

5.2.4. Mixture of Polarized Coherent and Polarized Thermal Light

In this section, we turn to the photoelectron statistics for a superposition of polarized coherent and polarized thermal light fields.

As has already been mentioned in Sec. 4.2.3, the stochastic signal that represents such a field, $V(t)$, is the sum of a complex Gaussian signal $V_{th}(t)$ and a deterministic signal $V_c(t) = A(t)\exp(j\Delta\omega t)$. From this, the statistics

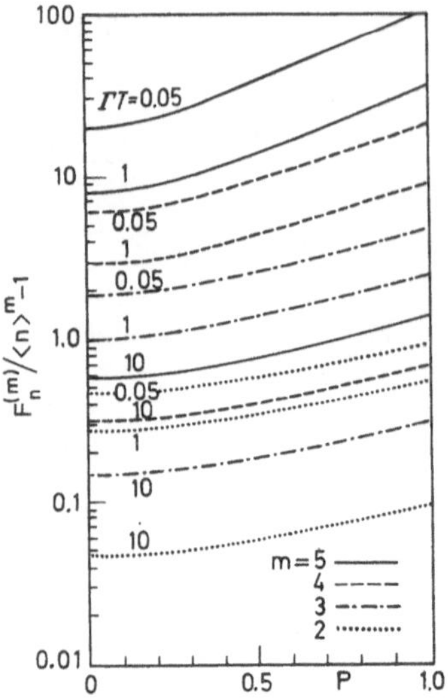

Fig.5.14. Dependence of the photoelectron counting normalized factorial moments of partially polarized thermal light that has a Lorentzian spectrum on the degree of polarization P and the time-bandwidth product γ (after JAISWAL and MEHTA [225])

of the process I(t) have been determined. Here we need the statistics of the integrated intensity W from which the photoelectron statistics can be determined. Let us first consider the limiting case of short integration time and a point detector. Moreover, let us assume that the integration time is shorter than the period of the slow envelope variation due to beating between the central frequencies, i.e., $T \ll \Delta\omega^{-1}$.

$T \ll \tau_c$, $A \ll A_c$, $\Delta\omega T \ll 1$

In this limit, the integrated intensity is proportional to the intensity itself, $W \sim I$. Hence, the statistical properties of W are identical to those of I (4.95-99). For convenience, we write these equations here again in the form

$$Q_W(s) = (1+s\langle W_{th}\rangle)^{-1}\exp[-sW_c/(1+s\langle W_{th}\rangle)] \tag{5.127}$$

$$P(W) = \frac{1}{\langle W_{th}\rangle}\exp\left(-\frac{W+W_c}{\langle W_{th}\rangle}\right)I_0\left(\frac{2\sqrt{W_cW}}{\langle W_{th}\rangle}\right) \tag{5.128}$$

$$\begin{aligned}\langle W^m\rangle &= m!\,\langle W_{th}\rangle^m L_m(-W_c/\langle W_{th}\rangle)\\ \langle \Delta W^2\rangle &= \langle W_{th}\rangle^2 + 2W_c\langle W_{th}\rangle = \langle W\rangle^2 - W_c^2\ .\end{aligned} \tag{5.129}$$

If we define the effective number of modes by $N_{eff} = \langle W\rangle^2/\langle\Delta W^2\rangle$, we get

$$N_{eff} = (1-W_c^2/\langle W\rangle^2)^{-1} \quad , \tag{5.130}$$

which is 1.0 for a pure thermal field and increases with W_c until it becomes infinite for a purely coherent field. The corresponding photoelectron-counting statistics can be determined by using the fundamental rules of DSP.PP processes. This gives [172, 277]

$$P(n) = \frac{\langle n_{th}\rangle^n}{(1+\langle n_{th}\rangle)^{n+1}} \exp\left(\frac{-\langle n_c\rangle}{1+\langle n_{th}\rangle}\right) L_n\left(\frac{-\langle n_c\rangle/\langle n_{th}\rangle}{1+\langle n_{th}\rangle}\right) \tag{5.131}$$

and

$$F_n^{(m)} = m!\ \langle n_{th}\rangle^m\ L_m(-\langle n_c\rangle/\langle n_{th}\rangle) \quad , \tag{5.132}$$

$$\langle\Delta n^2\rangle = (\langle n_{th}\rangle^2 + \langle n_{th}\rangle) + (\langle n_c\rangle) + (2\langle n_c\rangle\ \langle n_{th}\rangle) \quad .$$

Fig.5.15 illustrates the change of P(n) with the proportion of the coherent light in the mixture.

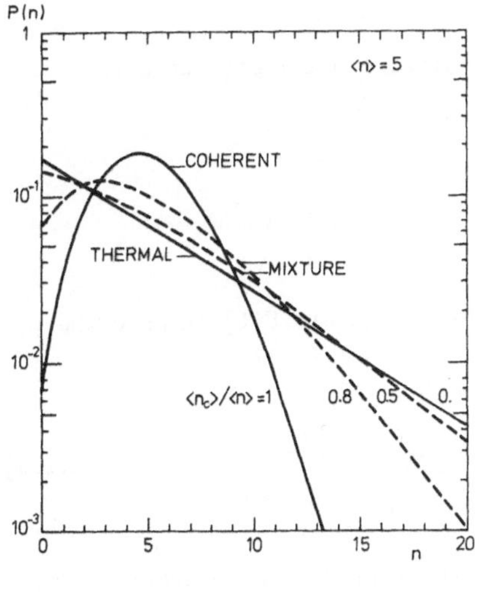

Fig.5.15. Photoelectron counting PD of a mixture of coherent and thermal light in the short-sampling-time limit

The limits of pure coherent and pure thermal fields are also shown. The change of $F_n^m/\langle n\rangle^m = \langle W^m\rangle/\langle W\rangle^m$ with the ratio $\langle n_{th}\rangle/\langle n_c\rangle$ has already been illustrated in Fig.4.11.

The statistics of times can be obtained from (5.127), (3.93), and (3.94) [415]

$$P_f(t) = T_s^{-1}Q(1)\,\frac{1+r_{th}^2 t/T_s}{(1+r_{th}t/T_s)^2} \quad ,$$

$$P_i(t) = T_s^{-1}Q(1)\,\frac{2(1+r_{th}^2 t/T_s)^2-r_c^2}{(1+r_{th}t/T_s)^4} \quad ,$$

where

$$Q(1) = (1+r_{th}t/T_s)^{-1}\exp\left[-\frac{r_c t/T_s}{(1+r_{th}t/T_s)}\right] .$$

Here, r_{th} is the ratio between the average rate of photoelectrons due to the thermal component alone and the average total rate, and $r_c=1-r_{th}$. As expected, the degree of bunching is less than that in the case of pure thermal light. For example,

$$P_i(0) = T_s^{-1}(2-r_c^2) \quad ,$$

instead of $2T_s^{-1}$ for thermal light.

Next, we allow the counting time to be arbitrary but limit ourselves to small detectors.

Arbitrary T/τ_c and $\Delta\omega T$, $A \ll A_c$

The Moment-Generating Function

As has been done in the case of thermal light, we expand $V(t)$ in an orthonormal basis

$$V(t) = \sum_k \alpha_k \Phi_k(t) = \sum_k (\alpha_{thk} + \alpha_{ck})\Phi_k(t) \quad , \tag{5.133}$$

where α_{thk} and α_{ck} are the contributions of the thermal and coherent parts, respectively, and are given by

$$\alpha_{thk} = \int_0^T V_{th}(t)\Phi_k^*(t)dt$$

$$\alpha_{ck} = \int_0^T V_c(t)\Phi_k^*(t)dt \quad . \tag{5.134}$$

Note that $\{\alpha_{ck}\}$ are deterministic complex quantities, whereas $\{\alpha_{thk}\}$ are joint complex Gaussian zero-mean random variables with covariance matrix Λ, whose elements are

$$\Lambda_{\ell k} = \int_0^T\int_0^T G_{th}(t_1,t_2)\phi_\ell^*(t_1)\phi_k(t_2)dt_1dt_2 \, , \; G_{th}(t,t') = \langle V_{th}^*(t)V_{th}(t')\rangle \; . \tag{5.135}$$

Now we can write the integrated intensity as a sum

$$W = \sum_k W_k \quad , \qquad W_k = |\alpha_{thk} + \alpha_{ck}|^2 \quad , \tag{5.136}$$

and

$$Q_W(s) = \langle \exp(-sW)\rangle = \left\langle \exp\left(-s\sum_k |\alpha_{thk} + \alpha_{ck}|^2\right)\right\rangle$$

$$= \left(\prod_k \int d^2\alpha_{thk}\right) P(\{\alpha_{thk}\}) \exp\left(-s\sum_k |\alpha_{thk} + \alpha_{ck}|^2\right) \quad .$$

The result of this multiple integration can be written in the matrix form

$$Q_W(s) = |\delta + T\Lambda|^{-1} \exp\left[[V_c]^\dagger(\Lambda^{-1}D^{-1} - \delta)\Lambda^{-1}[V_c]\right] \tag{5.137}$$

where

$$D = \Lambda^{-1} + sT\delta \quad , \tag{5.138}$$

δ is the unit matrix, and $[V_c]$ is the vector $\{\alpha_{ck}\}$.

The problem can be further simplified if the orthonormal basis $\{\phi_k(t)\}$ is the Karhunen-Loève basis that corresponds to the thermal part, which satisfies (19) or (5.23). In this case, $\{\alpha_{thk}\}$ are statistically independent; hence $W = \sum_k W_k$ is the sum of independent random variables, each having the distribution

$$P(W_k) = \frac{1}{m_k} \exp\left(-\frac{W_k+W_{ck}}{m_k}\right) I_0(2\sqrt{W_k W_{ck}}/m_k) \tag{5.139}$$

and a mgf

$$Q_{W_k}(s) = (1+sm_k)^{-1}\exp[-sW_{ck}/(1+sm_k)] \quad , \tag{5.140}$$

where m_k are the eigenvalues of

$$\int_0^T G_{th}(t,t')\Phi_k(t)dt = m_k\Phi_k(t') \quad , \tag{5.141}$$

and

$$W_{ck} = |\alpha_{ck}|^2 \quad , \quad W_{thk} = |\alpha_{thk}|^2 \quad . \tag{5.142}$$

The PD, P(W), is then the multiple convolution

$$P(W) = \circledast_k P_k(W) \tag{5.143}$$

and the mgf is the product

$$Q_W(s) = \prod_k Q_{W_k}(s) \quad . \tag{5.144}$$

The Moments

In order to find the moments of W, we write the cgf as a sum

$$\begin{aligned} Q_W^c(s) &= \sum_k Q_{W_k}^c(s) \\ &= -\sum_k \ln(1+sm_k) - s\sum_k W_{ck}(1+sm_k)^{-1} \quad , \end{aligned}$$

which can also be written in the form

$$Q_W^c(s) = -\sum_k \ln(1+sm_k) - s\sum_k W_{ck} + s^2 \sum_k W_{ck} m_k (1+sm_k)^{-1} \quad ,$$

from which we observe that the first and second terms are the cgf's of the thermal and coherent parts of the field, i.e.,

$$Q_W^c(s) = Q_{W_{th}}^c(s) + Q_{W_c}^c(s) + s^2 \sum_k W_{ck} m_k (1+sm_k)^{-1}$$

$$= Q_{W_{th}}^c(s) + Q_{W_c}^c(s) + \sum_{j=2}^{\infty} (-1)^j s^j \Big(\sum_k W_{ck} m_k^{j-1}\Big) \quad .$$

We can automatically use the rule for cumulant generation, (2.12), and obtain

$$K_W^{(1)} = \langle W_{th}\rangle + \langle W_c\rangle$$

$$K_W^{(m)} = K_{W_{th}}^{(m)} + m! \sum_k W_{ck} m_k^{m-1}, \quad m > 1 \quad .$$

Substituting from (5.142) and (5.134), we get

$$K_W^{(m)} = K_{W_{th}}^{(m)} + m! \int_0^T\!\!\int_0^T dt_1 dt_2 V_c^*(t_1) V_c(t_2) \Gamma^{(m-1)}(t_1,t_2) \quad , \tag{5.145}$$

where

$$\Gamma^{(1)}(t_1,t_2) = \sum_k \phi_k^*(t_1)\phi_k(t_2) = G_{th}(t_1-t_2)$$

$$\Gamma^{(j)}(t_1,t_2) = \sum_k m_k^j \phi_k^*(t_1)\phi_k(t_2) = \int_0^T \Gamma^{(1)}(t_1,t)\Gamma^{(j-1)}(t,t_2)dt \quad , \quad j > 1 \quad . \tag{5.146}$$

An expression for $K_{W_{th}}^{(m)}$ is given in Sec. 5.2.2.

If the coherent part is constant, i.e., $V_c(t) = A\, e^{j\Delta\omega t}$, then by writing

$$G_{th}(t_1,t_2) = \langle W_{th}\rangle\, g(t_1-t_2) \quad ,$$

we immediately get [226]

$$K_W^{(m)} = K_{W_{th}}^{(m)} + m!\, \bar{B}_m\, W_c\, \langle W_{th}\rangle^{m-1} \quad , \tag{5.147}$$

where

$$\bar{B}_m = \frac{\mathrm{Re}}{T^m} \int_0^T dt_1 \ldots \int_0^T dt_m \exp[j\Delta\omega(t_1-t_m)]g(t_1-t_2) \ldots g(t_{m-1}-t_m) \quad . \tag{5.148}$$

See also [350].

Counting Distribution

The factorial moments of n can be determined from the moments of W by using the rules of Sec. 3.5.2. The probability density P(n) can be written as a discrete convolution of the Poisson transform of $P(W_k)$, i.e.,

$$P(n) = \circledast_k \frac{m_k^n}{(1+m_k)^{n+1}} \exp\left(\frac{-W_{ck}}{1+m_k}\right) L_n\left(\frac{-W_{ck}/m_k}{1+m_k}\right) \quad . \tag{5.149}$$

The problem of determining closed-form expressions for P(W), P(n) and F_n cannot be pursued any further unless a mathematical law for the eigenvalues m_k is found, i.e., the spectrum is specified and the integral equation (5.141) solved. LACHS [279] computed the discrete convolutions in (5.149). LAXPATI and LACHS [295] developed recurrence relations for computing P(n) and $F_n^{(m)}$, which are well adapted to the computer.

Statistics of Times

General expressions for the probability distribution of photoelectron time of arrival and inter-event time can be obtained by using the expansion (5.133) and the fundamental equations (3.88) and (3.90), as has been done previously in the case of thermal light. This gives [287]

$$P_f(t) = T_s^{-1}Q(1)\eta(t,t)$$

and

$$P_i(t) = T_s^{-1}Q(1)\Big|[\eta(0,0)]^2+|\eta(0,t)|^2+|S(t)|^2|S(0)|^2+2\mathrm{Re}\{S(t)S^*(0)\eta(0,t)\}\Big| ,$$

where

$$S(t) = \sum_k \frac{\alpha_{ck}}{1+m_k}\Phi_k(t)$$

and as in (5.46)

$$\eta(t,t') = T_s \sum_k \frac{m_k}{1+m_k}\Phi_k^*(t)\Phi_k(t') \quad ,$$

and Q(1) is the mgf of the integrated intensity over the interval [0,t].

Note that if the coherent component of light is zero, $S(t) = 0$ and (5.44) and (5.45) are recovered.

Limit $T \ll \tau_c$, $A \ll A_c$, Arbitrary $\Delta\omega T$, Constant Coherent Component

In this limit, only one thermal mode exists and (5.134) gives

$$\begin{aligned} m_k &= \langle W_{th}\rangle \qquad k = 1 \\ &= 0 \qquad\qquad k > 1 \end{aligned}$$

$$\phi_k(t) = 1/\sqrt{T}$$

$$W_{ck} = W_c\Omega^2$$

$$\Omega = \sin(\Delta\omega T/2)/(\Delta\omega T/2) \quad . \tag{5.150}$$

Substituting in (5.140, 144), we get

$$\begin{aligned} Q_W(s) &= \frac{1}{1+s\langle W_{th}\rangle} \exp\left(\frac{\Omega^2 W_c\langle W_{th}\rangle s^2}{1+s\langle W_{th}\rangle}\right)\exp(-W_c s) \\ &= \frac{1}{1+s\langle W_{th}\rangle}\exp\left(\frac{-\Omega^2 W_c s}{1+s\langle W_{th}\rangle}\right)\exp[-(1-\Omega^2)W_c s] \quad . \end{aligned} \tag{5.151}$$

If $\Delta\omega T \ll 1$, then $\Omega = 1$ and the Rician-square statistics is reproduced. If $\Delta\omega T \gg 1$, then the beat signal between the two parts of the field is too fast to be detected and the mgf becomes the product of the mgf's of the thermal and coherent parts alone.

$$Q_W(s) = (1+s\langle W_{th}\rangle)^{-1}\exp(-sW_c) = Q_{W_{th}}(s) \cdot Q_{W_c}(s) \quad .$$

This can be seen from (5.151) by putting $\Omega = 0$. The same situation would occur if the thermal and coherent parts had orthogonal polarizations. Eq. (5.151) shows that W could be regarded as the sum of two components: I) coherent light with integrated intensity $(1-\Omega^2)W_c$, II) a mixture of thermal light with average integrated intensity $\langle W_{th}\rangle$ and coherent light with integrated intensity $\Omega^2 W_c$. For the PD, we get

$$P(W) = \frac{1}{\langle W_{th}\rangle} \exp\left[-\frac{W+(2\Omega^2-1)W_c}{\langle W_{th}\rangle}\right] I_0\left(\frac{2\Omega}{\langle W_{th}\rangle}\left\{W_c[W+(\Omega^2-1)W_c]\right\}^{1/2}\right) ,$$

$$\text{if} \quad W > (1-\Omega^2)W_c \quad , \qquad (5.152)$$

$$= 0 \quad \text{otherwise.}$$

For $\Omega = 1$, $(\Delta\omega T = 0)$, (5.152) gives (5.128) and for $\Omega = 0$, $(\Delta\omega T \gg 1)$, it becomes

$$P(W) = \langle W_{th}\rangle^{-1}\exp[-(W-W_c)/\langle W_{th}\rangle] \ , \qquad W > W_c \qquad (5.153)$$

$$= 0 \quad \text{otherwise.}$$

For the moments of W, we have

$$\langle W^m\rangle = \sum_{j=0}^{m} \frac{m!}{(m-j)!} [(1-\Omega^2)W_c]^{m-j} \langle W_{th}\rangle^j L_j(-\Omega^2 W_c/\langle W_{th}\rangle)$$

$$\langle \Delta W^2\rangle = \langle W_{th}\rangle^2 + 2\Omega^2 W_c\langle W_{th}\rangle \qquad (5.154)$$

corresponding to

$$N_{eff} = [1-W_c^2/\langle W\rangle^2-2(1-\Omega^2)W_c\langle W_{th}\rangle/\langle W\rangle^2]^{-1} \quad ,$$

which increases with increase of W_c and with decrease of Ω (i.e., increase of the separation frequency $\Delta\omega$).

The photoelectron-counting distribution can be determined by finding the Poisson transform of (5.152). The result is [370]

$$P(n) = (1+\langle n_{th}\rangle)^{-1}\exp\left[-\langle n_c\rangle \frac{1+\langle n_{th}\rangle(1-\Omega^2)}{1+\langle n_{th}\rangle}\right]$$
$$\sum_{j=0}^{n} \frac{1}{(n-j)!}\,[\langle n_c\rangle(1-\Omega^2)]^{n-j}\left(1+\frac{1}{\langle n_{th}\rangle}\right)^{-j}$$
$$L_j\left[-\frac{\langle n_c\rangle\,\Omega^2}{\langle n_{th}\rangle(1+\langle n_{th}\rangle)}\right]\,. \tag{5.155}$$

Note that with $\Delta\omega = 0$, $\Omega = 1$, (5.131) is reproduced. On the other hand, for $\Omega = 0$, the Poisson transform of (5.153) gives [150]

$$P(n) = \frac{1}{n!}\,\frac{\langle n_{th}\rangle^n}{(1+\langle n_{th}\rangle)^{n+1}}\,\exp(\langle n_c\rangle/\langle n_{th}\rangle)$$
$$\Gamma[n+1\,,\langle n_c\rangle(1+\langle n_{th}\rangle)/\langle n_{th}\rangle]$$

where $\Gamma(a,x) = \int_x^\infty e^{-t}t^{a-1}dt$

is the incomplete gamma function.

From (5.154),

$$F_n^{(m)} = \sum_{j=0}^{m}\frac{m!}{(m-j)!}[(1-\Omega^2)\langle n_c\rangle]^{m-j}\langle n_{th}\rangle^j L_j(-\Omega^2\langle n_c\rangle/\langle n_{th}\rangle)$$
$$\langle\Delta n^2\rangle = (\langle n_{th}\rangle^2 + \langle n_{th}\rangle) + (\langle n_c\rangle) + (2\Omega^2\langle n_c\rangle\langle n_{th}\rangle)\quad . \tag{5.156}$$

<u>Limit $T \gg \tau_c$, $A \ll A_c$, Arbitrary $\Delta\omega T$, Constant Coherent Component</u>

In this limit, $N=(T/\tau_c)$ equal thermal modes having $m_k = \langle W_{th}\rangle/N$ exist. By substituting in (5.140) and putting $W_c = \sum W_{ck}$, we get

$$Q_W(s) = (1+s\langle W_{th}\rangle/N)^{-N}\exp\left(\frac{\Omega^2 W_c\langle W_{th}\rangle s^2/N}{1+s\langle W_{th}\rangle/N}\right)\exp(-W_c s)\quad , \tag{5.157}$$

which reproduces (5.151) if $N = 1$.

If $A >> A_c$, then (5.157) holds with $N = (T/\tau_c)(A/A_c)$. The distribution $P(W)$ is obtained by an inverse Laplace transform of $Q_W(s)$

$$P(W) = \frac{1}{\langle W_{th}\rangle/N}\left[\frac{W-(1-\Omega^2)W_c}{\Omega^2 W_c}\right]^{N-\frac{1}{2}} \exp\left[-\frac{W+W_c(2\Omega^2-1)}{\langle W_{th}\rangle/N}\right]$$

$$I_{N-1}\left\{\frac{2\Omega\,[W_c W-(1-\Omega^2)W_c^2]^{1/2}}{\langle W_{th}\rangle/N}\right\}, \quad W > W_c(1-\Omega^2) \tag{5.158}$$

$$= 0 \quad \text{otherwise.}$$

The moments can be obtained from the properties of the generating function of Laguerre polynomials,

$$\langle W^m\rangle = \sum_{j=0}^{m} \frac{m!}{(m-j)!}\,[(1-\Omega^2)W_c]^{m-j}\left(\frac{\langle W_{th}\rangle}{N}\right)^j L_j^{N-1}\left(-\frac{W_c\Omega^2}{\langle W_{th}\rangle/N}\right), \tag{5.159}$$

where $L_j^N(x)$ is the Laguerre polynomial $[L_j^0(x) = L_j(x)]$, e.g.,

$$\langle \Delta W^2\rangle = \frac{1}{N}\left(\langle W_{th}\rangle^2 + 2\Omega^2 W_c\langle W_{th}\rangle\right) ,$$

which corresponds to

$$N_{eff} = N/[1-W_c^2/\langle W\rangle^2 - 2(1-\Omega^2)W_c\langle W_{th}\rangle/\langle W\rangle^2] .$$

If $\Delta\omega = 0$, (5.157) reduces to

$$Q_W(s) = (1+s\langle W_{th}\rangle/N)^{-N}\exp[-sW_c(1+s\langle W_{th}\rangle/N)] . \tag{5.160}$$

which is the mgf of the Bessel or the noncentral chi-square distribution. Eqs. (5.158) and (5.159) reduce to [368]

$$P(W) = \frac{1}{\langle W_{th}\rangle/N}\left(\frac{W}{W_c}\right)^{N-\frac{1}{2}} \exp\left(-\frac{W+W_c}{\langle W_{th}\rangle/N}\right) I_{N-1}\left(\frac{2(W_c W)^{1/2}}{\langle W_{th}\rangle/N}\right) \tag{5.161}$$

$$\langle W^m \rangle = m! \quad (\langle W_{th} \rangle / N)^m L_m^{N-1} (-N W_c / \langle W_{th} \rangle) \tag{5.162}$$

$$\begin{aligned} \langle \Delta W^2 \rangle &= (\langle W_{th} \rangle^2 + 2 W_c \langle W_{th} \rangle)/N \\ N_{eff} &= N/(1 - W_c^2/\langle W \rangle^2) \quad . \end{aligned} \tag{5.163}$$

It has been shown that this distribution can be approximated well by a log-normal distribution [354].

By taking the Poisson transform of (5.158), the photoelectron counting PD is obtained [278, 370]

$$\begin{aligned} P(n) &= \exp\left\{ - \frac{\langle n_c \rangle [N + \langle n_{th} \rangle (1-\Omega^2)]}{N + \langle n_{th} \rangle} \right\} \\ &\sum_{j=0}^{n} \frac{[\langle n_c \rangle (1-\Omega^2)]^{n-j} \; (\langle n_{th} \rangle / N)^j}{(n-j)! \; (1 + \langle n_{th} \rangle / N)^{N+j}} \\ &L_j^{N-1} \left[- \frac{\Omega^2 N^2 \langle n_c \rangle}{\langle n_{th} \rangle (\langle n_{th} \rangle + N)} \right] \quad . \end{aligned} \tag{5.164}$$

For $\Delta\omega = 0$, $\Omega = 1$, (5.164) simplifies to [368]

$$\begin{aligned} P(n) &= \frac{(\langle n_{th} \rangle / N)^n}{(1 + \langle n_{th} \rangle / N)^{n+N}} \exp\left(- \frac{\langle n_c \rangle}{1 + \langle n_{th} \rangle / N} \right) \\ &L_n^{N-1} \left[\frac{- \langle n_c \rangle}{(1 + \langle n_{th} \rangle / N) \langle n_{th} \rangle / N} \right] \quad , \end{aligned} \tag{5.165}$$

which is also the Poisson transform of (5.161). The factorial moments of n are obtained from (5.159) or (5.162) by putting $F_n^{(m)} = \langle W^m \rangle$ and replacing W_c and $\langle W_{th} \rangle$ by $\langle n_c \rangle$ and $\langle n_{th} \rangle$, respectively.

The Thermal Part has a Lorentzian Spectrum

If the thermal part has a Lorentzian spectrum, then the eigenvalues λ_k are as given in Sec. 2.4.4. Assuming also that the amplitude of the coherent part is constant, JAKEMAN and PIKE [243] calculated the product in (5.144) and (5.140). Their result can be written in the form

$$Q_W(s) = Q_{W_{th}}(s)\ \exp(-sW_cQ_2) \quad , \tag{5.166}$$

where $Q_{W_{th}}(s)$ is given by (5.54) and

$$Q_2 = -4Q_0/Q_1^2\ (\nu) \quad . \tag{5.167}$$

Here

$$Q_1^{-1}(\nu) = \cos(\tfrac{\nu}{2}) - \tfrac{\nu}{\gamma}\sin(\tfrac{\nu}{2}) \quad , \tag{5.168}$$

$$Q_0 = a\gamma(1-s\langle W_{th}\rangle)^{-1} - sb^2y^{-1}\langle W_{th}\rangle(\gamma^2+\nu^2)(1-s\langle W_{th}\rangle)^{-1}(\gamma^2+2s\gamma\langle W_{th}\rangle)d[\ln Q_1(y)]/ds$$

$$+ 2a^2b^2\nu^{-1}(\gamma^2+\nu^2)[(\gamma^2+\nu^2)^2(\nu^2-\gamma^2-2\gamma-2s\gamma\langle W_{th}\rangle)-4s\langle W_{th}\rangle\gamma^2(3\Omega^2+\gamma^2)]\ d[\ln Q_1(\nu)]/d\nu$$

$$- ab(\gamma^2+\nu^2)^2 d^2[\ln Q_1(\nu)]/d\nu^2 \quad ,$$

$$a = \tfrac{1}{2}(\nu^2+\gamma^2+2\gamma)^{-1} \quad , \qquad b = (\nu^2+\gamma^2+2s\langle W_{th}\rangle\gamma)^{-1} \quad ,$$

$$y^2 = -\gamma^2 - 2\gamma s\langle W_{th}\rangle \quad , \qquad \nu = \Delta\omega T \quad . \tag{5.169}$$

Notice that in the limit $\gamma \to 0$, (5.151) is reproduced as expected.

Limit $\nu \to 0$ ($\Delta\omega = 0$)

When the central frequencies of the coherent and thermal parts coincide (i.e., $\gamma \to 0$), (5.167-169) give

$$Q_W(s) = Q_{W_{th}}(s)\ \exp(-sW_cQ_2)$$

$$Q_2 = \frac{\gamma(\gamma+4)/(\gamma+2)}{(\gamma+2s\langle W_{th}\rangle)} - \frac{2/(\gamma+2)}{(1-s\langle W_{th}\rangle)} - \frac{4s d[\ln Q_1(y)]/ds}{(1-s\langle W_{th}\rangle)(\gamma+2s\langle W_{th}\rangle)} \tag{5.170}$$

In the limit $\gamma \to 0$, (5.127) is obtained.

The Moments

The moments of W can be obtained from the mgf given by (5.166-169) or (5.170), [243]. It is easier, however, to use the general equations (5.147, 148), and get

$$K_W^{(m)} = K_{W_{th}}^{(m)} + m!\; W_c \langle W_{th}\rangle^{m-1} \bar{B}_m \quad .$$

Substituting $g(t) = e^{-\Gamma|t|}$ in (5.148), we get [226]

$$\bar{B}_1 = 1$$

$$\bar{B}_2 = 2(\gamma^2+\nu^2)^{-2}\left|[(\gamma^2-\nu^2)\cos\nu - 2\gamma\nu\sin\nu]e^{-\gamma} - (\gamma^2-\nu^2) + \gamma(\gamma^2+\nu^2)\right| \qquad (5.171)$$

$$\bar{B}_3 = \frac{1}{\gamma(\gamma^2+\nu^2)}\,(2\gamma + e^{-\gamma}\cos\nu - e^{-2\gamma})$$

$$+ \frac{1}{\gamma(\gamma^2+\nu^2)^2}[2\gamma(\gamma^2-\nu^2)(1+e^{-\gamma}\cos\nu) - (3\gamma^2-\nu^2)(1-e^{-\gamma}\cos\nu) - 4\gamma e^{-\gamma}(1+\gamma)\nu\sin\nu]$$

$$+ \frac{4}{(\gamma^2+\nu^2)^3}[\gamma(\gamma^2-3\nu^2)(e^{-\gamma}\cos\nu - 1) + e^{-\gamma}\nu(\nu^2-3\gamma^2)\sin\nu] \quad .$$

If $\nu = 0$, these expressions simplify to

$$\bar{B}_2 = 2\gamma^{-2}(e^{-\gamma}+\gamma-1)$$

$$\bar{B}_3 = \gamma^{-3}[-e^{-2\gamma}+2(\gamma+4)e^{-\gamma}+4\gamma-7] \quad . \qquad (5.172)$$

Expressions for $K_{W_{th}}^{(m)}$ are given in (5.55), for m = 2 we get

$$\langle\Delta W^2\rangle = \xi(\gamma)\langle W_{th}\rangle^2 + \frac{4W_c\langle W_{th}\rangle}{(\gamma^2+\nu^2)^2}\left|[(\gamma^2-\nu^2)\cos\nu - 2\gamma\nu\sin\nu]e^{-\gamma} - (\gamma^2-\nu^2) + \gamma(\gamma^2+\nu^2)\right| \quad , \qquad (5.173)$$

where

$$\xi(\gamma) = (e^{-2\gamma}+2\gamma-1)/2\gamma^2 \qquad (5.174)$$

and for $\nu = \Delta\omega T = 0$,

$$\langle\Delta W^2\rangle = \xi(\gamma)\langle W_{th}\rangle^2 + 2\,\xi(\tfrac{\gamma}{2})W_c\langle W_{th}\rangle \quad . \tag{5.175}$$

The variance $\langle\Delta W^2\rangle$ has also been calculated [5] for light with a rectangular spectrum and a $(\Gamma^4+\omega^4)^{-1}$ spectrum.

From the moments of W, the photoelectron-counting moments can be simply obtained, for $\Delta\omega = 0$,

$$F_n^{(2)} = \langle n\rangle^2 + \xi(\gamma)\,\langle n_{th}\rangle^2 + 2\,\xi(\gamma/2)\,\langle n_c\rangle\,\langle n_{th}\rangle \quad .$$

JAKEMAN et al. [235] verified this relation experimentally.

Effect of the Detector Area

We consider now the effect of spatial integration over the detector surface on the moments of the integrated light intensity.

A straightforward generalization of the mathematical manipulations used to study the effect of temporal integration (cf. page 209) gives

$$K_W^{(m)} = K_{W_{th}}^{(m)} + m!\ \mathrm{Re}\left\{\iint_{OA}^{T} d\underline{x}_1 \iint_{OA}^{T} d\underline{x}_2 V_c^*(\underline{x}_1)V_c(\underline{x}_2)\Gamma^{(m-1)}(\underline{x}_1,\underline{x}_2)\right\} \quad , \tag{5.176}$$

where

$$\Gamma^{(j)}(\underline{x}_1,\underline{x}_2) = \iint_{OA}^{T} d\underline{x}\,\Gamma^{(1)}(\underline{x}_1,\underline{x})\Gamma^{(j-1)}(\underline{x},\underline{x}_2) \ , \quad \Gamma^{(1)}(\underline{x}_1,\underline{x}_2) = G_{th}(\underline{x}_1,\underline{x}_2) \quad .$$

For example, m = 2 gives

$$\langle\Delta W^2\rangle = \langle\Delta W_{th}^2\rangle + 2\,\eta\,W_c\langle W_{th}\rangle \quad , \tag{5.177}$$

where

$$\eta = \frac{1}{W_c\langle W_{th}\rangle}\ \mathrm{Re}\left\{\iint_{OA}^{T} d\underline{x}_1 \iint_{OA}^{T} d\underline{x}_2 V_c^*(\underline{x}_1)V_c(\underline{x}_2)G_{th}(\underline{x}_1,\underline{x}_2)\right\} \tag{5.178}$$

represents the mixing efficiency.

For a cross-spectrally pure stationary thermal field and a constant coherent field, (5.176) gives

$$K_W^{(m)} = K_{W_{th}}^{(m)} + m!\ \bar{B}_m \bar{D}_m W_c \langle W_{th}\rangle^{m-1} \tag{5.179}$$

where $\bar{B}_m$ are given by (5.148) and

$$\bar{D}_m = \frac{1}{A^m} \mathrm{Re}\left\{\int_A d\underline{r}_1 \ldots \int_A d\underline{r}_m\ g(\underline{r}_1,\underline{r}_2) \ldots g(\underline{r}_{m-1},\underline{r}_m)\right\} .$$

In particular, for m = 2, $\eta = \bar{B}_2\bar{D}_2$ and by using (5.95-98) we can write

$$\langle \Delta W^2\rangle = \frac{1}{N_t}\frac{1}{N_s} \langle W_{th}\rangle^2 + 2\eta_t\eta_s W_c \langle W_{th}\rangle \quad ,$$

where

$$\eta_t = \bar{B}_2 = \frac{2}{T^2}\int_0^T (T-\tau)|g(\tau)|d\tau$$

$$\eta_s = \bar{D}_2 = \frac{1}{A^2} \mathrm{Re}\left\{\int_A d\underline{r}_1 \int_A d\underline{r}_2\ g(\underline{r}_1,\underline{r}_2)\right\}$$

and N_t and N_s are the number of temporal and spatial modes in the thermal light given by (5.97) and (5.98).

We shall denote N_s^{-1} by the symbol f(A) and η_s by the symbol $f_D(A)$ and write

$$\langle \Delta W^2\rangle = \frac{1}{N_t} f(A)\langle W_{th}\rangle^2 + 2\eta_t f_D(A) W_c \langle W_{th}\rangle \ . \tag{5.180}$$

Now we take an example to illustrate the dependence of $\eta_s = f_D(A)$ on the coherence area of the thermal field and the area of the detector. As in Sec. 5.2.2, page 195, we take a circular detector of radius a and a spatial coherence function as in (5.100). Direct substitution gives [229, 388]

$$f_D(A) = \eta_s = \frac{A_c}{A}\left[1 - J_0^2(2\sqrt{A/A_c}) - J_1^2(2\sqrt{A/A_c})\right] . \tag{5.181}$$

This is plotted in Fig.5.16 as a function of A/A_c. The function $f_D(A)$ drops from a value of 1 at $A \ll A_c$ and becomes approximately equal to A_c/A (where $A_c = \pi r_c^2$) when $A \gg A_c$. For the purpose of comparison, we plot $f(A) = N_s^{-1}$ on the same graph. This shows that the contribution of spatial integration to the mixing efficiency η_s drops sharply as the detector area exceeds the coherence area of the fields.

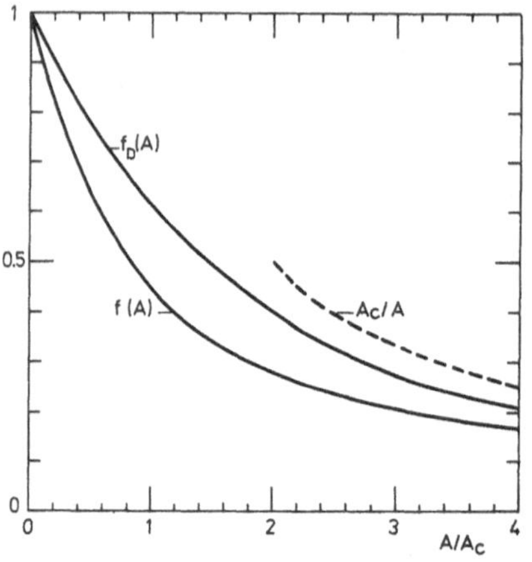

Fig.5.16. Dependence of the spatial integration factors $f(A)$ and $f_D(A)$ on the ratio between the area of the detector and the coherence area of the field

In deriving (5.180), we have assumed that the coherent field is constant and that the thermal field is homogeneous over the detector surface. This ideal situation is never practically realized. We should therefore assume a geometrical model for the mixed fields and substitute in (5.178) for the mixing efficiency. ANDRADE and RYE [6] have studied the effect of phase-front-curvature mismatch, tilt, and axis displacement between the mixed beams. Previous studies based on antenna properties [288, 445] show that for a significant mixing efficiency, the beating beams should interset at the detector within a solid angle $\Omega \lesssim \lambda^2/A$.

The Approximate Formula

As we have seen in Sec. 5.2.2, page 193, the integrated intensity of a purely thermal light can be approximated very well by a sum of N independent exponentially distributed random variables, if N is properly chosen.

Generalizing this to the case of superposition of coherent and thermal light, we can also assume that the integrated intensity is the sum of N independent modes and obtain the statistics of (5.157-165). The number of modes N should be chosen such that the second moment is exactly generated. Eq. (5.159) gives

$$N = (\langle W_{th}\rangle^2 + 2\Omega^2 W_c \langle W_{th}\rangle)/\langle \Delta W^2\rangle \quad , \tag{5.182}$$

which, for $\Delta\omega = 0$ ($\Omega = 1$), becomes

$$N = (\langle W\rangle^2 - W_c^2)/\langle \Delta W^2\rangle \ . \tag{5.183}$$

The variance $\langle \Delta W^2\rangle$ should be calculated exactly from (5.147, 148). For a Lorentzian spectrum, (5.173-175) could be used. For a large detector (5.180) must be used. HORÁK et al. [218] examined this approximation when the thermal part has a Lorentzian spectrum. They compared the third moment calculated from the exact and approximate formulas and found that the accuracy is very good over all ranges of the parameters involved. The accuracy increases as the proportion of the coherent light increases.

PEŘINA et al. [373] compared the approximate expressions of the counting distribution with those calculated by using the closed-form recursion relations of LAXPATI and LACHS [295] for several spectral distributions. The approximation turns out to be best for rectangular and Gaussian spectra and worst for Lorentzian spectra, but excellent in practically all cases.

PEŘINA and MISTA [371] proposed the use of the parameter N to define an effective coherence time, coherence area, and coherence volume for superposed coherent and thermal radiation. For example, if a point detector is used, the effective coherence time is equal to T/N. If a fast detector is used, the effective coherence area is taken equal to A/N. In general, the coherence volume equals cTA/N where c is the velocity of light. A consequence of this is that the coherence volume does not exactly factorize into a product of the coherence length and coherence area even if the radiation is cross-spectrally pure.

5.2.5 Mixture of Coherent Light and Partially Polarized Thermal Light

Referring to Sec. 4.2.4, we see that in this case the detected energy is the sum of two components, $W = W_1 + W_2$, where W_1 is the detected energy due to a sum of coherent light-carrying energy $W_{c1} = W_c\cos^2(\theta)$ and thermal light having an average energy $\langle W_{th1}\rangle = (1/2)(1+P)\langle W_{th}\rangle$, whereas, W_2 corresponds to a mixed coherent and thermal light with average energies $W_{c2} = W_c\sin^2(\theta)$ and $\langle W_{th2}\rangle = (1/2)(1-P)\langle W_{th}\rangle$, respectively.

The problem of determining the statistics of W and its corresponding photoelectron-counting statistics becomes a straightforward use of the relations

$$
\begin{aligned}
Q_W(s) &= Q_{W_1}(s) \cdot Q_{W_2}(s) \\
P(W) &= P_1(W) \circledast P_2(W) \\
K_W^{(m)} &= K_{W_1}^{(m)} + K_{W_2}^{(m)} \\
P(n) &= P_1(n) \circledast P_2(n) \\
F_n^{(m)} &= \sum_{\ell=0}^{m} \binom{m}{\ell} F_{n_1}^{(\ell)} F_{n_2}^{(m-\ell)} \quad .
\end{aligned}
\tag{5.184}
$$

Because we already know the statistics of W_1 and W_2 separately under various conditions (Sec. 5.2.5), we can consider the problem solved. The only computational difficulty lies in determining the convolution in P(W). The discrete convolution in P(n) can be very easily calculated by use of a computer. Therefore, we write here explicit formulas for P(W) under some of the cases discussed in Sec. 5.2.5 [374].

Limit of $T << \tau_c$, $A << A_c$, $T\Delta\omega << 1$

$$
\begin{aligned}
P(W) = {} & \frac{1}{\langle W_{th1}\rangle}\left(\frac{W}{W_{c2}}\right)^{1/2} \exp\left(-\frac{W}{\langle W_{th2}\rangle} - \frac{W_{c1}}{\langle W_{th1}\rangle} - \frac{W_{c2}}{\langle W_{th2}\rangle}\right) \\
& \sum_{k=0}^{\infty} \frac{1}{K!}\left[\left(1-\frac{\langle W_{th2}\rangle}{\langle W_{th1}\rangle}\right)\left(\frac{W}{W_{c2}}\right)^{1/2}\right]^k L_k\left(\frac{W_{c1}\langle W_{th2}\rangle/\langle W_{th1}\rangle}{\langle W_{th2}\rangle-\langle W_{th1}\rangle}\right) \\
& I_{k+1}[2(W_{c2}W)^{1/2}/W_{th1}] \, , \qquad W > 0 \, .
\end{aligned}
\tag{5.185}
$$

N-Modes

$$
P(W) = N\frac{\langle W_{th2}\rangle^{N-1}}{\langle W_{th1}\rangle^{N}}\left[\frac{(W-B)}{\Omega^2 W_{c2}}\right]^{N-\frac{1}{2}} \exp\left[-N\frac{(W-B)}{\langle W_{th2}\rangle} - N\Omega^2\frac{W_{c1}}{\langle W_{th1}\rangle} - N\Omega^2\frac{W_{c2}}{\langle W_{th2}\rangle}\right]
$$

$$\sum_{k=0}^{\infty}\left[\left(1-\frac{\langle W_{th2}\rangle}{\langle W_{th1}\rangle}\right)^2\frac{(W-B)}{\Omega^2 W_{c2}}\right]^{k/2} L_k^{N-1}\left[\frac{N\Omega^2 W_{c1}\langle W_{th2}\rangle}{\langle W_{th1}\rangle(\langle W_{th2}\rangle-\langle W_{th1}\rangle)}\right]$$

$$I_{k+2N-1}\left[2N\Omega\frac{[W_{c2}(W-B)]^{\frac{1}{2}}}{\langle W_{th2}\rangle}\right], \qquad W \geq B \quad , \tag{5.186}$$

$$= 0 \quad , \quad W < B \quad ,$$

where

$$B = (1-\Omega^2)W_c \tag{5.187}$$

and

$$\Omega^2 = \sum_{k=1}^{N} \Omega_k^2 W_{ck}/W_c \quad , \tag{5.188}$$

where Ω_k^2 characterizes the frequency shift of the coherent component from the kth mode (as in (5.150)).

5.2.6 Quasi-Stationary Gaussian Light

The statistical properties of quasi-stationary Gaussian light have been examined in Sec. 4.2.5. Here we study the properties of the photoelectron process due to such a field. We start with a field with real amplitude, and then discuss the general case.

Real Amplitude

If the counting time is much shorter than the coherence time $T \ll \tau_c$, and the detector is small, $A \ll A_c$, then the integrated intensity W has the same distribution as the intensity I. Thus, W has a chi-square distribution with one degree of freedom characterized by

$$Q_W(s) = (1+2\langle W\rangle s)^{-1/2} \tag{5.189}$$

$$P(W) = (2\pi W\langle W\rangle)^{-1/2}\exp(-W/2\langle W\rangle) \tag{5.190}$$

$$\left.\begin{aligned} K_W^{(m)} &= 2^{(m-1)}(m-1)!\langle W\rangle^m \\ \langle W^m\rangle &= (2m-1)!!\ \langle W\rangle^m \\ \langle \Delta W^2\rangle &= 2\langle W\rangle^2 \quad . \end{aligned}\right\} \tag{5.191}$$

The photoelectron counting statistics can be obtained by using usual methods,

$$P(n) = \frac{(2n-1)!!}{n!}\ \frac{\langle n\rangle^n}{(1+2\langle n\rangle)^{n+\frac{1}{2}}} \tag{5.192}$$

$$F_n^{(m)} = (2m-1)!!\ \ \langle n\rangle^m \tag{5.193}$$

$$\langle \Delta n^2\rangle = \langle n\rangle + 2\ \langle n\rangle^2 \quad . \tag{5.194}$$

Examining (5.189-194) and comparing them to (5.33-35, 41, 42) which describe an N-mode detected light we immediately find that if we put $N = 1/2$ in (5.33-35, 41, 42) we obtain (5.189-194). Because we have already found that as N increases the uncertainties of W and of n decrease, we see that the quasi-stationary Gaussian light with real amplitude produces more uncertainty of n than a thermal field. The graphs of P(n) and $F_n^{(m)}$ in Fig.5.17 and Fig.5.18 illustrate these effects.

The probability distributions of the time of arrival and the inter-event times are obtained by differentiating (5.189),

$$P_f(t) = T_s^{-1}(1 + 2t/T_s)^{-3/2} \tag{5.195}$$

$$P_i(\tau) = 3T_s^{-1}(1 + 2\tau/T_s)^{-5/2} \quad , \tag{5.196}$$

where T_s^{-1} is the average number of counts per second. Eqs. (5.195) and (5.196) are valid for a large degeneracy parameter. Fig.5.19 compares these distributions with the corresponding distributions for thermal light.

In the limit $T \gg \tau_c$, we can easily see that the statistics are described by (5.33-35, 41, 42) with $N = (1/2)T/\tau_c$. Moreover, (5.33-35, 41, 42) describe the case of arbitrary T/τ_c and A/A_c approximately, provided that N is chosen such that the second moment is exact.

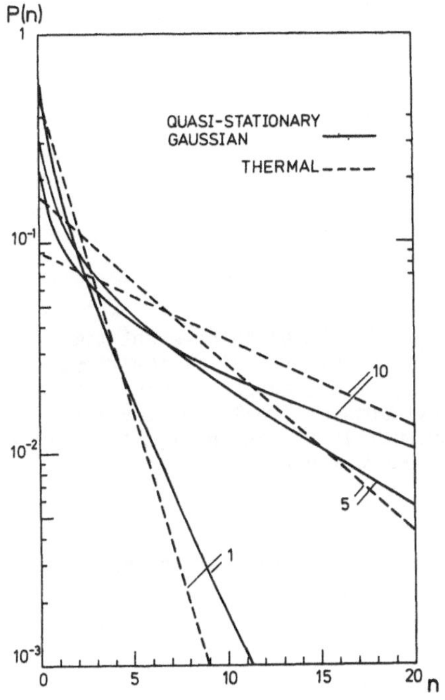

Fig.5.17. Comparison between the photoelectron counting PD of quasi-stationary Gaussian (real amplitude) and thermal light, for the indicated values of $\langle n \rangle$, in the limit of short sampling time

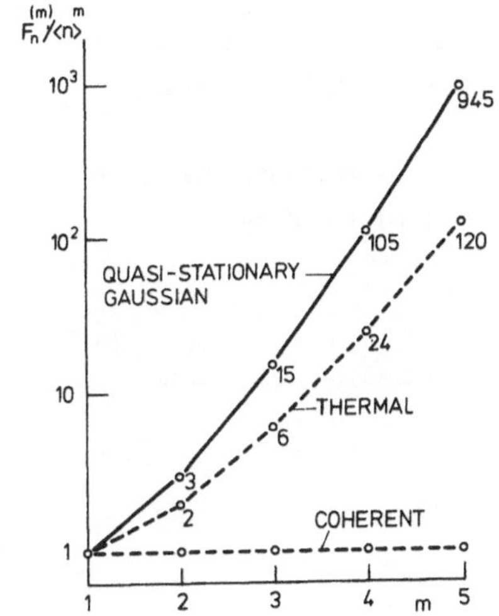

Fig.5.18. Comparison between the photoelectron counting normalized factorial moments of quasi-stationary Gaussian (real amplitude), thermal, and coherent light, in the limit of short sampling time

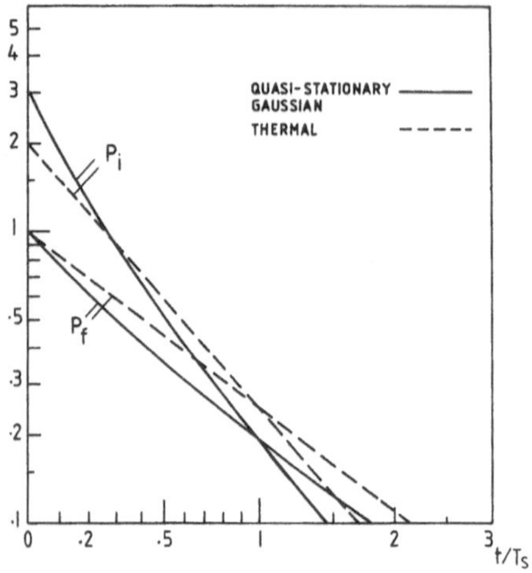

Fig.5.19. PD of inter-event time and time of arrival of the first event in the cases of quasi-stationary Gaussian (real amplitude) and thermal light, in the limit of large degeneracy parameter

For a field with a Lorentzian spectrum, an analysis similar to that in Sec. 5.2.2, page 181, gives [379]

$$Q_W(s) = \left\{e^{\gamma}\left[\cosh z + (\gamma/2z + z/2\gamma)\sinh z\right]^{-1}\right\}^{1/2}$$
$$z^2 = \gamma^2 + 4\gamma\langle W\rangle s \quad , \tag{5.197}$$

which can be obtained from (5.54) by taking the square root and doubling the average value $\langle W\rangle$. It can also be obtained from (5.87) by putting $N_s = 1/2$.

In general, we can relate the cgf of a real quasi-stationary Gaussian (RQSG) field to the cgf of an equivalent thermal (TH) field by

$$Q_W^c(s)\Big|_{RQSG} = \frac{1}{2}\, Q_{2W}^c(s)\Big|_{TH} \quad , \tag{5.198}$$

and the cumulants by

$$K_W^{(m)}\Big|_{RQSG} = \frac{1}{2}\, K_{2W}^{(m)}\Big|_{TH} \tag{5.199}$$

$$= \frac{1}{2}\,(m-1)!B_m\langle 2W\rangle^m = 2^{m-1}(m-1)!B_m\langle W\rangle^m \quad , \tag{5.200}$$

where B_m are given by (5.29) and for a Lorentzian spectrum by (5.55).

General Complex Amplitude

In the limit $T << \tau_c$ and $A << A_c$, if we assume that the real and imaginary components of the complex amplitude have the same variance and a correlation coefficient ρ, we get

$$P(W) = \frac{1}{a\langle W\rangle} I_0\left(\frac{\rho}{a^2}\frac{W}{\langle W\rangle}\right)\exp\left(\frac{-W}{a^2\langle W\rangle}\right) , \quad a^2 = 1 - \rho^2 \tag{5.201}$$

$$Q_W(s) = a\,[\,(1+a^2\langle W\rangle s)^2 - \rho^2]^{-1/2} \tag{5.202}$$

$$\langle\Delta W^2\rangle = (1 + \rho^2)\langle W\rangle^2 \quad , \tag{5.203}$$

which corresponds to [379]

$$P(n) = \frac{(a^2\langle n\rangle)^n}{(1+a^2\langle n\rangle)^{n+1}} \frac{a}{[1-\rho^2/(1+a^2\langle n\rangle)^2]^{n+\frac{1}{2}}}$$

$$\sum_{\ell=0}^{n} \binom{n}{\ell}\binom{n+\ell}{\ell}\left(-\frac{1}{2}\right)^{\ell} \left\{1+[1-\rho^2/(1+a^2\langle n\rangle)^2]^{-\frac{1}{2}}\right\}^{\ell} \tag{5.204}$$

and to the time statistics

$$P_f(t) = T_s^{-1}a^3 \frac{1+a^2t/T_s}{[(1+a^2t/T_s)^2-\rho^2]^{3/2}} \quad ,$$

$$P_i(\tau) = T_s^{-1}a^3 \frac{\frac{3}{3[1-\rho^2(1+a^2\tau/T_s)^{-2}]^{-1}-1}}{[(1+a^2\tau/T_s)^2-\rho^2]^{3/2}} \tag{5.205}$$

If we put $\rho = 0$ in the above equations, we reproduce the equations that correspond to a thermal field. On the other hand, the limit $\rho = 1$ reproduces (5.189-194), which correspond to a quasi-stationary Gaussian field with real amplitude. Other values of ρ, $0 < \rho < 1$, correspond to properties intermediate between these two cases.

5.2.7 Transient Thermal Light

In Sec. 4.2.6, we have presented a possible model of light that relaxes to a steady-state thermal light with a Lorentzian power-spectrum. The optical field is characterized by a time constant $\tau_c = 1/\Gamma$ that describes the decay of its average intensity and also the coherence time of its fluctuations in the equilibrium state. What would be the photoelectron statistics that correspond to this light?

If the sampling time T is short compared to τ_c ($\gamma = \Gamma T << 1$), then the collected light energy is proportional to the light intensity and is therefore exponentially distributed with a decaying average as in (4.124). This corresponds to a Bose-Einstein-distributed number of counts, with decaying mean

$$\langle n(t)\rangle = \langle n(\infty)\rangle + [\langle n(0)\rangle - \langle n(\infty)\rangle]\exp(-2\Gamma t) \quad .$$

The effect of the sampling time can be considered by using a Karhunen-Loève expansion in the interval [t,t+T] as we have done in Sec. 5.2.2 [420, 516, 520].Here, it is necessary to find the eigenvalues of the integral equation

$$\int_t^{t+T} G_V(t,t')\Phi_j(t')dt' = m_j\Phi_j(t) \quad ,$$

$$G_V(t,t') = P_e\exp(-\Gamma|t_1-t_2|)+(P_i-P_e)\exp[-\Gamma(t_1+t_2)]$$

and determine the mgf of $W = \int_t^{t+T} |V(t)|^2dt$ from

$$Q_W(s) = \prod_j(1+sm_j)^{-1} \quad .$$

Manipulations similar to those in Sec. 5.2.2 lead to the expression

$$Q_W(s) = e^{\gamma}\left\{\cosh z + \sinh z\left[\frac{\gamma}{z} + \frac{b}{2}\left(\frac{z}{\gamma} - \frac{\gamma}{z}\right)\right]\right\}^{-1} \tag{5.206}$$

$$z = \gamma^2+2\gamma\langle I(\infty)\rangle T \quad , \quad b = \langle I(t)\rangle/\langle I(\infty)\rangle \quad , \quad \gamma = \Gamma T \quad .$$

Notice that if we put b = 1, we reproduce (5.54). The photoelectron statistics $P(n)$, $F_n^{(m)}$, $P_f(t)$, $P_i(\tau)$ can be obtained from (5.206) by expressions similar to (5.60, 61, 66, 67) (cf. [420, 516, 520]. See also [16]).

ZARDECKI et al. [520] used calculations based on this model to account for the effect of counting time on measurements of photoelectron statistics of laser light, following the moment of its switching on.

5.2.8 Laser Light Described by a Van der Pol Oscillator Model

Referring to Sec. 4.2.7, in which the statistics of the intensity fluctuation of a laser (Van der Pol, or Risken model) was described, we can apply the by-now familiar rules and determine the photoelectron statistics.

Assuming that the counting time T is much shorter than the coherence time and that the detector is small, we see that the detected energy W is proportional to the instantaneous local intensity I. Taking the Poisson transform of P(I) as given by (4.138), we get [37, 40]

$$P(n) = \frac{(\sqrt{\pi}\langle n_0\rangle/2)^n}{n!} \; \frac{R_n(a-\sqrt{\pi}\langle n_0\rangle)}{R_0(a)} \quad , \tag{5.207}$$

where $R_n(x)$ is determined by (4.145) and (4.147).

It satisfies the recurrence relation

$$nP(n) = \frac{\pi}{2}\langle n_0\rangle^2 \left[P(n-2) - \left(1 - \frac{a}{\sqrt{\pi}\langle n_0\rangle}\right)P(n-1)\right] \quad . \tag{5.208}$$

As is mentioned in Sec. 4.2.7, the parameter a is the pumping parameter that determines whether the laser is excited above ($a > 0$), below ($a < 0$) or at ($a = 0$) threshold. The parameter $\langle n_0\rangle$ determined by

$$\langle n_0\rangle = \langle I_0\rangle T \tag{5.209}$$

is the average count at threshold. The average number of counts is given by

$$\langle n\rangle = \langle n_0\rangle \left\{\frac{\sqrt{\pi}}{2}\, a + e^{-a^2/4}/[1 + \Phi(a/2)]\right\} \quad . \tag{5.210}$$

From (5.207-210), we can write P(n) as a function of $\langle n\rangle$ and a.

For large negative values of a (laser well below threshold), we can approximate the distribution by a Bose-Einstein distribution. For large positive values of a, the distribution is approximated by a Poisson distribution that corresponds to an almost constant light intensity. For $a = 0$, an intermediate distribution is obtained.

The factorial moments of this distribution can be determined from (4.144) and (3.68)

$$F_n^{(m)} = \left(\frac{\sqrt{\pi}}{2} \langle n_0 \rangle\right)^m R_m(a)/R_0(a) \quad . \tag{5.211}$$

The values of $F_n^{(m)}/\langle n \rangle^m$ for $a = 0$ are the same as in Table 4.1, page 143.

Statistics of Times

From (4.144) and in the limit of a large degeneracy parameter, the mgf of the integrated intensity is given by

$$Q_W(s) = R_0(a - s\sqrt{\pi} I_0 T)/R_0(a) \quad ,$$

where $R_0(x)$ is defined by (4.145).

By using the differentiation rules (3.93) and (3.94), we can get the distribution of the forward recurrence time and the inter-event time

$$P_f(t) = T_s^{-1} \frac{\sqrt{\pi}}{2} R_1(a - \sqrt{\pi}\tau/T_s)/R_0(a) \quad , \tag{5.212}$$

and

$$P_i(\tau) = T_s^{-1} \frac{\pi}{4} R_2(a - \sqrt{\pi} t/T_s)/R_0(a) \quad . \tag{5.213}$$

These distributions are plotted in Figs.5.20 and 5.21 (for $a = 0$) together with the corresponding distributions for a = a very large positive number (coherent light) and a = a very large negative number (thermal light). This clearly shows the effect of the pumping parameter a on the degree of bunching of the photoelectron events.

The photoelectron statistics of laser light has been the subject of extensive experimental work (e.g., [1, 18, 20, 21, 86, 92, 93, 116, 121, 151, 237, 305, 342, 344, 450]).

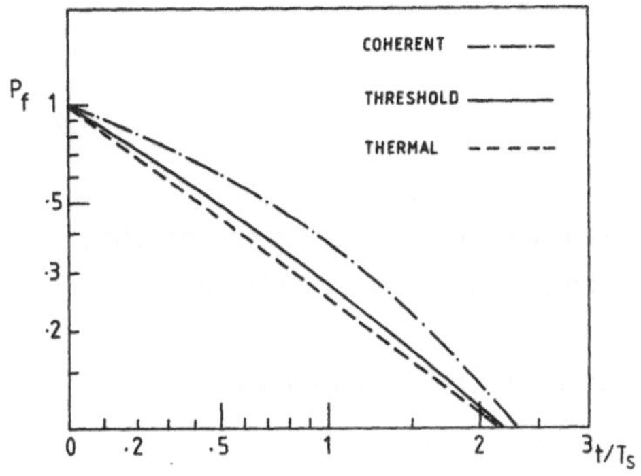

Fig.5.20. Comparison between the PD of time of arrival of the first photoelectron due to coherent light, thermal light, and laser light at threshold

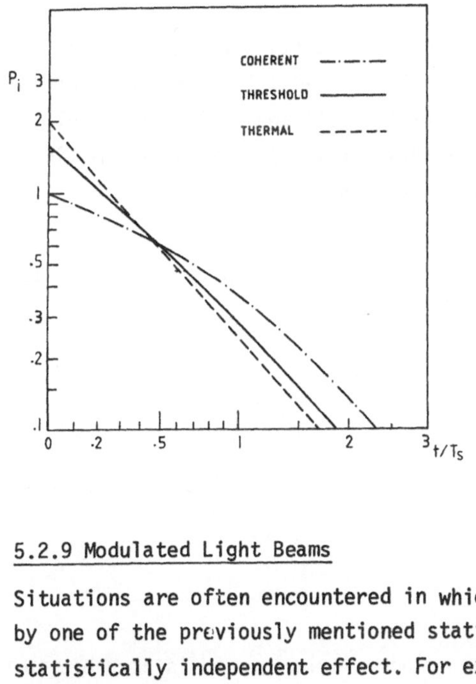

Fig.5.21. Comparison between the PD of inter-event time in the cases of coherent light, thermal light, and laser light at threshold

5.2.9 Modulated Light Beams

Situations are often encountered in which a light beam (described for example by one of the previously mentioned statistical models) is modulated by a statistically independent effect. For example, this happens if the beam propagates through a fluctuating medium or if it is scattered from a fluctuating target. In an optical communication system, modulation is essential for the transfer of information. Modulation has also been introduced as a means of testing the theory of photoelectron counting; modulation can enhance the distinction between sources that have different statistics.

We first derive some general relations for the photoelectron statistics of modulated light, and then turn to several specific examples.

Let the intensity of a light beam be the product of two effects,

$$I(t) = \alpha(t) I_0(t) \quad , \tag{5.214}$$

where $I_0(t)$ and $\alpha(t)$ are statistically independent random processes that represent the intensity of the original beam and the modulation effect, respectively.

The integrated intensity over an interval $[t,t+T]$ is given by

$$W = \int_t^{t+T} \alpha(\theta) I_0(\theta) d\theta \quad . \tag{5.215}$$

In general, it is very difficult to relate its statistics to the statistics of $I_0(t)$ and $\alpha(t)$. Simplifications may be obtained if T is short compared to one or both the time scales of fluctuations in $I_0(t)$ and $\alpha(t)$. For example, if $\alpha(t)$ varies slowly with time in the interval $[t,t+T]$ or is a stochastic process with a correlation time $\tau_{c\alpha}$ such that $T \ll \tau_{c\alpha}$, then

$$W = W_1 W_0 \quad , \tag{5.216}$$

where

$$W_0 = \int_t^{t+T} I(\theta) d\theta \quad ,$$

and

$$W_1 = \alpha(t) \quad .$$

W is then the product of two independent random variables and is amenable to simpler statistical analysis.

If the unmodulated light is coherent, then W_0 is deterministic and the statistics of the modulated light W exactly copy the statistics of the modulation $\alpha(t)$. Coherence of the unmodulated light is however not necessary for this. If the unmodulated light has a coherence time τ_{co} that is much shorter than the detection time T ($T \gg \tau_{co}$), then (as we have already seen in Sec. 5.2.2) W_0 is deterministic and W has the same statistics as α [489, 490]. This, of course, means that $\tau_{co} \ll \tau_{c\alpha}$.

On the other hand, if the unmodulated beam itself has a correlation time (coherence time) τ_{CO} much larger than T, then

$$W = W_1 W_0 \quad ,$$

where, this time,

$$W_0 = I_0(t)\,T \; ,$$

and

$$W_1 = \frac{1}{T}\int_t^{t+T} \alpha(\theta)d\theta \quad .$$

If $T \ll \tau_{CO}$ and at the same time $T \ll \tau_{C\alpha}$, then

$$W = W_1 W_0$$

$$W_0 = I_0(t)T$$

$$W_1 = \alpha(t) \quad .$$

In any of the above three cases, W can be written as a product of two independent random variables. Before specific cases are studied, it is expedient to find how the statistics of W are related to those of W_0 and W_1. A straightforward exercise in transformation of random variables shows that

$$P(W) = \int_0^\infty W_0^{-1} P_0(W_0) P_1(W/W_0) dW_0 \tag{5.217}$$

$$Q_W(s) = \left< Q_{W_0}(sW_1) \right>_{W_1} = \left< Q_{W_1}(sW_0) \right>_{W_0}$$

$$= \int_0^\infty Q_{W_1}(sW_0) P_0(W_0) dW_0 \tag{5.218}$$

$$\langle W^m \rangle = \langle W_0^m \rangle \langle W_1^m \rangle \quad , \tag{5.219}$$

and

$$P(n) = \frac{1}{n!}\int_0^\infty dW_1 W_1^n P_1(W_1) \int_0^\infty e^{-W_0W_1}\, W_0^n P_0(W_0) dW_0 \quad , \tag{5.220}$$

or

$$P(n) = \int_0^\infty P(n|W_1)P(W_1)dW_1 \quad ,$$

where

$$P(n|W_1) = \int_0^\infty \frac{(W_0W_1)^n}{n!}\, e^{-W_0W_1} dW_0 \quad . \tag{5.221}$$

Now we consider some specific examples.

Coherent Light Modulated by Thermal Noise

If a coherent beam is modulated by thermal noise (e.g., when a coherent beam is scattered from a target that has a thermally fluctuating susceptibility), then the resultant modulated beam is itself thermal and its statistics are described as in Sec. 5.2.2.

Coherent Light Modulated by Real Gaussian Noise

If a coherent light beam passes through an intensity modulator (e.g., one that uses the electrooptic effect), and if the modulating voltage is Gaussian noise, then the modulation $\alpha(t)$ is a chi-square process of one degree of freedom and the modulated field is a quasi-stationary Gaussian field with real amplitude. The statistics of such a field have already been studied in Sec. 5.2.6. These statistics have been verified experimentally by BENDJABALLAH and PERROT [46-48].

Thermal Light Modulated by Thermal Noise (Gaussian-Gaussian Scattering)

The problem of a thermal light beam scattered by thermal fluctuations in a target is more difficult. In this case, $I = I_0\alpha$, where both I_0 and α are squares of complex Gaussian stationary stochastic processes. Let us consider the statistics of the integrated intensity W in the case $T \ll \tau_{co}$ and

$T \ll \tau_{c\alpha}$. This means that $W = W_0 W_1$, where both W_0 and W_1 are exponentially distributed. Using (5.218), we get

$$\begin{aligned} Q_W(s) &= \int_0^\infty \frac{1}{1+s\langle W_0\rangle W_1} \frac{1}{\langle W_1\rangle} \exp\left(-\frac{W_1}{\langle W_1\rangle}\right) dW_1 \\ &= \frac{-1}{s\langle W\rangle} \exp\left(\frac{1}{s\langle W\rangle}\right) E_i\left(-\frac{1}{s\langle W\rangle}\right) , \qquad s \neq 0 \end{aligned} \tag{5.222}$$

where $E_i(x) = \int_{-\infty}^{x} e^t/t \, dt$ is the exponential-integral function. The moments of W are given by

$$\begin{aligned} \langle W^m\rangle &= (m!)^2 \langle W\rangle^m \\ \langle \Delta W^2\rangle &= 3 \langle W\rangle^2 \quad , \end{aligned} \tag{5.223}$$

which should be compared to $\langle W^m\rangle = m!\langle W\rangle^m$ and $\langle \Delta W^2\rangle = \langle W\rangle^2$ for an exponential distribution. The effective number of modes is

$$N_{eff} = 1/3 \quad ,$$

showing that the field is actually more chaotic than a thermal field. The photoelectron-counting distribution P(n) can be obtained by performing the integration in (5.220), which gives [54]

$$P(n) = \frac{n!}{\langle n\rangle^{\frac{1}{2}}} \exp\left(\frac{1}{2\langle n\rangle}\right) W_{-n-\frac{1}{2},0}\left(\frac{1}{\langle n\rangle}\right) , \tag{5.224}$$

where $W_{m,n}(.)$ is the Whittaker function. The factorial moments are, of course, given by

$$F_n^{(m)} = (m!)^2 \langle n\rangle^m \quad . \tag{5.225}$$

In Fig.5.22, we compare the factorial moments $F_n^{(m)}$ for coherent, thermal, and modulated thermal light.

The time statistics can be obtained by differentiating the mgf (5.222) according to (3.93,94). This gives

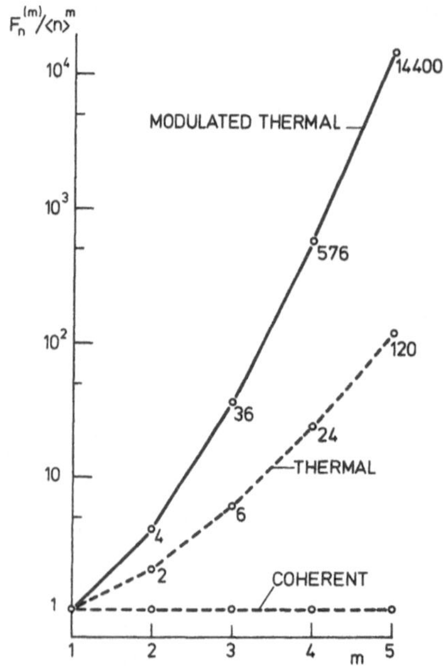

Fig.5.22. Comparison between the photoelectron counting normalized factorial moments of coherent light, thermal light, and thermal light modulated by thermal noise

$$P_f(t) = -T_s^{-1}(T_s/t)^2\left[1+(1+T_s/t)e^{T_s/t}E_i(-T_s/t)\right]\ , \quad t \neq 0\ ,$$
$$= 1\ , \quad t = 0\ , \qquad (5.226)$$

$$P_i(\tau) = -T_s^{-1}(T_s/\tau)^3\left\{(3+\tau/T_s)+\left[2+4T_s/\tau+(T_s/\tau)^2\right]e^{T_s/\tau}E_i(-T_s/\tau)\right\}\ , \quad \tau \neq 0\ , \qquad (5.227)$$
$$= 4 \quad \tau = 0\ .$$

These distributions are plotted in Fig.5.23 and are compared with the distributions due to unmodulated thermal light.

Gaussian-Gaussian scattering can be observed by doubly scattering laser light from a suspension of particles in Brownian motion [497].

Thermal Light Whose Intensity is Modulated by Real Gaussian Noise

In this case, I_0 is exponentially distributed and α has a chi-square distribution with one degree of freedom. If $T \ll \tau_{co}, \tau_{c\alpha}$, then $W = W_0W_1$ where W_0 is exponential and W_1 is chi_1^2. Then the integral of (5.217) gives

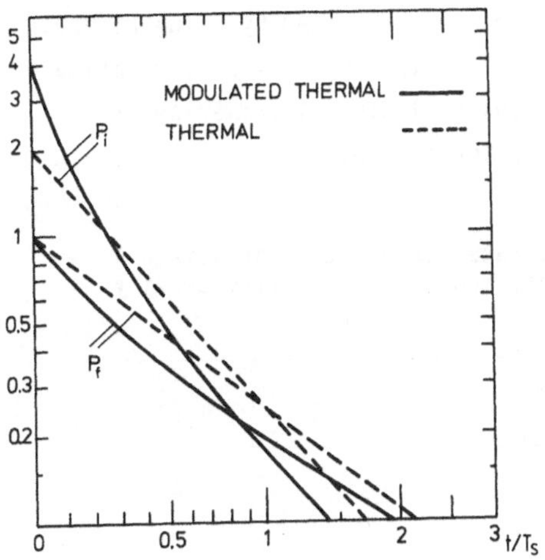

Fig.5.23. Time statistics of photoelectrons generated by thermal light, and by thermal light modulated by thermal noise

$$P(W) = \frac{1}{\sqrt{\pi}\langle W\rangle}\left(\frac{2\langle W\rangle}{W}\right)^{\frac{1}{4}} K_{\frac{1}{2}}\left(\frac{2W}{\langle W\rangle}\right) \quad , \tag{5.228}$$

where $K_\nu(.)$ is the Bessel function of imaginary argument. Eq. (5.218) yields

$$Q_W(s) = \left(\frac{\pi}{2s\langle W\rangle}\right)^{\frac{1}{2}} \exp\left(\frac{1}{2s\langle W\rangle}\right)\left[1-\Phi\left(\frac{1}{\sqrt{2s\langle W\rangle}}\right)\right] \quad , \tag{5.229}$$

where $\Phi(x)$ is the error function.

The photoelectron counting PD is obtained by performing the integral in (5.221), which gives

$$P(n) = \frac{\Gamma(n+\frac{1}{2})}{\sqrt{\pi/2}}\left(\frac{\langle n\rangle}{2}\right)^{3/4} \exp\left(\frac{1}{4\langle n\rangle}\right) W_{-n-\frac{1}{4},\,-\frac{1}{4}}\left(\frac{1}{2\langle n\rangle}\right) \quad . \tag{5.230}$$

The time statistics are represented by the distributions [48]

$$P_f(t) = T_s^{-1} 6(T_s/t)^{5/2} \exp\left(\frac{T_s}{4t}\right) D_{-5}(\sqrt{T_s/t}) \tag{5.231}$$

and

$$P_i(\tau) = T_s^{-1}(T_s/\tau)^{3/2} \exp\left(\frac{T_s}{4\tau}\right) D_{-3}(\sqrt{T_s/\tau}) \quad , \tag{5.232}$$

where $D_n(.)$ is the parabolic cylinder function, obtained by using (5.229) and the differentiation rule (3.93, 94). Note that $P_i(0) = 6$, indicating a very high degree of photoelectron bunching. Table 5.1 summarizes this for several of the fields that have been considered.

Table 5.1. Normalized variance of integrated intensity, normalized photoelectron counting factorial moment, and the probability density of zero inter-event time interval

	$\langle\Delta W^2\rangle/\langle W\rangle^2 = F_n^{(2)}/\langle n\rangle^2 - 1$	$P_i(0)$
Coherent	0	1
Thermal (polarized)	1	2
Quasi-stationary Gaussian	2	3
Thermal modulated by thermal noise	3	4
Thermal modulated by real Gaussian noise	5	6

Intensity Modulation with a Periodic Deterministic Signal Unsynchronized with Sampling

Let the modulation $\alpha(t)$ be a periodic deterministic signal with period T_α. If the sampling is not synchronized with this periodic signal, then an extra randomness is introduced in the counting distribution. This randomness arises because some samples are taken when $\alpha(t)$ is at its peak, whereas others are taken when $\alpha(t)$ is at its trough, in a random fashion.

If we assume that the counting time T is much shorter than the modulation period T_α and that the center of the sampling time t is a uniformly distributed random variable over the period $[0,T_\alpha]$, then the probability of obtaining a value of α between α and $\alpha + d\alpha$ is given by

$$P(\alpha)d\alpha = P(t)dt = (1/T_\alpha)dt \quad ,$$

i.e.,

$$P(\alpha) = (1/T_\alpha)\,\frac{d\alpha}{dt} \quad .$$

Hence, $P(\alpha)$ is completely determined by the modulation waveform. We can now write

$$W_1 = \alpha$$

and

$$W_0 = \int_0^T I_0(t)dt$$

and use the general equations (5.217-221) to determine the resulting photoelectron statistics. The results are summarized in Tables 5.2,3, which include square, triangular, and sinusoidal modulation waveforms for cases of

Table 5.2. Modulation waveforms and their effects on the mgf on the integrated intensity

MODULATION	$P(\alpha)$	$Q_W(S)$
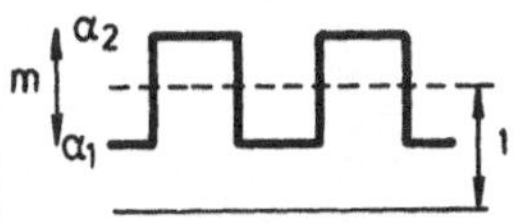 SQUARE	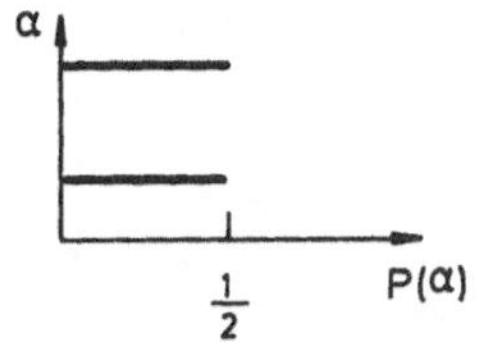	$Q_W(S)=\frac{1}{2}\left[Q_{W_0}(S\alpha_1)+Q_{W_0}(S\alpha_2)\right]$
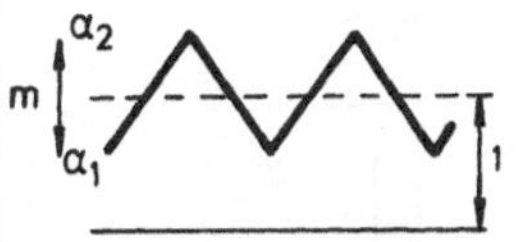 TRIANGULAR	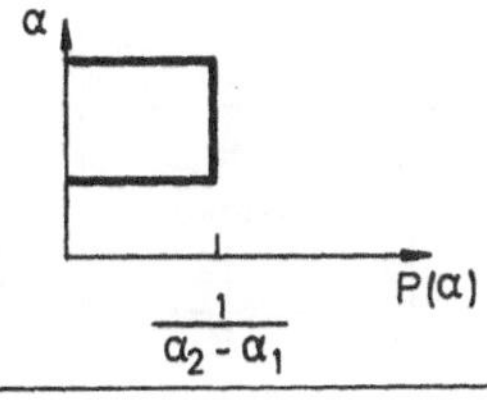	$Q_W(S)=\frac{1}{m}\int_{\alpha_1}^{\alpha_2} Q_{W_0}(S\alpha)\,d\alpha$
 $\alpha(t)=1+0.5m\cos(\Omega t)$ SINUSOIDAL	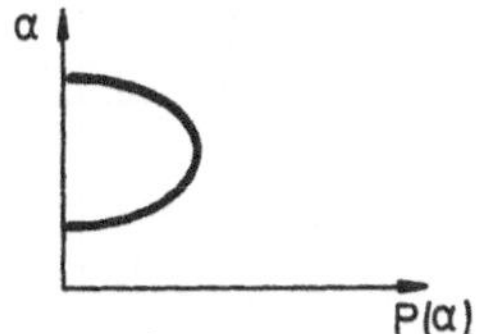 $P(\alpha)=\frac{1}{\pi}\left[(\alpha_2-\alpha)(\alpha-\alpha_1)\right]^{-1/2}$	$Q_W(S)=\frac{1}{\pi}\int_{\alpha_1}^{\alpha_2}\frac{Q_{W_0}(S\alpha)\,d\alpha}{\sqrt{(\alpha_2-\alpha)(\alpha-\alpha_1)}}$

Table 5.3 Basic equations describing the statistics of modulated light

Primary Light Modulation	$Q_{W_0}(s) = e^{-sW_0}$	Thermal $(T \ll \tau_{co})$ $Q_{W_0}(s) = (1+sW_0)^{-1}$
Square	$Q_W(s) = \frac{1}{2}\sum_{i=1}^{2} e^{-s\langle W_i\rangle}$ $\quad \langle W_i\rangle = \alpha_i \langle W_0\rangle$ (5.233)	$Q_W(s) = \frac{1}{2}\sum_{i=1}^{2}(1+s\langle W_i\rangle)^{-1}$ (5.235)
	$P(n) = \frac{1}{2}\sum_{i=1}^{2}\frac{\langle n_i\rangle^n e^{-\langle n_i\rangle}}{n!}$ $\quad \langle n_i\rangle = \alpha_i\langle n\rangle$ (5.234)	$P(n) = \frac{1}{2}\sum_{i=1}^{2}\frac{\langle n_i\rangle^n}{(1+\langle n_i\rangle)^{n+1}}$ (5.236)
Triangular	$Q_W(s) = \frac{1}{sm\langle W\rangle}[e^{-s\langle W_1\rangle} - e^{-s\langle W_2\rangle}]$ (5.237)	$Q_W(s) = \frac{1}{sm\langle W\rangle}\ln\left(\frac{1+s\langle W_2\rangle}{1+s\langle W_1\rangle}\right)$ (5.239)
	$P(n) = \frac{-1}{2m\langle n\rangle}\sum_{i=1}^{2}(-1)^i e^{-\langle n_i\rangle}\sum_{k=0}^{n}\frac{\langle n_i\rangle^k}{k!}$ (5.238)	$P(n)\ \frac{1}{2m\langle n\rangle}\left\{\ln\left(\frac{1+\langle n_2\rangle}{1+\langle n_1\rangle}\right) - \sum_{k=1}^{n}\frac{1}{k}\sum_{i=1}^{2}(-1)^i\left(\frac{\langle n_i\rangle}{1+\langle n_i\rangle}\right)^k\right\}$ (5.240)
Sinusoidal	$Q_W(s) = e^{-s\langle W\rangle} I_0(m\langle W\rangle s)$ (5.241)	$Q_W(s) = [(1+\langle W_1\rangle s)(1+\langle W_2\rangle s)]^{-1/2}$ (5.243)
	$P(n) = \frac{\langle n\rangle^n e^{-\langle n\rangle}}{n!}\sum_{\ell=0}^{n}\binom{n}{\ell}\left(\frac{-m}{2}\right)^{\ell}\sum_{k=0}^{\ell}\binom{\ell}{k} I_{\lvert\ell-2k\rvert}(m\langle n\rangle)$ (5.242)	$P(n) = \frac{\langle n\rangle^n \mu^n}{R^{n+1}} P_n\left(\frac{\mu^2\langle n\rangle+1}{\mu R}\right)$ $\quad m < 1$ (5.244) $= \frac{(2n)!}{2^n(n!)^2}\frac{\langle n\rangle^n}{(1+2\langle n\rangle)^{n+\frac{1}{2}}}$ $\quad m = 1$ $\mu^2 = 1-m^2 \qquad R^2 = 1+2\langle n\rangle+\langle n\rangle^2\mu^2$

both coherent and thermal unmodulated beams, under the conditions $T \ll \tau_{co}$ [132]. The same results could be obtained by assuming $\alpha(t)$ to have a deterministic waveform with uniformly distributed random phase [45].

The resulting photoelectron-counting distributions are plotted in Fig.5.24. These distributions can be understood if they are regarded as averages over a Poisson distribution (coherent case) or a Bose-Einstein distribution (thermal case) with random means. In the case of square-wave modulation, the averaging is done over two means with equal weights. In the triangular-wave-modulation case, we average over a uniform distribution of the mean (see Table 5.2). Thus, for the coherent case (and if the average number of counts $\langle n \rangle$ is large), the unmodulated beam produces a counting distribution that is a narrow pulse around the mean. With square-wave modulation, we average over two means and could end by getting a double-peaked distribution (Fig.5.24a).

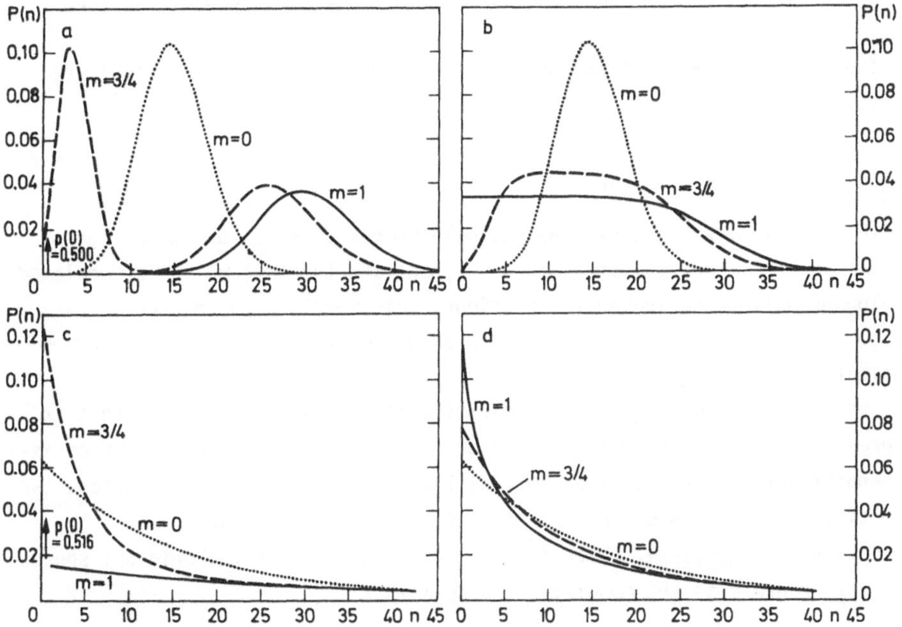

Fig.5.24a-d. Photoelectron counting PD for: a) square-wave modulated coherent light, b) triangularly modulated coherent light, c) square-wave modulated coherent light, d) triangularly modulated thermal light; for $\langle n \rangle = 15$, and for the indicated modulation depths (after DIAMENT and TEICH [132])

With triangular-wave modulation, we average over a uniform distribution of means and we obtain the interesting flat counting distribution of Fig.5.24b [132, 473, 474].

The effect of modulation in the case of a thermal source is not as drastic as in the case of a coherent source (Fig.5.24c,d) because the original counting distribution is broad and the averaging of broad distributions with different means would not produce a distribution with a different shape.

These distributions have been experimentally studied by FRAY et al. [149], PEARL and TROUP [365], BENDJABALLAH and PERROT [45] and others. The conditions that the modulation period and the sampling time are short were dropped in the analysis by CLARK and O'NEILL [97] , who considered the general case of periodic modulation with arbitrary shape. Their results were verified experimentally by KITAZIMA [267].

If we compare (5.241) to (4.188), we observe that sinusoidally modulated coherent light results in the same photoelectron counting statistics as a mixture of coherent light and phase fluctuating light. Therefore, (5.242) also describes the counting statistics for heterodyned phase-fluctuating fields [136].

The time statistics of photoelectrons generated by modulated light are discussed by TROUP [486].

Log-Normally Modulated Light

In this section, we consider the photoelectron statistics due to a light beam (coherent, thermal, or a mixed coherent and thermal) which is modulated by log-normally distributed fluctuations. This model has been used to describe the modulation effect of light propagation through a turbulent atmosphere (see, e.g., [293]).

The importance of a log-normal distribution stems from the fact that it represents the limiting distribution of the product of a large number of independent effects in the same sense as the normal distribution describes the limiting distribution of the sum of many random contributions.

Let $I(x) = I_0(\underline{x})\alpha(\underline{x})$, where $I_0(\underline{x})$ is the intensity of the original source and $\alpha(\underline{x})$ represents the modulation. Assume that

$$\alpha(\underline{x}) = \exp [\psi(\underline{x})] \tag{5.245}$$

where $\psi(\underline{x})$ is a stationary Gaussian stochastic process with mean μ and variance σ^2. At a space-time point $\underline{x}$, α has a log-normal distribution (cf. Table 2.1, see also [4]); then

$$P(\alpha) = \frac{1}{\sqrt{2\pi}\sigma\alpha} \exp[-(\ln\alpha-\mu)^2/2\sigma^2] \ .$$

Assuming also that $\langle\alpha\rangle = 1$ (conservation of energy), we have

$$\langle\alpha\rangle = \exp(\mu + \tfrac{1}{2}\sigma^2) = 1$$

from which

$$\mu = -\sigma^2/2 \ .$$

Thus, the probability distribution

$$P(\alpha) = \frac{1}{\sqrt{2\pi}\sigma\alpha} \exp\left[-(\ln\alpha + \tfrac{1}{2}\sigma^2)^2/2\sigma^2\right] \tag{5.246}$$

becomes a function of only one parameter, σ, which describes the degree of fluctuations of the modulation effect (the degree of turbulence of the atmosphere for example).

Assuming that the counting time T is much shorter than both the coherence time of the source τ_{co} and the correlation time of the modulation $\tau_{c\alpha}$, and that the area of the detector is much smaller than the coherence area of the process $\psi(\underline{x})$, we can write $W = W_0\alpha$. From Table 2.1

$$\langle\Delta W^2\rangle/\langle W\rangle^2 = (\langle\Delta W_0^2\rangle/\langle W_0\rangle^2) \cdot \left[\exp(\sigma^2) - 1\right] \ ,$$

i.e., log-normal modulation increases the uncertainty by a factor $[\exp(\sigma^2)-1]$.

Using (5.221), we write

$$P(n) = \int_0^\infty P_0(n|\alpha)P(\alpha)d\alpha \ , \tag{5.247}$$

where $P_0(n|\alpha)$ is the counting distribution due to the source with mean $\alpha\langle n\rangle$. Now the problems is to perform the integration in (5.247) for the cases described by (5.250), (5.251), or (5.252).

No closed-form solution of the integral in (5.247) is available. BLUEMEL et al. [61] used numerical methods to determine P(n). CLARK and KARP [96] approximated the log-normal distribution by a non-central chi-square distribution [354] and obtained an analytic expression for P(n). DIAMENT and TEICH [133] provided an excellent approximate solution, by use of the method of steepest descent, as described in the following.

By putting $k = \alpha\langle n\rangle$, we can write (5.247) in the form

$$P(n) = \int_0^\infty P_0(n|k)P(k)dk \quad , \tag{5.248}$$

where

$$P(k) = \frac{1}{\sqrt{2\pi}\sigma k} \exp\left\{-[\ln (k/\langle n\rangle) + \tfrac{1}{2}\sigma^2]^2/2\sigma^2\right\} \quad , \tag{5.249}$$

and $p_0(n|k)$ represents the counting distribution due to an unmodulated beam with average counts k. For a *coherent* beam

$$P_0(n|k) = \frac{k^n e^{-k}}{n!} \quad , \tag{5.250}$$

and for a *thermal* beam,

$$P_0(n|k) = \frac{k^n}{(1+k)^{n+1}} \tag{5.251}$$

For a *mixed coherent and thermal light*,

$$P_0(n|k) = \frac{(ak)^n}{(1+ak)^{n+1}} \exp\left(\frac{-bk}{1+ak}\right)L_n\left[-\frac{b}{a(1+k)}\right] \quad , \tag{5.252}$$

$$a = \langle n_{th}\rangle/\langle n\rangle \quad , \quad b = \langle n_c\rangle/\langle n\rangle \quad .$$

Changing the variable of integration to $x = [\ln (k/M)]/2\sigma$ and using (5.249), we can write

$$P(n) = \frac{1}{\sqrt{2\pi}} \int_{-\infty}^{\infty} dx \exp\left[\ln P_0(n|k) - \frac{1}{2}\left(x + \ln M - \ln\langle n\rangle + \frac{1}{2}\right)^2\right] \quad .$$

The parameter M can be determined from the stationarity condition,

$$\ln M = \ln \langle n\rangle - \frac{1}{2}\sigma^2 + \sigma^2 q_1(n,M) \quad , \tag{5.253}$$

where

$$q_m(n,k) = \frac{\partial^m}{\partial(\ln k)^m} \ln P_0(n|k) \quad . \tag{5.254}$$

A series expansion in the variable x gives

$$P(n) = \frac{P_0(n|M)\exp[-\tfrac{1}{2}\sigma^2 q_1^2(n,M)]}{[1-\sigma^2 q_2(n,M)]^{\frac{1}{2}}} \tag{5.255}$$

The approximation applies to any reasonable single-peaked distribution $P_0(n|k)$.

Coherent Case

Using (5.250), we get

$$q_1 = n - M \quad , \quad q_2 = - M \tag{5.256}$$

from which M is determined by the nonlinear equation,

$$M = \langle n\rangle \exp[\sigma^2(-\tfrac{1}{2} + n - M)] \quad , \tag{5.257}$$

and

$$P(n) = \frac{M^n e^{-M}}{n!} \frac{\exp[-\tfrac{1}{2}\sigma^2(M-n)^2]}{(1+\sigma^2 M)^{\frac{1}{2}}} \tag{5.258}$$

Thermal Case

$$q_1 = (n-M)/(M+1) \, , \quad q_2 = - M(n+1)/(M+1)^2 \tag{5.259}$$

$$M = \langle n\rangle \exp\left[\sigma^2\left(\frac{n-M}{M+1} - \frac{1}{2}\right)\right] \tag{5.260}$$

and

$$P(n) = \frac{M^n}{(1+M)^{n+1}} \frac{\exp\{-\frac{1}{2}\sigma^2[(M-n)/(M+1)]^2\}}{[1+\sigma^2 M(n+1)(M+1)^{-2}]^{\frac{1}{2}}} \tag{5.261}$$

Mixed Coherent and Thermal [134]

$$\begin{aligned}
q_1 &= \frac{n-aM}{1+aM} - \frac{(1-a)M}{(1+aM)^2} A \\
q_2 &= -\frac{(1+n)aM}{(1+aM)^2} - \frac{(1-aM)(1-a)M}{(1+aM)^3} A - \frac{(1-a)^2M^2}{(1+aM)^4} B \\
A &= 1-L_n'(x)/L_n(x) \\
B &= [L_n'(x)/L_n(x)]^2 - L_n''(x)/L_n(x) \\
x &= \frac{a-1}{1+aM} ,
\end{aligned} \tag{5.262}$$

where a is the ratio of the thermal part to the total intensity. M and $P(n)$ can be obtained by direct substitution in (5.253) and (5.255).

Samples of the photoelectron-counting distributions obtained in the above cases are illustrated in Fig.5.25. For $\sigma = 0$, i.e., the modulation has no fluctuations (quiescent atmosphere), the Poisson and Bose-Einstein distributions for coherent and thermal light are obtained (Fig.5.25a). For mixed unmodulated light, an intermediate distribution is obtained. As the fluctuations of the modulation increase (σ increases), the counting distribution is broadened. The Poisson distribution characterized by a narrow shape around the mean is affected most by modulation. It is considerably broadened and its peak shifts to a decreasing count. At a high value of $\sigma (\sigma > 1.5)$, the most likely count becomes $n = 0$. The Bose-Einstein distribution, which is a straight line in a semilogarithmic plot, is curved so that the probability of small counts is higher and that of large counts is lower. This corresponds to clustering of photoelectron events. The mixed-light distribution suffers an effect intermediate between the effects of coherent and thermal light. Fig.5.25c shows that for very large fluctuations of the modulation ($\sigma \gtrsim 1.5$) the count distribution becomes relatively insensitive to the source statistics, which is overshadowed by the strong fluctuations of the modulation.

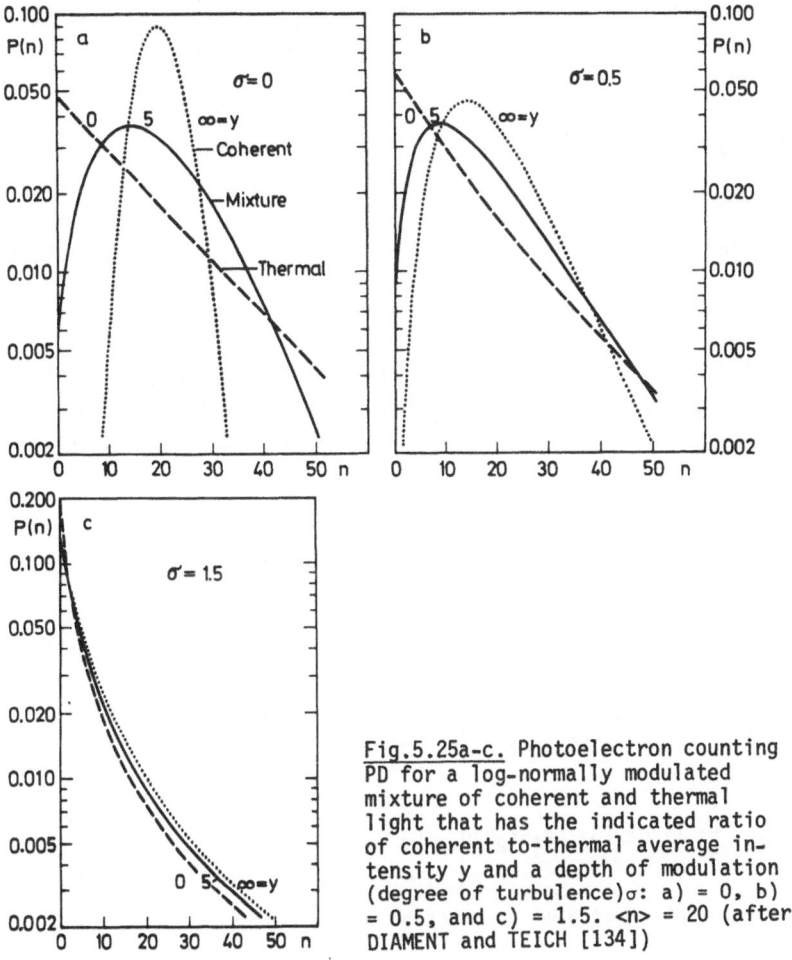

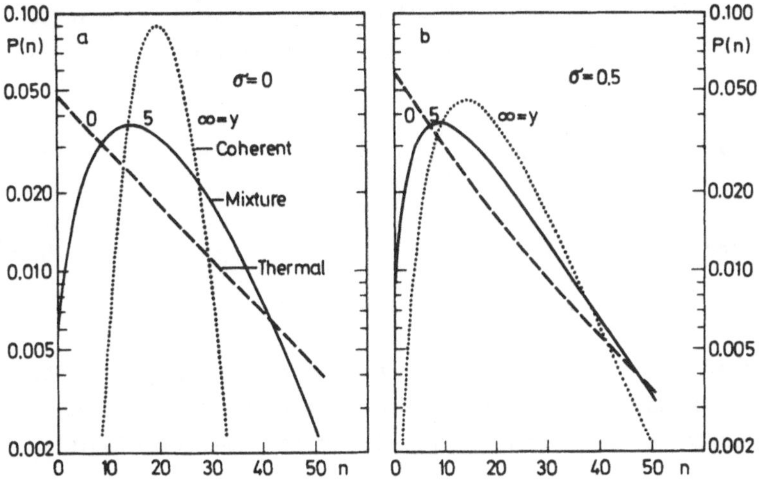

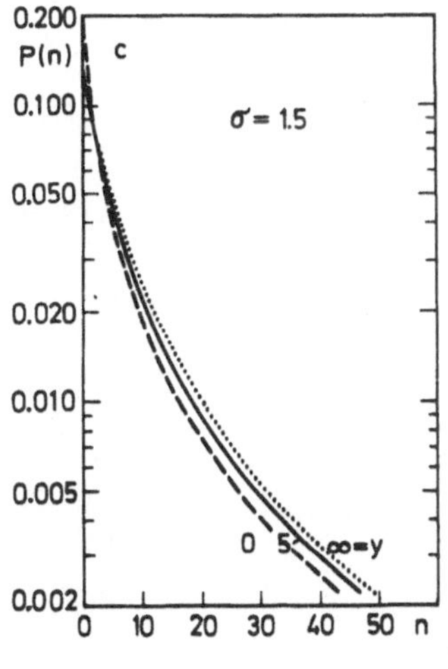

Fig.5.25a-c. Photoelectron counting PD for a log-normally modulated mixture of coherent and thermal light that has the indicated ratio of coherent to-thermal average intensity y and a depth of modulation (degree of turbulence)σ: a) = 0, b) = 0.5, and c) = 1.5. $\langle n \rangle = 20$ (after DIAMENT and TEICH [134])

In spite of the analytic difficulties, the photoelectron statistics of log-normally modulated light have been studied further. ROSENBERG and TEICH [405] derived exact expressions for the first-, second-, and third-order photoelectron-counting cumulants. YEN et al. [517] considered the effect of a non-interfering background. LACHS and LAXPATI [282] included a chaotic interfering background and used numerical techniques to calculate the distributions.

Effect of the Detector Area

When the area of the detector is not much smaller than the coherence area of the field and that of the modulation effect, the previous analysis has to be significantly modified. If we assume that the unmodulated beam has a large coherence area, then the integrated intensity can be written as

$$W = I_0 \int_A \alpha(\underline{r})d\underline{r} = W_0\bar{\alpha} \quad , \tag{5.263}$$

$$\bar{\alpha} = \frac{1}{A}\int_A \alpha(\underline{r})d\underline{r} \quad , \quad W_0 = I_0 A \quad . \tag{5.264}$$

The modulation effect is therefore smoothed by integration over the detector area. The smoothed modulation factor $\bar{\alpha}$ is not log-normally distributed. Its moments can be obtained by using the properties of the underlying Gaussian process $\psi(\underline{r})$. If the primary beam is coherent, then

$$\langle W \rangle = \langle W_0 \rangle \langle \bar{\alpha} \rangle = \langle W_0 \rangle \tag{5.265}$$

$$\langle W^2 \rangle = \langle W_0 \rangle^2 \frac{1}{A^2} \iint_{AA} \langle \alpha(\underline{r}_1)\alpha(\underline{r}_2)\rangle d\underline{r}_1 d\underline{r}_2 \quad . \tag{5.266}$$

Substituting from (5.245), we obtain (cf. also Sec. 4.2.9)

$$\langle W^2 \rangle = \langle W_0 \rangle^2 \frac{1}{A^2} \iint_{AA} \exp[G_{\Delta\psi}(\underline{r}_1,\underline{r}_2)]d\underline{r}_1 d\underline{r}_2 \ , \tag{5.267}$$

$$\Delta\psi = \psi(\underline{r}) - \langle \psi(\underline{r}) \rangle \quad ,$$

from which

$$\langle \Delta W^2 \rangle = \langle W \rangle^2 \left(e^{\sigma^2} - 1\right)/N_\sigma \quad , \tag{5.268}$$

where

$$N_\sigma = \left(e^{\sigma^2} - 1\right)/\frac{1}{A^2}\iint_{AA} \exp[G_{\Delta\psi}(\underline{r}_1,\underline{r}_2) - 1]d\underline{r}_1 d\underline{r}_2 \quad . \tag{5.269}$$

The factor $\left(e^{\sigma^2} - 1\right)$ represents the broadening effect of log-normal modulation and the factor N_σ represents the narrowing effect of averaging over the detector area. In the limit of small A, $N_\sigma = 1$. In the limit of large A, $N_\sigma = A_\sigma/A$, where A_σ represents the coherence area of the modulation effect [152].

As A increases, N_σ increases and the variance of W decreases. Thus, the effect of atmospheric turbulence can be reduced by increasing the receiver aperture. Experimental evidence, however, indicates that saturation is reached when N_σ reaches the value 10 [265].

A good approximation for the distribution of $\bar{\alpha}$ is a log-normal distribution with an effective variance $\bar{\sigma}$ determined from

$$\left(e^{\bar{\sigma}^2} - 1\right) = \left(e^{\sigma^2} - 1\right) / N_\sigma \quad . \tag{5.270}$$

This can be justified by the permanence property of the log-normal distribution [349].

5.3 Multifold Photoelectron Statistics

The previous section, 5.2, dealt with the statistics of the integrated light intensity, the number of photoelectron counts in a single time interval, the interval between photoelectron events and the time of occurrence of the mth event. The analysis covered several models of optical fields that have different statistical properties. Each model corresponds to a characteristic photoelectron-counting distribution, which distinguishes it. It was also shown that such distributions are affected by the shape of the power spectrum of the field, if the counting time is of the same order of magnitude as the coherence time of the field. Moreover, the profile of the spatial-coherence function of the field affects the photoelectron-counting distribution if the detector has an area comparable to the coherence area of the field.

The spectral- and spatial-coherence properties of the field manifest themselves more clearly in the joint statistics of photoelectron counts of two short intervals separated by a time delay comparable to a coherence time or of two point detectors separated by a distance comparable to a coherence length. This makes the study of joint or multifold photoelectron statistics of obvious importance.

Also, in Chap. 4, it was emphasized that coherence functions of higher order must be considered if a complete characterization of non-Gaussian light is desired. These functions can be measured by a complicated set of interference experiments (as in [62]) but the most direct approach is multifold photoelectron counting.

In the present section, we reconsider some of the models of optical fields introduced in Sec. 4.2 and studied in Sec. 5.2, in an attempt to determine the joint photoelectron statistics in each case. Thus, Sec. 5.3 is a formal generalization of Sec. 5.2. The quantities of interest in this section are $[W] = (W_1, W_2, \dots W_N)$, where W_j is the light energy collected by detector j in an interval $[t_j, t_j+T]$ and $[n] = (n_1, n_2, \dots n_N)$, where n_j is the number of photoelectrons counted in the interval j by detector j. The statistics of the set of times $\{t_i\}$ at which photoelectrons occur in one detector will also be studied. Because the photoelectron events at each detector are governed by a DSP.PP, we shall draw freely from the general rules of Sec. 3.5.

5.3.1 Polarized Thermal Light

The Limit $T \ll \tau_c$ and $A \ll A_c$

In this limit, the detected energies [W] are proportional to the local intensities [I], thus [W] have the same statistics as [I]. These have been derived in Sec. 4.2.1; they are summarized here

$$Q_{[W]}([s]) = |\Delta|^{-1} \quad , \quad \Delta_{ij} = \delta_{ij} + s_i G(\underline{x}_i, \underline{x}_j) \quad , \tag{5.271}$$

e.g.,

$$Q_{W_1, W_2}(s_1, s_2) = \left[(1+s_1\langle W_1\rangle)(1+s_2\langle W_W\rangle) - s_1 s_2 \langle W_1\rangle\langle W_2\rangle |g(\underline{x}_1, \underline{x}_2)|^2\right]^{-1} \tag{5.272}$$

$$\langle W_1 W_2 \dots W_N\rangle = \sum_\pi \prod_{j=1}^N \langle W_j\rangle\langle W_{\pi j}\rangle g(\underline{x}_j, \underline{x}_{\pi j}) \quad , \tag{5.273}$$

e.g.,

$$\langle W_1 W_2\rangle = \langle W_1\rangle\langle W_2\rangle[1+|g(\underline{x}_1, \underline{x}_2)|^2] \quad , \tag{5.274}$$

and

$$K^{[I]}_{[W]} = \sum_c \prod_{j=1}^N \langle W_j\rangle\langle W_{cj}\rangle g(\underline{x}_j, \underline{x}_{cj}) \quad . \tag{5.275}$$

These expressions cover two special cases I) single-point detector and several counting times, in which case $\underline{x}_j = \underline{r}, t_j$; II) several detectors opened at the same time, in which case $\underline{x}_j = \underline{r}_j, t$.

Counting Statistics

The JPD P([n]) and the factorial moments $F_{[n]}^{[m]}$ can be determined from (5.271) by use of (3.70) and (3.81), respectively. It is easier, though, to obtain a recurrence relation [38] by writing

$$Q_{[W]}([s]) \; |\Delta| = 1$$

and by using the Leibnitz differentiation rule to get

$$\sum_{[\ell]=[0]}^{[1]} [-1]^{[\ell]} \frac{\partial^{[\ell]}}{\partial[s]^{[\ell]}} |\Delta| _{[s]=[1]} P([n-\ell]) = 0 \quad , \tag{5.276}$$

and

$$\sum_{[\ell]=[0]}^{[1]} [-1]^{[\ell]} \frac{1}{[m-\ell]!} \frac{\partial^{[\ell]}}{\partial[s]^{[\ell]}} |\Delta| _{[s]=[0]} F_{[n]}^{[m-\ell]} = 0 \quad ; \tag{5.277}$$

e.g., $N = 1$ gives $P(n) = [\langle n\rangle/(\langle n\rangle+1)]P(n-1)$, the recurrence relation of a Bose-Einstein distribution (5.8), and $N = 2$ gives

$$\sum_{\ell_1=0}^{1} \sum_{\ell_2=0}^{1} A_{\ell_1,\ell_2} P(n_1-\ell_1, n_2-\ell_2) = 0 \quad , \tag{5.278}$$

where

$$A_{\ell,m} = (-1)^{\ell+m} \frac{\partial^\ell}{\partial s_1^\ell} \frac{\partial^m}{\partial s_2^m} |\Delta| _{s_1 = s_2 = 1} \quad .$$

Substitution from (5.271) gives

$$A_{00} = 1 + \langle n_1\rangle + \langle n_2\rangle + \langle n_1\rangle\langle n_2\rangle\beta$$

$$A_{10} = \langle n_1\rangle(1 + \langle n_2\rangle \beta)$$

$$A_{01} = \langle n_2\rangle(1 + \langle n_1\rangle \beta)$$

$$A_{11} = \langle n_1\rangle\langle n_2\rangle\ \beta$$

$$\beta = 1 - |g(\underline{x}_1,\underline{x}_2)|^2 \quad . \tag{5.279}$$

Equations (5.278) and (5.279), together with the initial conditions

$$P(n_1,0) = A_{10}^{n_1} / A_{00}^{n_1+1}$$
$$P(0,n_2) = A_{01}^{n_2} / A_{00}^{n_2+1} \tag{5.280}$$

completely determine $P(n_1,n_2)$. Algebraic manipulation of these recurrence relations leads to the explicit formula [13]

$$P(n_1,n_2) = \frac{(n_1+n_2)!}{n_1!n_2!}\frac{A_{10}^{n_1+n_2}}{A_{00}^{n_1+n_2+1}}\ {}_2F_1\left(-n_1,-n_2;-n_1-n_2;\frac{A_{00}A_{11}}{A_{10}^2}\right) , \tag{5.281}$$

where we assume that $T_1 = T_2$ and $\langle n_1\rangle = \langle n_2\rangle$, i.e., $A_{10} = A_{01}$ and ${}_2F_1$ is the Gaussian hypergeometric function.

Notice that for $t_1-t_2 \to \infty$ or $\underline{r}_1-\underline{r}_2 \to \infty$, $|g| \to 0$ and we get $P(n_1,n_2) = P(n_1)P(n_2)$ as we expect. On the other hand, if $\underline{x}_1 = \underline{x}_2$ we get $|g| = 1$ and

$$P(n_1,n_2) = \binom{n}{n_1}\frac{\langle n\rangle^n}{(1+2\langle n\rangle)^{n+1}} , \quad n = n_1+n_2 \quad , \tag{5.282}$$

which is the fully correlated case, corresponding to

$$Q_{W_1,W_2}(s_1,s_2) = [1+(s_1+s_2)\langle W\rangle]^{-1} = Q_W(s_1+s_2),$$
$$W = W_1 + W_2 \quad . \tag{5.283}$$

Of interest also is the conditional probability $P(n_1|n_2) = P(n_1,n_2)/P(n_2)$. By using (5.281) and (5.7), we obtain the distribution sketched in Fig.5.26, [13].

The *factorial moments of* $[n]$ can be determined from (5.277) or directly from the equality between the factorial moments of [n] and the ordinary

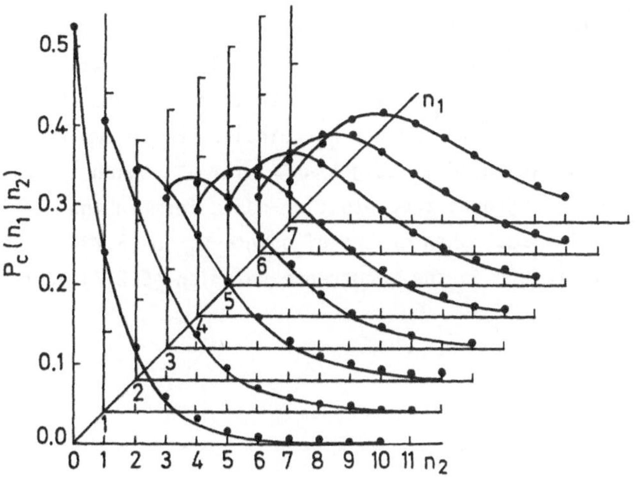

Fig.5.26. Conditional photoelectron counting PD for thermal light. The sampling times are equal, small and separated by a time delay such that $|g|^2 = 0.65$. $\langle n\rangle = 2$. The dots represent experimental data (after ARECCHI et al. [13])

moments of [W], [80]. Using the moment-expansion property of a complex Gaussian field (4.72), we can write

$$\langle n_1 n_2 \ldots n_N\rangle = \sum_{\pi} \prod_{j=1}^{N} (\langle n_j\rangle\langle n_{\pi j}\rangle)^{1/2} g_{j\pi_j} \quad , \tag{5.284}$$

where $g_{ij} = g(\underline{x}_i, \underline{x}_j)$, e.g., for N = 2

$$\langle n_1 n_2\rangle = \langle n_1\rangle\langle n_2\rangle(1+|g_{12}|^2) \quad , \tag{5.285}$$

whereas the cases N = 3 and 4 can be written in the form

$$\langle \Delta n_1 \Delta n_2 \Delta n_3\rangle = 2\langle n_1\rangle\langle n_2\rangle\langle n_3\rangle \operatorname{Re}\{g_{12}g_{23}g_{32}\} \quad , \tag{5.286}$$

$$\begin{aligned}\langle \Delta n_1 \Delta n_2 \Delta n_3 \Delta n_4\rangle = \langle n_1\rangle\langle n_2\rangle\langle n_3\rangle\langle n_4\rangle \Big[&|g_{12}|^2 |g_{34}|^2 \\ &+ |g_{13}|^2 |g_{24}|^2 + |g_{14}|^2 |g_{23}|^2 + 2\operatorname{Re}\{g_{12}g_{23}g_{34}g_{41} \\ &+ g_{12}g_{24}g_{43}g_{31} + g_{13}g_{32}g_{24}g_{41}\}\Big] \quad .\end{aligned} \tag{5.287}$$

Equation (5.285) is of extreme importance because it indicates how $|g(\underline{x}_1,\underline{x}_2)|$ can be determined simply by measuring $\langle n_1 n_2\rangle$ and $\langle n_i\rangle$, $i = 1,2$. This will be discussed in detail in Chap. 7. The reader is warned that (5.284) and (5.285) should not be used if two or more of the points $\underline{x}_j$ coincide. For example, if we put $\underline{x}_1 = \underline{x}_2$ (i.e., $|g| = 1$) in (5.285), we get $\langle n^2\rangle = 2\langle n\rangle^2$, which is not correct (see (5.10)) because it ignores the detection-noise term in the correct relation $\langle n^2\rangle = 2\langle n\rangle^2 + \langle n\rangle$. When groups of points $\underline{x}_j$ in (5.284) coincide, it is necessary to return to the recurrence relation (5.277) or to the basic equation

$$F_{[n]}^{[m]} = \langle [W]^{[m]}\rangle$$

and to use it together with (4.72). For example,

$$F_{n_1 n_2}^{2,1} = \langle n_1(n_1-1)n_2\rangle = \langle W_1^2 W_2\rangle$$

gives

$$\langle \Delta n_1^2 \Delta n_2\rangle = \langle n_1\rangle^2 \langle n_2\rangle |g_{12}|^2 + \langle n_1 n_2\rangle \ . \tag{5.288}$$

Similarly

$$\begin{aligned}\langle \Delta n_1^2 \Delta n_2^2\rangle = {} & \langle n_1\rangle^2 \langle n_2\rangle^2 \left(1+4|g_{12}|^2+4|g_{12}|^4\right) \\ & + \langle n_1^2 n_2\rangle + \langle n_1 n_2^2\rangle - \langle n_1 n_2\rangle \ ,\end{aligned} \tag{5.289}$$

and

$$\langle \Delta n_1^3 \Delta n_2\rangle = 9\langle n_1\rangle^3 \langle n_2\rangle |g_{12}|^2 + 3\langle n_1^2 n_2\rangle - 2\langle n_1 n_2\rangle \ . \tag{5.290}$$

Single Small Detector ($A \ll A_c$), Arbitrary Sampling Time

Statistics of [W]

We are first interested in finding the statistics of the set of random variables $[W] = (W_1, \ldots W_N)$, where W_j is the integrated intensity in an interval T centered around the time t_j. For simplicity, equal nonoverlapping intervals are assumed.

mgf of $[W]$. In order to find

$$Q_{[W]}([s]) = \left\langle \exp\left(-\sum_{j=1}^{N} s_j W_j\right)\right\rangle$$

it is logical to try to generalize the approach used in Sec. 5.2.2 by expanding the field in a Karhunen-Loève series. This can be done in two ways.

Method 1 [227, 460]. The field $V(t)$ is expanded in an orthonormal basis defined by the integral equation,

$$\left(\sum_{j=1}^{N} s_j \int_{I_j} dt'\right) g(t-t')\Phi_\ell(t') = \lambda_\ell \Phi_\ell(t) \quad , \tag{5.291}$$

where I_j represents the time interval $(t_j-T/2, t_j+T/2)$. Note that λ_ℓ is a function of $(s_1, s_2, \ldots s_N)$, i.e., $\lambda_\ell = \lambda_\ell([s])$.

The orthonormality condition in this case becomes

$$\left(\sum_{j=1}^{N} s_j \int_{I_j} dt\right)\Phi_\ell(t)\Phi_k^*(t) = \delta_{\ell k} \quad . \tag{5.292}$$

This enables us to write

$$\sum_{j=1}^{N} s_j W_j = \sum_{\ell=0}^{\infty} |\alpha_\ell|^2 \quad ,$$

where α_ℓ are circularly symmetric complex random variables with variances equal to $\langle W\rangle \lambda_\ell([s])/T$. This permits us to find the expectation value of the mgf and to obtain

$$Q_{[W]}([s]) = \prod_{\ell=0}^{\infty}\left[1+\frac{\langle W\rangle}{T}\lambda_\ell([s])\right]^{-1} \quad . \tag{5.293}$$

The problem is now to find the eigenvalues $\lambda_\ell([s])$ or, if possible, to find directly the infinite product of (5.293).

The easiest example in which (5.291) can be solved is the case N = 2 in the limit $T \ll \tau_c$. This example is discussed only for the purpose of demonstration, because we already know the answer. In this case, the integral in (5.291) reduces to an algebraic equation valid for all values of t. By choosing any two different values of t, we obtain two algebraic equations that can be solved together to yield two independent eigenvalues λ_ℓ. Inserting these two values in (5.293), we obtain $Q_{W_1,W_2}(s_1,s_2)$. The answer is, of course, given by (5.272). The case N = 3 can similarly be treated [60].

A second example is the case of light with a Lorentzian spectrum $g(\tau) = \exp(-\Gamma|\tau|)$. JAKEMAN [227] solved the eigenvalue problem in the two-fold case, N = 2, and obtained

$$Q(s_1,s_2) = [Q^{-1}(s_1)Q^{-1}(s_2)-e^{-2\Gamma|\tau|}y(s_1)y(s_2)]^{-1} \quad , \tag{5.294}$$

where

$$y(s_j) = \tfrac{1}{2}(\gamma/z_j-z_j/\gamma)\sinh(z_j) \quad ,$$

$$z_j^2 = \gamma^2+2\gamma s_j\langle W\rangle \, , \quad j = 1,2 \quad ,$$

and Q(s) is the single-fold mgf given by (5.54).

SRINIVASAN et al. [460] found a generalized technique for obtaining an explicit formula for the infinite product of (5.293).

Method 2 [131]. We can alternatively use a Karhunen-Loève expansion in a basis that is orthogonal over a long time interval $[0,T_0]$ that covers all of the intervals under consideration

$$V(t) = \sum_k \alpha_k\Phi_k(t) \tag{5.295}$$

$$\int_0^{T_0} g(t-t')\Phi_k(t')dt' = \lambda_k\Phi_k(t) \quad . \tag{5.296}$$

The integrated intensity in interval j can be written as

$$W_j = \sum_{\ell,m} \alpha_\ell^*\alpha_m \int_{I_j} \Phi_\ell^*(t)\Phi_m(t)dt \quad ,$$

and the multifold mgf takes the form

$$Q_{[W]}([s]) = \left\langle \exp\left(-\sum_{\ell,m} \alpha_\ell^* \alpha_m F_{\ell m}\right)\right\rangle, \tag{5.297}$$

where

$$F_{\ell,m} = \sum_{j=1}^{N} s_j \int_{I_j} \Phi_\ell^*(t)\Phi_m(t)dt \quad . \tag{5.298}$$

The variables $\{\alpha_\ell\}$ are statistically independent, circular complex Gaussian variables with variances proportional to $\{\lambda_\ell\}$. Hence, the expectation value in (5.297) can be determined and we get

$$Q_{[W]}([s]) = |\Delta|^{-1} \quad , \quad \Delta_{\ell m} = \delta_{\ell m} + \lambda_m F_{\ell m} \quad . \tag{5.299}$$

Obviously, solution of the integral equation (5.296) is easier than (5.291), but calculating the infinite product in (5.293) is easier than calculating the determinant of the infinite matrix in (5.299).

The infinite matrix Δ can of course be diagonalized, but this will lead to another homogeneous or inhomogeneous Fredholm equation of the second kind. Because this is in no way easier than method 1, we stop here and refer the reader to [131] for details.

The Moments of [*W*]. The moments of [W] can be most directly obtained by integrating over (4.73). This gives the general expression

$$K_{[W]}^{[1]} = \sum_c \int_{t_1}^{t_1+T_1} dt_1' \ldots \int_{t_N}^{t_N+T_N} dt_N' \prod_{j=1}^{N} G(t_j'-t_{cj}') \quad , \tag{5.300}$$

where c represents cyclic permutations. If T_i are euqal, then (5.300) can be written in the form

$$K_{[W]}^{[1]} = B_N \langle W\rangle^N \quad , \tag{5.301}$$

where

$$B_N = \frac{1}{T^N}\sum_c \int_{t_1}^{t_1+T} dt_1' \ldots \int_{t_N}^{t_N+T} dt_N' \prod_{j=1}^{N} g(t_j'-t_{cj}') \quad , \tag{5.302}$$

e.g.,

$$B_2 = \frac{1}{T^2}\int_{t_1}^{t_1+T} dt_1 \int_{t_2}^{t_2+T} dt_2 |g(t_1-t_2)|^2 = \frac{1}{T^2}\int_{-T}^{T} dt(T- t\ )|g(t_1-t_2+t)|^2$$

$$B_3 = \frac{2}{T^3}\int_{t_1}^{t_1+T} dt_1' \int_{t_2}^{t_2+T} dt_3' \left|\mathrm{Re}\ g(t_1'-t_2')g(t_2'-t_3')g(t_3'-t_1')\right| \quad . \tag{5.303}$$

Notice that $K_W^{(m)}$ follows as a special case from (5.301) (See. (5.28,29)).

Example. Lorentzian Spectrum. By putting $|g(t)| = \exp(-\Gamma|t|)$ in (5.303) and assuming that $\tau \geq T$, we get

$$B_2 = \mathrm{sh}^2(\gamma)e^{-2\Gamma\tau} \quad , \quad \tau = t_2-t_1 \quad , \tag{5.304}$$

$$B_3 = \mathrm{sh}^2(\gamma)\left[e^{-2\Gamma\tau_1} + e^{-2\Gamma\tau_2} + 3e^{-2\Gamma(\tau_1+\tau_2)}\right] \quad ,$$

$$\tau_1 = t_2-t_1 \quad , \quad \tau_2 = t_3-t_2 \quad , \tag{5.305}$$

where

$$\mathrm{sh}(\gamma) = \sinh(\gamma)/\gamma \quad , \quad \gamma = \Gamma T \quad .$$

From this, we can write the important equation

$$\langle W_1W_2\rangle/\langle W\rangle^2 = g_W(\tau) = 1 + \mathrm{sh}^2(\gamma)\exp(-2\Gamma|\tau|) \quad . \tag{5.306}$$

Statistics of [*n*]. The moments of [n] can be determined directly from those of [W] as we have done in Sec. 5.2.2, e.g., for Lorentzian light

$$\langle n_1n_2\rangle/\langle n_1\rangle\langle n_2\rangle = g_n(\tau) = 1 + \mathrm{sh}^2(\gamma)\exp(-2\Gamma|\tau|)$$

$$|\tau| \geq T \quad . \tag{5.307}$$

In order to visualize the effect of the counting time on the correlation between n_1 and n_2, we plot $g_n(\tau)$ as function of $\Gamma|\tau|$ for several values of γ (Fig.5.27).

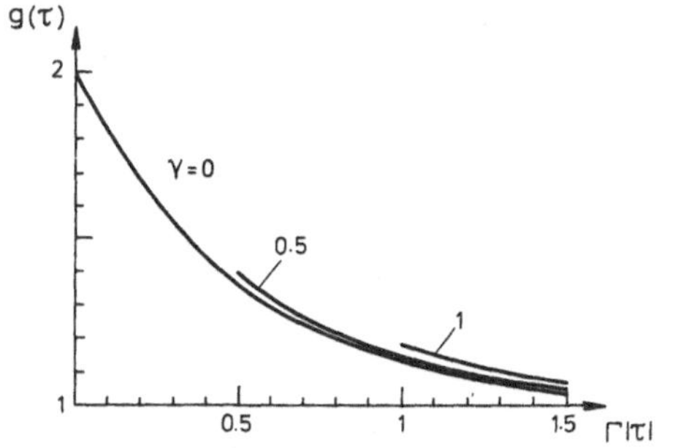

Fig.5.27. Effect of the sampling time on the photoelectron counting correlation function for thermal light with Lorentzian spectrum

The JPD of [n] is more difficult to determine, because we have not found a general explicit formula for the multifold mgf of [W]. Even in the double-fold case of a field with a Lorentzian spectrum, the explicit formula (5.294) cannot be used to determine $P(n_1,n_2)$, because it would be difficult to determine the higher-order derivatives.

An approximate formula for the mgf (discussed in this subsection, page 263) may be helpful in determining P([n]).

Single Detector of Arbitrary Area. Arbitrary Sampling Time

This section is a generalization of Sec. 5.2.2, page 189, to the multifold case. Because of the cross-spectral purity, the field can be expanded in separate temporal and spatial modes by using eigenfunctions of the integral equations

$$\left(\sum_{i=1}^{N} s_i \int_{t_i}^{t_i+T} dt'\right) g(t-t')\chi_\ell(t') = \lambda_\ell \chi_\ell(t) \tag{5.308}$$

$$\int_A g(\underline{r},\underline{r}')\xi_j(\underline{r})d\underline{r} = \mu_j \xi_j(r') \ , \qquad \sum_j \mu_j = 1 \ . \tag{5.309}$$

By an analysis similar to that of Sec. 5.2.2, page 189, we see that the mgf of the integrated intensities [W] is related to the eigenvalues of (5.308) and (5.309) by

$$Q_{[W]}([s]) = \prod_j \prod_\ell \left[1+\mu_j \frac{\langle W\rangle}{T} \lambda_\ell([s])\right]^{-1}$$

$$= \prod_j Q_{[\mu_j W]}([s]) \quad , \tag{5.310}$$

where $Q_{\mu_j W}([s])$ is the mgf of a small detector that detects an energy $\mu_j W$. This equation is the generalization of (5.82).

Example. $N = 2$, $T \ll \tau_c$, Arbitrary A/A_c

Here we have

$$Q(s_1,s_2) = \prod_j \left[(1+s_1\mu_j\langle W_1\rangle)(1+s_2\mu_j\langle W_2\rangle)-s_1s_2\mu_j^2\langle W_1\rangle\langle W_2\rangle |g(t_1-t_2)|^2\right]^{-1} \quad . \tag{5.311}$$

If $A \gg A_c$, then (5.310) simplifies to

$$Q_{[W]}([s]) = \left[Q_{[W/N]}([s])\right]^N \quad , \tag{5.312}$$

where $N = A/A_c$.

Moments of [W]

The moments of [W] can best be obtained by integrating (4.73) over the time intervals and detector areas. This gives the general relation

$$K^{[1]}_{[W]} = \sum_c \prod_{j=1}^N \frac{1}{T_j} \int_{I_j} dt'_j \frac{1}{A} \int_A d\underline{r}'_j \langle W_j\rangle\langle W_{cj}\rangle g(\underline{x}'_j,\underline{x}'_{cj}) \quad . \tag{5.313}$$

Eqs. (5.89) and (5.300) follow as special cases of the general equation (5.313).

For N = 2 and cross-spectrally pure light, (5.313) gives

$$\langle \Delta W_1 \Delta W_2\rangle = \langle W_1\rangle\langle W_2\rangle \frac{1}{T_1T_2} \int_0^{T_1} dt_1 \int_\tau^{\tau+T_2} dt_2 |g(t_1-t_2)|^2$$

$$\cdot \frac{1}{A^2} \int_A d\underline{r}_1 \int_A d\underline{r}_2 \; |g(\underline{r}_1,\underline{r}_2)|^2 \quad . \tag{5.314}$$

As an *example*, let us take *one circular detector* with radius a and two equal time intervals $T_1 = T_2$ and assume the light to have a *Lorentzian* spectrum, $g(t) = exp(-\Gamma|t|)$ and a spatial coherence function given by $g(\underline{r}_1,\underline{r}_2) = 2J_1(2|\underline{r}_1-\underline{r}_2|/r_c)/(2|\underline{r}_1-\underline{r}_2|/r_c)$. Eq. (5.314) gives [233, 429, 430]

$$\langle W_1W_2\rangle/\langle W_1\rangle\langle W_2\rangle = g_W(\tau) = 1 + sh^2(\gamma)f(A)exp(-2\Gamma|\tau|) \quad , \tag{5.315}$$

where $f(A) = N_s^{-1}$ is given by (5.98) and (5.101), and is illustrated in Figs.5.12 and 5.16.

Recall that (5.315) is not valid for $\tau = 0$. For $\tau = 0$, we have a single-fold second-order moment, which, according to (5.55) and (5.89), takes the form

$$\langle W^2\rangle/\langle W\rangle^2 = g_W(0) = 1 + \left(\frac{e^{-2\gamma}+2\gamma-1}{2\gamma^2}\right) f(A) \quad . \tag{5.316}$$

Other examples of higher-order moments for single and several detectors are treated by CANTRELL and FIELDS [83].

Several Photodetectors Simultaneously Counting. Cross-Spectrally Pure Light

In this case, we perform the opposite of (5.308) and (5.309), namely, we expand our fields in a basis determined by the integral equations

$$\int_0^T g(t'-t)\chi_\ell(t')dt' = \lambda_\ell\chi_\ell(t) \tag{5.317}$$

and

$$\sum_j s_j \int_{A_j} g(\underline{r},\underline{r}')\xi_j(\underline{r})d\underline{r} = \mu_j\xi_j(\underline{r}') \quad , \tag{5.318}$$

where $\{\mu_j\}$ are now functions of $(s_1,s_2, \ldots s_N)$.

The moment-generating function can now be written in the form

$$Q_{[W]}([s]) = \prod_j \prod_\ell \left[1 + \lambda_\ell \frac{<W>}{T} \mu_j([s])\right]^{-1}$$
$$= \prod_j Q_W\left[\mu_j([s])\right] \quad , \tag{5.319}$$

where $Q_W(s)$ is the single-fold mgf for one detector.

Thus, (5.319) relates the multifold mgf to the single-fold mgf provided that the eigenvalues of the spatial integral equation (5.318) have been determined (see also [81]).

Let us take the case in which $A << A_c$ as an example. In this case, the integral equation (5.318) is reduced to a matrix equation,

$$\sum_{i=1}^{N} s_i g_{ki} \xi_{ij} = \mu_j \xi_{kj} \quad , \quad g_{ki} = g(\underline{r}_k, \underline{r}_j) \quad , \; \xi_{ij} = \xi_i(\underline{r}_j) \quad , \tag{5.320}$$

which gives N values of μ_j, each being a function of $s_1, s_2, \ldots s_N$. For example, for $N = 2$, this gives

$$\mu_j = \frac{1}{2}(s_1+s_2) \pm \frac{1}{2}\left[(s_1-s_2)^2 + 4s_1 s_2 |g(\underline{r}_1, \underline{r}_2)|^2\right]^{1/2} \quad .$$

These two values have to be substituted in (5.319) and we get

$$Q_{W_1, W_2}(s_1, s_2) = Q_{\mu_1 W}(s_1) Q_{\mu_2 W}(s_2) \quad ,$$

where $Q_W(s)$ is determined as described in Sec. 5.2.2. If $T << \tau_c$, then

$$Q_W(s) = (1+s<W>)^{-1} \quad ,$$

which gives

$$Q_{W_1, W_2}(s_1, s_2) = \left[1+s_1<W>)(1+s_2<W>) - <W>^2 s_1 s_2 |g(\underline{r}_1, \underline{r}_2)|^2\right]^{-1} \quad ,$$

which is what we expect from (5.272).

An Approximate Formula

The formula of the mgf for $T \ll \tau_c$ and $A \ll A_c$ (5.271) can be generalized to describe approximately the case of arbitrary T/τ_c and A/A by following an approach similar to that of Sec. 5.2.2, page 193 [419].

Writing

$$Q(s_1, s_2, \dots s_N) = |\Delta|^{-N} \quad , \tag{5.321}$$

where

$$\Delta_{ij} = \delta_{ij} + s_i(\langle \Delta W_i \Delta W_j \rangle / N)^{1/2} \quad , \tag{5.322}$$

we immediately see that this mgf generates the exact $\langle W_i W_j \rangle$ for all i,j = 1,2, ... N if N is determined from

$$N = \langle W \rangle^2 \; / \; \langle \Delta W^2 \rangle \; . \tag{5.323}$$

Notice that for N = 1 we obtain the approximate formula of Sec. 5.2.2. For $T \ll \tau_c$ and $A \ll A_c$, $N = 1$ and we again obtain (5.271). This formula ensures that if any two detectors are separated by distances longer than a coherence length or if two time intervals are separated by a time delay longer than a coherence time, then $\langle \Delta W_\ell \Delta W_k \rangle = \langle \Delta W_\ell^2 \rangle \, \delta_{\ell k}$ and the mgf is factorized properly.

It should be noted that the spectral and coherence properties of the field are imbedded in $\{\langle \Delta W_i \Delta W_j \rangle\}$, which have to be determined exactly.

SALEH [419] compared the two-fold moments $\langle W_1^2 W_2 \rangle$ and $\langle W_1^2 W_2^2 \rangle$ generated by the approximate formula to their corresponding exact values for spatially coherent light with Lorentzian spectrum and found the approximation satisfactory. Moreover, comparison of the three-fold moment $\langle W_1 W_2 W_3 \rangle$ with its exact value gives satisfactory results. There is no guarantee, however, that the approximate formula holds for higher-order moments.

Statistics of Times.
Probability Density of Multicoincidence PDC

From the discussion in Sec. 3.2.3, the PDC of the photoelectron events is related to the intensity correlation function by

$$\lambda(t_1,t_2, \ldots t_N) = \int_A d\underline{r}_1 \ldots \int_A d\underline{r}_N \; G_I(\underline{r}_1,t_1; \ldots ; \underline{r}_N,t_N) \quad .$$

For a small detector, and for thermal light

$$\lambda(t_1, \ldots t_N) = A^N \sum_{\pi} \prod_{j=1}^{N} G_V(t_j,t_{\pi_j}) \quad ,$$

e.g.,

$$\lambda(0,\tau) = A^2 \langle I(0)\rangle^2 \left[1 + |g_V(0,\tau)|^2\right] \quad .$$

When $\tau \gg \tau_c$, $|g_V(0,\tau)| = 0$ and the PDC takes its lowest value, $A^2\langle I(0)\rangle^2$, which corresponds to independent random emissions. When $\tau \ll \tau_c$, $|g_V(0,\tau)| \simeq 1$; the PDC is higher by a factor of 2. This is another manifestation of the bunching phenomenon discussed in Sec. 5.2.2. Compare this to the case of coherent light in which $\lambda(0,\tau) = A^2\langle I(0)\rangle^2$ for all τ, i.e., photoelectrons are emitted independently (Poisson process). Experimental verification of this fact was very important to the early development of the theory of photoelectron statistics [17, 351].

JPD of Instants of Occurrence in a Fixed Interval

From (3.92), the JPD of obtaining n photoelectrons at times $t_1,t_2, \ldots t_n$ in an interval [0,T] is given by

$$P_c(n;t_1, \ldots t_n) = \left\langle \exp\left[-\int_0^T \lambda(t)dt\right] \prod_{i=1}^{n} \lambda(t_i) \right\rangle , \qquad (5.324)$$

where we assume that the area of the detector is small ($A \ll A_c$).

In order to evaluate this expectation value, we resort again to the Karhunen-Loeve expansion over the time interval [0,T], (5.15,19), from which

$$P_c(n;t_1, \ldots t_n) = \left\langle \exp\left(-\sum_k |\alpha_k|^2\right) \prod_{i=1}^{n} \sum_{k_1} \sum_{k_2} \alpha_{k_1}^* \alpha_{k_2} \Phi_{k_1}^*(t_i)\Phi_{k_2}(t_i) \right\rangle (5.325)$$

where α_k are independent complex circularly symmetric Gaussian variables with variances equal to the eigenvalues m_k. Evaluating the expectation, we obtain the general expression [303, 304, 418].

$$P_c(n;t_1, \dots t_n) = Q(1)\langle n\rangle^n \sum_{\pi_n} \prod_{i=1} \eta(t_i, t_{\pi_i}) \quad , \tag{5.326}$$

where

$$\eta(t,t') = \frac{1}{\langle n\rangle} \sum_k \frac{m_k}{1+m_k} \Phi_k^*(t)\Phi_k(t') \quad . \tag{5.327}$$

Here π_n represents permutations of $(\pi_1, \pi_2, \dots \pi_n)$ over $(1,2, \dots n)$ and

$$Q(1) = \prod_k (1+m_k)^{-1} \tag{5.328}$$

is the probability of observing no events in the interval [0,T]. Expressions for Q(1) have been previously studied under different conditions. For $n = 1$,

$$P_c(1;t) = Q(1)\langle n\rangle\eta(t,t) \quad ,$$

and for $n = 2$,

$$P_c(2;t_1,t_2) = Q(1)\langle n\rangle^2[\eta(t_1,t_1)\eta(t_2,t_2) + |\eta(t_1,t_2|^2] \quad , \tag{5.329}$$

and so on.

For stationary fields, $\eta(t_1,t_2)$ is a function of the time difference, i.e.,

$$\eta(t_1,t_2) = \eta(t_1-t_2) \quad .$$

Consequently, $P_c(1,t)$ is independent of t, i.e., if one electron is detected in the interval [0,T], it is equally likely to be anywhere in this interval.

In order to understand the behaviour of the function $\eta(t,t')$, we compare it to the normalized field-correlation function,

$$g(t,t') = \frac{1}{\langle n\rangle} \sum_k m_k \Phi_k^*(t)\Phi_k(t') \quad .$$

For $m_k \ll 1$ (very small average number of counts in [0,T]), $\eta(t,t') \simeq g(t,t')$. For light with a rectangular spectrum, m_k has a rectangular behaviour and $\eta(t,t') \sim g(t,t')$. In general [304], the width of the function η is of the same order of magnitude as that of the function g (the coherence time τ_c).

This means that many of the permutations in (5.326) have negligible contributions to the sum. Only the terms that have $|t_i - t_{\pi i}| < \tau_c \;\forall\; i$ would be nonnegligible. For example, if *the average number of counts per coherence time is very small*, then it is unlikely that two photoelectrons would have a time difference less than τ_c and consequently,

$$P_c(n;t_1, \ldots t_n) = Q(1)\langle n\rangle^n \prod_{i=1}^{n} \eta(t_i, t_i) \quad ,$$

which is independent of t_i for a stationary field.

The evaluation of $\eta(t,t')$ is analytically difficult and computer techniques are usually necessary. Consideration of the following limiting cases is therefore important.

The Limit $T \ll \tau_c$. In this limit, only one mode exists and (5.327) gives

$$\eta(t,t') = \frac{1}{T}\,\frac{1}{1+\langle n\rangle} \quad .$$

This corresponds to

$$P_c(n;t_1, \ldots t_n) = \langle n\rangle^n/(1+\langle n\rangle)^{n+1} \quad ,$$

independently of the times of occurrence.

The Limit $T \gg \tau_c$. When the observation time T is much longer than the field's coherence time τ_c (which is certainly the case when natural light is detected), we can use the values,

$$m_k \approx \langle n\rangle g(\omega_k) \quad , \quad \omega_k = k(2\pi/T)$$
$$\phi_k(t) \approx \frac{1}{\sqrt{T}} \exp(j\omega_k t) \quad , \quad k = 0, \pm 1, \pm 2, \ldots , \tag{5.330}$$

as our eigenvalues and eigenfunction (cf. Sec. 2.6.6), where $g(\omega)$ is the power spectrum of the field. We then write the summations in (5.327) in the form of integrations by using the relation [498]

$$\sum_k f(m_k) \approx \frac{T}{2\pi} \int_{-\infty}^{\infty} f[g(\omega)]d\omega \quad . \tag{5.331}$$

This gives

$$\sum_k \frac{m_k}{1+m_k} \exp(j\omega_k t) \approx \frac{T}{2\pi} \int_{-\infty}^{\infty} \frac{\langle n\rangle g(\omega)}{1+\langle n\rangle g(\omega)} \, d\omega \quad ,$$

which when used in (5.327) gives

$$\eta(t_1-t_2) = \int_{-\infty}^{\infty} \frac{g(\omega)}{1+\langle n\rangle g(\omega)} \exp[j\omega(t_1-t_2)]d\omega \quad . \tag{5.332}$$

This gives a simple relation for evaluating η in the stationary case. Because

$$g(t_1-t_2) = \int_{-\infty}^{\infty} g(\omega)\exp[j\omega(t_1-t_2)]d\omega$$

and because $g(\omega)/[1+\langle n\rangle g(\omega)]$ has approximately the same width as $g(\omega)$, we conclude that $\eta(t_1-t_2)$ has approximately the same width as $g(t_1-t_2)$. It is clear that if $g(\omega)$ is rectangular, then

$$\eta(t_1-t_2) = \frac{1}{1+\langle n\rangle\tau_c/T} \, g(t_1-t_2) \quad .$$

If the field is Lorentzian, i.e.,

$$G(\tau) = \langle n\rangle \exp(-\Gamma|\tau|) \quad ,$$

then

$$\eta(\tau) = (\alpha T)^{-1}\exp(-\alpha\Gamma|\tau|) \quad ,$$
$$\alpha = (1+2\langle n\rangle/\Gamma T)^{1/2} \quad . \tag{5.333}$$

The width of $g(\tau)$ is the coherence time $\tau_c = \Gamma^{-1}$. The width of $\eta(\tau)$ is $\tau_\eta = \tau_c/(1+2\langle n\rangle/\Gamma T)^{\frac{1}{2}}$. Whereas τ_c represents the coherence time of the field, τ_η represents the effective coherence time of the detected photoelectrons. When the number of photoelectrons per coherence time is small, $\tau_\eta = \tau_c$. As this number increases, τ_η decreases, i.e., as the density of events increases, the memory time becomes shorter.

We finally remark that the approximation $\eta(\tau) \approx g(\tau)$ is equivalent to factorizing the expectation in (5.324) in the form

$$\left\langle \exp\left[-\int_0^T \lambda(t)dt\right] \prod_{i=1}^{n} \lambda(t_i)\right\rangle \approx \left\langle \exp\left[-\int_0^T \lambda(t)dt\right]\right\rangle \left\langle \prod_{i=1}^{n} \lambda(t_i)\right\rangle .$$

The first factor is the probability of zero events and the second is proportional to the intensity coherence function,

$$\left\langle \prod_{i=1}^{n} \lambda(t_i)\right\rangle = \langle n\rangle^n g_I(t_1, \ldots t_n) = \langle n\rangle^n \sum_{\pi} \prod_{i=1}^{n} g_V(t_i - t_{\pi_i}) .$$

Under this approximation, the joint probability of instants of occurrence can be factorized as a product of a function of n and a function of the times $t_1, t_2, \ldots t_n$.

5.3.2 Mixture of Coherent and Polarized Thermal Light

This section is a generalization of Sec. 5.2.4, which uses the results of Sec. 4.2.3.

The Limit $T \ll \tau_c$, $A \ll A_c$, and $T\Delta\omega \ll 1$

In this limit, the light energies [W] collected by the detectors are proportional to the local irradiances [I]. Therefore the statistics of [W] are given by the equations of Sec. 4.2.3, (4.100-110) with [W] replacing [I].

Because the joint factorial moments of the number of counts [n] are proportional to the moments of [W] that have the same order, they can be straightforwardly determined. For example, for light with symmetric spectrum, we get

$$\langle n_1 n_2\rangle/\langle n\rangle^2 = g_n(\underline{r}_1,\underline{r}_2,\tau)$$
$$= 1 + \left(\frac{\langle n_{th}\rangle}{\langle n\rangle}\right)^2 |g_V(\underline{r}_1,\underline{r}_2,\tau)|^2 + \frac{2\langle n_{th}\rangle\langle n_c\rangle}{\langle n\rangle^2}|g_V(\underline{r}_1,\underline{r}_2,\tau)|\cos(\Delta\omega\tau) \quad . \tag{5.334}$$

Similarly, $\langle n_1 n_2 n_3\rangle$, $\langle n_1^2 n_2\rangle$, $\langle n_1 n_2 n_3 n_4\rangle$, $\langle n_1^2 n_2^2\rangle$, ... , etc., can be determined.

The joint probability distribution P([n]) can, in principle, be obtained from the mgf (4.100-103) by differentiation, but explicit formulas are extremely difficult to find. By directly performing enough differentiations to recognize a pattern, FILLMORE [140] found a lengthy expression for $P(n_1,n_2)$. The reader is referred to the original paper for this expression.

Arbitrary T/τ_c, A/A_c, and $T\Delta\omega$

An expression can be written for the multifold mgf in terms of the eigenvalues of

$$\left[\sum_j s_j \int_{t_j}^{t_j+T} dt'\right]\int_{A_j} d\underline{r}'G(\underline{r}',\underline{r},t-t')\Phi_k(\underline{r}',t') = m_k([s])\Phi_k(\underline{r},t) \quad , \tag{5.335}$$

namely

$$Q([s]) = \prod_k (1+m_k)^{-1}\exp(-W_{ck}s_k)\exp\left(W_{ck}\frac{m_k}{1+m_k}\right) , \tag{5.336}$$

where

$$W_{ck} = \left|\sum_j s_j \int_{t_j}^{t_j+T} dt \int_{A_j} d\underline{r}V_c(\underline{r},t)\Phi_k(\underline{r},t)\right|^2 \quad . \tag{5.337}$$

But, owing to its complexity, we should expect that such an expression has very little utility.

The moments of [W] can be determined from Q([s]), as has been done in Sec. 5.2.4, page 209 [338]; but they can also be determined by integrating (4.104-110) over time and space. This gives

$$K_{[W]}^{[1]} = K_{[W_{th}]}^{[1]} + K_{[W_c]}^{[1]} + \int_{t_1}^{t_1+T_1} dt_1' \int_{A_1} d\underline{r}_1' \cdots \int_{t_N}^{t_N+T_N} dt_N' \int_{A_N} d\underline{r}_N' \sum_{\pi}$$

$$V_c^*(\underline{x}_{k_1}')V_c(\underline{x}_{k_N}')G(\underline{x}_{k_1}',\underline{x}_{k_2}')G(\underline{x}_{k_2}',\underline{x}_{k_3}') \cdots G(\underline{x}_{k_{N-1}}',\underline{x}_{k_N}') \quad , \tag{5.338}$$

where $\sum_{\pi}$ is the summation over N! permutations of $(K_1, \ldots K_N)$ over the integers (1,2, .. N).

Moments of order other than [1] can be obtained from (5.338) by lumping variables together, e.g.,

$$K_W^{(m)} = K_{W,W,\ \ldots\ W}^{1,1,\ \ldots\ 1}$$

Example

For the case $V_c(t) = A\exp(j\Delta\omega t)$ and when the thermal light has a symmetric spectrum, we get [350]

$$\langle \Delta W_1 \Delta W_2 \rangle = \frac{1}{T^2} \iint_{0\tau}^{T\tau+T} dt_1 dt_2 \frac{1}{A^2} \int_{A_1} d\underline{r}_1 \int_{A_2} d\underline{r}_2$$

$$\langle W_{th}\rangle^2 |g_{th}(\underline{r}_1,\underline{r}_2,t_1-t_2)|^2 + 2\langle W_{th}\rangle W_c\ |g_{th}(\underline{r}_1,\underline{r}_2,t_1-t_2)| \tag{5.339}$$

$$\cos[\Delta\omega(t_1-t_2)] \quad .$$

In the special case of a single circular detector with radius a and when

$$|g_{th}(\underline{r}_1,\underline{r}_2,\tau)| = J(2|\underline{r}_1-\underline{r}_2|/r_c)\exp(-\Gamma|\tau|) \quad , \qquad J(x) = 2J_1(x)/x \tag{5.340}$$

and $\quad \Delta\omega = 0 \quad ,$

$$\langle W_1 W_2\rangle = \langle W\rangle^2 + \langle W_{th}\rangle^2 \xi(\gamma) f(A) e^{-2\Gamma|\tau|} + 2\langle W_{th}\rangle W_c \xi\left(\frac{\gamma}{2}\right) f_D(A) e^{-\Gamma|\tau|} \quad , \tag{5.341}$$

where

$$\xi(\gamma) = \mathrm{sh}^2(\gamma) \qquad \tau \neq 0 \tag{5.342}$$

$$= (e^{-2\gamma}+2\gamma-1)/2\gamma^2 \qquad \tau = 0 \quad , \qquad \gamma = \Gamma T \quad . \tag{5.343}$$

The factor $f_D(A) = \eta_s$ has already been determined (cf. (5.181) and Fig.5.16).

An expression for $\langle W_1^2 W_2\rangle$ that applies to this same example can be found in [421]. Other moments can be similarly determined but obviously they become mathematically much more complicated.

The photoelectron-counting moment that corresponds to the above example is

$$\begin{aligned} g_n(\tau) &= \langle n_1 n_2\rangle / \langle n_1\rangle \langle n_2\rangle \\ &= 1 + \epsilon + (\langle n_{th}\rangle/\langle n\rangle)^2 \xi(\gamma) f(A) e^{-2\Gamma|\tau|} \\ &\quad + 2\langle n_{th}\rangle\langle n_c\rangle \xi(\gamma/2) f_D(A) e^{-\Gamma|\tau|} \quad , \end{aligned} \tag{5.344}$$

where

$$\xi(\gamma) = \mathrm{sh}^2(\gamma) \quad , \quad \epsilon = 0 \quad , \quad \text{for} \quad \tau \neq 0 \tag{5.345}$$

$$\xi(\gamma) = (e^{-2\gamma}+2\gamma-1)/2\gamma^2 \quad , \quad \epsilon = \langle n\rangle^{-1} \quad , \quad \text{for} \quad \tau = 0 \quad . \tag{5.346}$$

An approximate formula for the double-fold statistics was found by SALEH [421].

5.4 Nonideal Effects in Photodetectors

The analysis presented in the previous sections of this chapter was based on an ideal mathematical model that describes photoelectron statistics. Photoelectron events follow a doubly stochastic Poisson point process whose rate is proportional to the instantaneous light power collected by the detector. In this section, we examine some of the effects that may lead to modifications or deviations from this ideal model.

5.4.1 Dark Current

It is well known that in addition to the photoelectrons, a photodetector produces a small number of spurious pulses known as dark pulses. These pulses are present whether the detector is illuminated or not. They act as a background which adds to the uncertainty of the observed optical signal and could considerably modify the measured statistics. When weak light is involved, dark current could be the limiting factor in determination of the minimum detectable signal (cf. Sec. 6.2.1).

The most important process that leads to dark pulses is thermionic emission from the photocathode. The rate of thermionic emission depends on temperature and can be reduced by cooling. Dark pulses can also be produced by ions of gases adsorbed on the envelope of the tube or by cosmic rays.

It can usually be assumed that the dark pulses are produced randomly at a constant rate. They follow therefore a homogeneous Poisson process. If this rate is known, the statistics of the combination of photoelectron pulses and dark pulses can easily be determined by use of the rules of addition of two statistically independent point processes (5.104-111). Dark pulses can also be regarded as if they originate from a coherent constant background light that does not interfere with the optical signal. When the measured optical signal alone corresponds to Poisson photoelectron pulses, the statistics of the combination with dark pulses remains Poisson. Otherwise, the photoelectron statistics are modified by the presence of dark pulses.

In some cases, departure from the Poisson statistics of dark pulses has been observed [492]. This might be caused by the generation of groups of after-pulses, following a strong cosmic-ray pulse, or by ion feedback in photomultipliers.

Photomultipliers that exhibit such effects should not be used for measurement of photoelectron statistics [352].

5.4.2 Dead-Time Effect

Most photodetectors require some characteristic time after each registration of a pulse, during which the detector does not respond to any external field and no further emissions can be registered. This time is called the recovery time or the dead time of the detector.

One obvious consequence of the dead-time effect is a reduction of the mean number of observed pulses. Modifications of the photoelectron statistics should also be expected, especially when short sampling times are used. This would set a limit on the maximum speed at which a detector could be used to observe optical signals. In this subsection, we discuss possible modificaitons of the photoelectron statistics due to the dead-time effect and methods of calculating them.

We assume that after each photoelectron pulse the detector is blocked for a time τ_D, then released to behave as an ideal detector until the next pulse arrives. We first consider the case of detecting coherent constant light and then generalize the results to fluctuating light.

In the absence of a dead-time effect, the time interval between two pulses, τ, follows an exponential distribution (3.30). By definition, the dead-time effect modifies this distribution to

$$\begin{aligned} P_i(\tau) &= \lambda\exp(-\lambda\tau') \quad , \qquad \tau' \geq 0 \quad , \\ &= 0 \qquad\qquad\qquad\quad \tau' < 0 \quad , \qquad \tau' = \tau - \tau_D \quad , \end{aligned} \tag{5.347}$$

where λ is the rate of occurrence of photoelectrons. The time interval between two consecutive events could not be less than τ_D.

If we assume that the detector is unblocked at $t = 0$, then the probability distribution of the time of arrival of the first photoelectron should not be modified by the dead-time effect. It therefore remains exponential

$$\begin{aligned} P_f(t_{10}) &= \lambda\exp(-\lambda t_{10}) \quad , \qquad t_{10} \geq 0 \quad , \\ &= 0 \quad , \qquad\qquad\qquad\quad t_{10} < 0 \quad . \end{aligned} \tag{5.348}$$

Because the underlying photoelectron process is Poisson, the times t_{10}, τ_1, τ_2, ... are statistically independent (cf. Fig.3.1).

The joint distribution of t_{10}, τ_1, τ_2, ... completely defines the point process of photoelectrons. Determining the other statistical properties of the process is a straightforward mathematical exercise. For example, the PD of the time of arrival of the nth photoelectron can be obtained by noting that (see Fig.3.1)

$$t_{no} = t_{10} + \tau_1 + \ldots + \tau_{n-1} \quad .$$

Because the elements of this sum are statistically independent, the mgf's are related by

$$Q_{t_{n0}}(s) = Q_{t_{10}}(s) \prod_{i=1}^{n-1} Q_{\tau_i}(s) = \frac{\lambda}{\lambda+s}\left[\frac{\lambda}{\lambda+s}\exp(-s\tau_D)\right]^{n-1} \quad ,$$

from which the PD of t_{no} is

$$P_f(t_{no}) = \frac{\lambda^{n+1}}{n!}(t'_{no})^n \exp(-\lambda t'_{no}) \quad , \quad t'_{no} \geq 0 \quad ,$$
$$= 0 \quad , \qquad t'_{no} < 0 \quad , \quad t'_{no} = t_{no} - n\tau_D \quad . \qquad (5.349)$$

As expected, it is impossible to obtain n photoelectrons in a time interval shorter than $n\tau_D$.

The PD of the number of counts P(n) in an interval [0,T] can be obtained by noting that it is equal to the probability that $t_{n-1} \leq T$ and $t_n \geq T$. This enables us to write

$$P(n) = \int_0^T P_f(t_{n-1})dt_{n-1} - \int_0^T P_f(t_n)dt_n$$
$$= \int_0^{T'_{n-1}} dt\ \lambda^n\ t^{n-1}\ e^{-\lambda t}/(n-1)! - \int_0^{T'_n} dt\ \lambda^{n+1}\ t^n\ e^{-\lambda t}/n! \quad ,$$

where $T'_n = T - n\tau_D$.

By using the identity $\int_0^x dt\ t^n\ e^{-t}/n! = 1 - e^{-x} \sum_{m=0}^{n} x^m/m!$ ([184]page 940), we can write

$$P(n) = \sum_{k=0}^{n} (\lambda T'_n)^k e^{-\lambda T'_n}/k! - \sum_{k=0}^{n} (\lambda T'_{n-1})^k e^{-\lambda T'_{n-1}}/k! \quad ,$$
$$T'_n = T - n\tau_D \quad . \qquad (5.350)$$

If P(n,T) denotes the PD of obtaining n counts in the interval [0,T], then we can write

$$P(n,T) = \sum_{k=0}^{n} P_0(k,T'_n) - \sum_{k=0}^{n-1} P_0(k,T'_{n-1}) \quad , \quad T'_n = T - n\tau_D \quad , \qquad (5.351)$$

or

$$P(n,T) = P_0(n,T'_n) + \sum_{k=0}^{n-1} [P_0(k,T'_n) - P_0(k,T'_{n-1})] \quad ,$$

where the subscript 0 stands for the distribution in the absence of the dead-time effect. Note that if $\tau_D = 0$, then $T'_n = T$ and $P(n,T) = P_0(n,T)$. Also, for $n = 0$, $P(0,T) = P_0(0,T)$ as expected from the assumption that the detector is unblocked at $t = 0$.

Now we turn to the case when λ is a fluctuating function of time. If we assume that T is much shorter than τ_c, the coherence time of fluctuations in λ, then all we need is to average the right-hand side of (5.350) over fluctuations in the random variable λ. This reproduces (5.351) provided that $P_0(n,T)$ includes the effect of fluctuations of λ, Eq. (5.351) therefore relates the dead-time-corrected photoelectron-counting distribution to the uncorrected distribution, for all light fields such that $T \ll \tau_c$.

When the dead time τ_D is much less than the counting time T, we can expand the right-hand side of (5.350) in a power series of τ_D/T and retain the lower-order terms. This gives

$$P(n) \simeq \langle[(\lambda T)^n e^{-\lambda T}/n!][1+n(\lambda T-n+1)\tau_D/T]\rangle \quad , \tag{5.352}$$

or

$$P(n) \simeq P_0(n)[1-n(n-1)\tau_D/T] + P_0(n+1)[n(n+1)\tau_D/T] \quad . \tag{5.353}$$

The approximate relation (5.352) was obtained by DE LOTTO et al. [130] using methods due to FELLER [137]. The exact formula (5.350) was obtained by BEDARD [41] and was put in the form (5.351) by CANTOR and TEICH [78]. These expressions were applied to the coherent and the partially polarized case [41], to the Van der Pol oscillator case [78] and to the case of mixed coherent and thermal light [475].

The Moments

By using (5.351), we can determine the moments of n, e.g.,

$$\langle n\rangle = \sum_{n=0}^{\infty} \sum_{k=0}^{n} P_0(k,T'_n) \quad .$$

It is difficult, however, to obtain a general expression for higher-order moments.

If we use the approximate formula (5.353), then we can relate the factorial moments $F_n^{(m)}$ to the factorial moments in the absence of the dead-time effect $F_n^m|_0$ by the formula [229]

$$F_n^{(m)} \simeq F_n^{(m)}\big|_0 - m\left[(m-1)F_n^{(m)}\big|_0 + F_n^{(m+1)}\big|_0\right]\tau_D/T \quad . \tag{5.354}$$

Note that the equality $F_n^{(m)} = \langle W^m\rangle$ no longer holds. The process is no longer a doubly stochastic Poisson process.

For *coherent light*, the exact formula gives

$$\begin{aligned}\langle n\rangle &= \langle n\rangle_0\,(1+\bar{n}_D)^{-1} + \tfrac{1}{2}\,\bar{n}_D^2\,(1+\bar{n}_D)^{-2}\\ \langle\Delta n^2\rangle &= \langle n\rangle_0\,(1+\bar{n}_D)^{-3} \quad ,\end{aligned} \tag{5.355}$$

where $\bar{n}_D = \langle n\rangle_0\tau_D/T$ is the average number of counts in a dead time. The approximate formula gives

$$\begin{aligned}\langle n\rangle &= \langle n\rangle_0(1-\bar{n}_D)\\ \langle\Delta n^2\rangle &= \langle n\rangle_0(1-3\bar{n}_D) \quad .\end{aligned} \tag{5.356}$$

The mean is reduced by a factor of approximately $\bar{n}_D$ and the variance by a factor $3\bar{n}_D$. The normalized variance $\langle\Delta n^2\rangle/\langle n\rangle^2 \approx (1-\bar{n}_D)/\langle n\rangle_0$ is reduced by a factor $\bar{n}_D$, i.e., the distribution becomes narrower than a Poisson distribution. This is a manifestation of a debunching effect.

For *thermal light*, the approximate formula gives

$$\begin{aligned}\langle n\rangle &\approx \langle n\rangle_0\,(1-2\bar{n}_D)\\ \langle\Delta n^2\rangle &\approx \langle n\rangle_0(1-6\bar{n}_D) + \langle n\rangle_0^2\,(1-8\bar{n}_D)\\ \langle\Delta n^2\rangle/\langle n\rangle^2 &\approx (1-4\bar{n}_D) + (1-2\bar{n}_D)/\langle n\rangle_0 \quad .\end{aligned} \tag{5.357}$$

The mean is reduced by a factor $2\bar{n}_D$. The distribution narrowing is considerably more than in the Poisson case. The Bose-Einstein distribution of thermal light is originally broader than the Poisson distribution. But, with the dead-time effect, it could become narrower than a dead-time-corrected Poisson distribution. For example, if $\tau_D/T = (2+4\langle n\rangle_0)^{-1}$, both distributions lead to $\langle\Delta n^2\rangle/\langle n\rangle^2 \simeq 1/\langle n\rangle_0$. Fig.5.28 illustrates these effects.

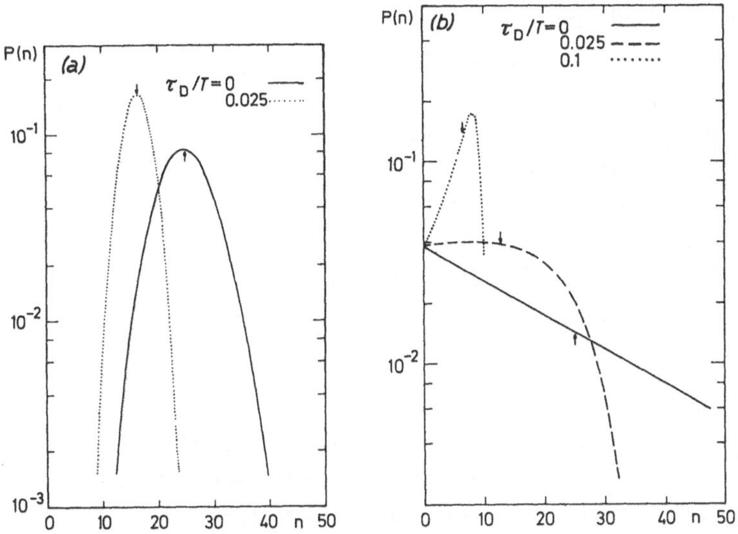

Fig.5.28a and b. Effect of the dead time on the phòtoelectron counting statistics for: a) coherent light, b) thermal light in the limit of short sampling time. The arrows indicate the means of their corresponding curves (after CANTOR and TEICH [78])

CANTOR and TEICH [78] compared the exact distributions to the approximate distribution obtained from (5.352). They note that for $\tau_D/T = 0.001$ and $\langle n \rangle_0 = 29$ the distributions cannot be distinguished graphically. For $\tau_D/T = 0.005$, the distributions are sufficiently different, particularly for count numbers greater than the mean.

These distributions were verified experimentally by JOHNSON et al. [253], who used a tungsten lamp in the limit $T \gg \tau_c$. A Poisson distribution corrected to the first order in τ_D/T appeared to fit the experimental results. CHANG et al. [86] observed the dead-time effect by using laser light above and below the threshold. SERRALLACH and ZULAUF [438] used the light scattered from particles undergoing Brownian motion and verified experimentally the effect of dead time on photoelectron statistics of thermal light with a Lorentzian spectrum. The dead-time phenomenon was observed earlier by KOLLIN [271] for ultraviolet radiation.

Joint Probability Distributions

A straightforward generalization of (5.351) gives the exact relation

$$P(n,m;T,T) = \sum_{k=0}^{n}\sum_{\ell=0}^{m} P_0(k,\ell;T'_n,T'_m) + \sum_{k=0}^{n-1}\sum_{\ell=0}^{m-1} P_0(k,\ell;T'_{n-1},T'_{m-1})$$

$$-2\sum_{k=0}^{n}\sum_{\ell=0}^{m-1} P_0(k,\ell;T'_n,T'_{m-1}), \quad n > 0, \quad m > 0 \quad , \qquad (5.358)$$

$$T'_n = T - n\tau_D \quad ,$$

where $P(n,m;T,T')$ is the JPD of obtaining n counts in interval T and m counts in interval T'. Here we assume that dead times do not overlap the samples.

A first-order approximation valid for small τ_D/T is [234]

$$P(n,m) \approx \left\langle \frac{(\lambda_1 T)^n e^{-\lambda_1 T}}{n!} \, \frac{(\lambda_2 T)^m e^{-\lambda_2 T}}{m!} \left\{1+[n(\lambda_1 T-n+1)+m(\lambda_2 T-m+1)]\tau_D/T\right\}\right\rangle \qquad (5.359)$$

or

$$P(n,m) \approx P_0(n,m)-\Big\{[n(n-1)+m(m-1)]P_0(n,m)-n(n+1)P_0(n+1,m)$$
$$-m(m+1)P_0(n,m+1)\Big\}\tau_D/T \quad . \qquad (5.360)$$

Equations (5.359) and (5.360) are generalizations of (5.352) and (5.353). The joint moments can be obtained similarly, e.g.,

$$\langle nm\rangle = \langle nm\rangle_0 - 2\langle n(n-1)m\rangle_0\tau_D/T \quad . \qquad (5.361)$$

For thermal light, this takes the form

$$\langle nm\rangle \simeq \langle n\rangle^2\left[1+(1-4\bar{n}_D)|g_V(\tau)|^2\right] \quad , \qquad (5.362)$$

where τ is the time delay between the intervals and $g_V(\tau)$ is the normalized coherence function of the optical field. Note that the second term, which contains information about the field correlation, is reduced by a factor $4\bar{n}_D$.

One final remark is appropriate. All of the equations of this subsection were based on the assumption that the dead time τ_D is constant. Effects of fluctuations of τ_D could be included by averaging over these fluctuations. This is discussed by CANTOR et al. [77].

5.4.3 Photoelectric-Current Relaxation and Photomultiplier-Gain Fluctuations

The emission of a photoelectron at a time t_i causes an instantaneous flow of an electric charge e in the anode circuit. So far, we have assumed that the charge is highly localized in time and have treated it as an impulse of infinitesimal duration. This is obviously an idealization that requires some clarification. Actually, an electron emitted at time t_i produces an electric current

$$i(t) = h(t-t_i) \quad , \quad t > t_i \quad ,$$

where h(t) is the current response function. The function h(t) has a finite width determined by the transit time between the cathode and the anode of the detector and by the anode circuit. The electric current due to a collection of photoelectron emissions is therefore given by

$$i(t) = \sum_i h(t-t_i) \quad .$$

If a photomultiplier is used, then the charge of each emitted electron is multiplied by the photomultiplier gain G_m, and

$$i(t) = \sum_i G_m h(t-t_i) \quad .$$

Two methods of processing the observed photoelectric current are possible: the analog method and the digital method.

In the *analog method*, an integrator circuit, corresponding to a broad h(t), is used to smooth the current i(t) as shown in Fig.5.29. The total electric charge accumulated by the anode is thus recorded. The observation i(t) is the datum to be used for information processing. Because the emission times $\{t_i\}$ are random, the function i(t) is a stochastic function. This is known as the shot-noise process; many aspects of its properties were discussed by PAPOULIS [358], BAR-DAVID [31], PICINBONO et al. [378], and KARP and GAGLIARDI [258].

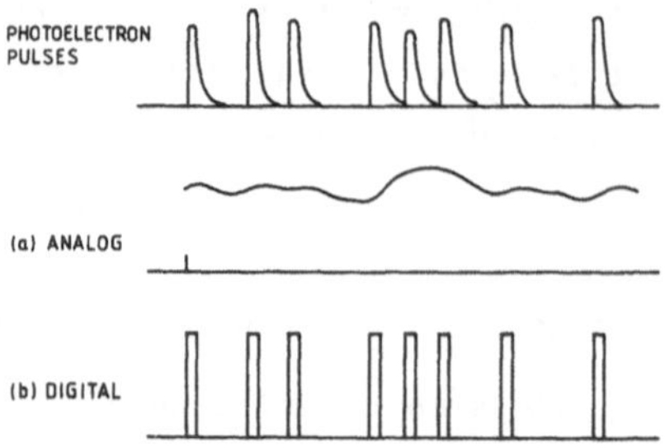

Fig.5.29. Analog and digital processing of photoelectron pulses

Additional fluctuations of the current i(t) result from fluctuation of the photomultiplier gain G_m (see, e.g., [280]) and of the shape of the function h(t). Moreover, thermal noise produced by the elements of the circuit is added to i(t).

In the *digital method*, the photoelectron pulses are used to trigger the generation of standardized very narrow electric pulses. This eliminates the effects of fluctuations of G_m and h(t) and gives a record of photoelectron emissions $\{t_i\}$. Moreover, the effect of thermal noise is considerably reduced. The only remaining source of randomness is the fluctuation of $\{t_i\}$ due to the uncertainty produced by the phenomenon of photodetection and the effect of field fluctuations.

In this book, we limit ourselves to digital methods. We are therefore not interested in the properties of i(t).

Part III

Applications

In Part II, we investigated the statistical properties of light fields of various origins and their corresponding photoelectron processes. Part III is concerned with the inverse problem-determination of the properties of an optical field from the results of some measurements on photoelectrons detected by a photodetector that intercepts the field. An optical field illuminates the surface of

I) one photodetector
or II) an array of photodetectors.

The generated photoelectron events are observed in one of various possible ways

I) counts in fixed intervals of time,
II) intervals between photoelectron events,
III) coincidence between photoelectron events,
IV) instants of occurrence of photoelectron events.

From such measurements, information about the light field is sought. Examples of the kind of information that may be of interest are

I) the average light intensity,
II) the degree of polarization,
III) the light power spectrum,
IV) the spatial coherence functions,
V) the statistical model that describes the fluctuation of the optical field.

For an estimation problem to be meaningful, some *a priori* information must be assumed. For example, we may desire an estimate of the power spectrum of an optical field, given that the fluctuations of the field are *a priori* known to be Gaussian. The profile of the power spectrum may be *a priori* known except for one or several parameters that are to be estimated. Another problem is to estimate the average intensity of an optical beam, given the *a priori* information that it may take only one of M known values. The problem is then a test of hypotheses. With the above classifications of possible observed data, possible parameters to be estimated, and possible given *a priori* information, we can form, from their combinations, quite a large number of interesting problems that have applications in different fields of science and engineering. For example, estimation of the power spectrum has obvious applications to spectroscopy. Estimation of light intensity applies to optical communication and imaging. Methods of estimating statistical parameters of a light beam are important in study of the physics of light sources.

For the purpose of classification, we divide this part of the book into two chapters. Chapters 6 and 7 contain problems pertaining mostly to optical communication and spectroscopy, respectively. The classification is not exclusive; these problems may be of interest in many other fields.

6. Applications to Optical Communication

A propagating light wave can be used to transfer information from a "transmitter" plane to a "receiver" plane. This can be accomplished by utilizing variations of its intensity, phase, frequency, or polarization. These variations could be functions of time or space or both. The best example is that of light emitted from or scattered by the surface of an object. As the light propagates, it carries information about the object to the viewer in the form of a spatial intensity distribution - an image. Along with the image, the viewer always receives unwanted light coming from other sources - the background. Because of the uncertain nature of the photodetection process, the viewer ends with a noisy replica of the image and the background. Techniques of statistical estimation and detection theory can be used very effectively to reduce the background and combat the uncertainties.

Another example is a beam of light whose intensity is made to vary with time in accordance with a signal (i.e., is modulated by the signal). At the receiver, the intensity of the beam is measured and the signal is recovered (Fig.6.1). Again, background radiation unavoidably enters the detector and noise is produced by the process of detection as well as by extraneous sources.

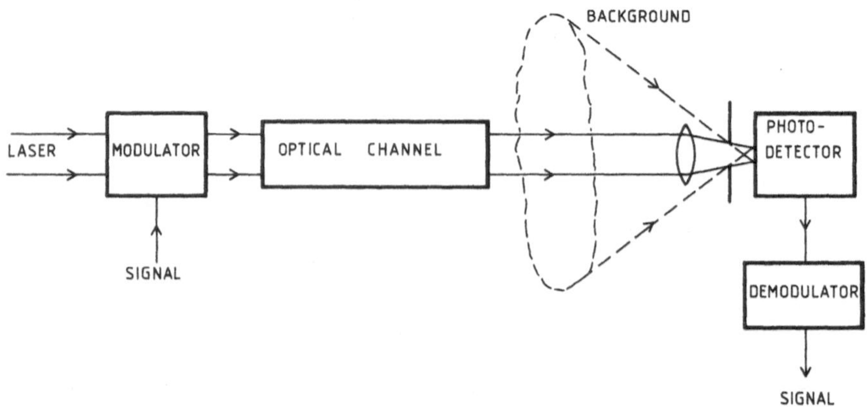

Fig.6.1. Optical communication system

In order to transmit optical information over long distances, the source has to be extremely intense (as in the case of a star) or otherwise the radiated energy should be concentrated in the form of a very narrow beam. This ensures the arrival of sufficient energy to the receiver and the identification of the information-carrying light from the ever-present background light. If the usual techniques of modulation, multiplexing, and information retrieval are to be used, and if the very wide bandwidth that is available at the frequency of light is to be efficiently utilized. the information-carrying light beam has to be coherent. Also, a carrier that has a small spectral width permits the filtering of the wideband background noise.

With the advent of the laser, it was immediately realized that such an intense, narrow, highly coherent, and narrowband beam of light could be used for communication purposes.

An optical communication system is not just a simple generalization of a radio frequency or a microwave communication system. The optical components (sources, modulators, filters, detectors, etc.) are completely different from their low-frequency counterparts. Propagation in the atmosphere at the relatively small optical wavelength makes effects such as clouds, fog, smog, haze, and even turbulence in the clear atmosphere very important. Moreover, the effect of background-diffused radiation from the sun and the stars, which fill the whole hemisphere, cannot be ignored in an optical communication system. Photon detection at optical frequencies is limited by quantum effects which must be properly accounted for if the design of the detectors and decoders is to be optimized. Efficient multiplexing systems that are to make full use of the available enormous bandwidth must be expected to be more complicated than previous systems.

All of these peculiar aspects of communication at optical frequencies have stimulated extensive research. This chapter is not intended to provide a complete description of the results of these investigations. Some books and several review articles cover many aspects of optical communication [128, 165, 219, 220, 259, 266, 387, 408].

In this chapter we examine, in general terms, the problem of estimating optical parameters and detecting optical signals from photoelectron measurements. We apply the analysis to analog and digital optical communication systems. Throughout this chapter, we study the role played by photoelectron statistics in the design and performance of basic modulation systems.

Section 6.1 is a short review of the different classifications of optical communication systems. Sections 6.2 and 6.3 present methods of estimating stationary and time-varying optical signals. They also cover applications to analog modulation systems. In Sec. 6.4 the problem of detecting the presence of an optical signal is discussed and basic digital communication systems are analyzed. Throughout Sec. 6.2-4, the receivers are assumed to be photoelectron counters. Receivers based on time measurements are considered in Sec. 6.5.

6.1 Classification of Optical Communication Systems

1) Optical communication systems can be classified according to the parameter of the optical wave that is used to carry the signal: amplitude, phase, frequency, intensity, or polarization.

According to the modulation format, or coding, communication systems are usually divided into two major classes: analog and digital. In the *analog modulation* system, a carrier field $V(t) = \mathcal{V}(t)\exp(j\omega_s t)$ can be modulated by a signal $S(t)$ in one of the following ways:

Amplitude Modulation (AM),	if	$\mathcal{V}(t) \sim 1 + S(t)$
Phase Modulation (PM),	if	$\mathcal{V}(t) \sim \exp[jS(t)]$
Frequency Modulation (FM),	if	$\mathcal{V}(t) \sim \exp[j\int_0^t S(t')dt']$
Intensity Modulation (IM),	if	$I(t) \sim \lvert\mathcal{V}(t)\rvert^2 \sim 1 + S(t)$.

Composite systems could also be used. For instance, in Intensity Modulation-PM coding (IM/PM), $I(t) \sim \cos[\omega_c t + S(t)]$. Analog polarization modulation (PL) can be obtained by using linear polarization and taking the angle of polarization proportional to $S(t)$. Alternatively, the ratio between the intensity of the right-to-left circularly polarized components in the carrier wave could be taken proportional to $S(t)$. A pulsed wave of light may also be used as a carrier. The amplitude, duration, or position of the pulse could be made proportional to the signal, in which case we obtain Pulse-Amplitude Modulation (PAM), Pulse-Duration Modulation (PDM), and Pulse-Position Modulation (PPM).

In all analog modulation systems, the receiver is to recover the signal $S(t)$. Methods of estimation theory can be used to determine the optimum estimators and assess their accuracies.

In the *digital systems*, the signal is sampled at equal intervals of time. The samples are then quantized, and coded into binary or M-ary codes.

Binary Coding. During each interval, the transmitter sends one of two possible symbols denoted bit 1 and bit 0. The receiver is to decide which symbol was sent, by testing both hypotheses. Examples are:

Pulse-Gated Binary Modulation (PGBM) or On-Off Keying. Bit 1 is represented by the presence of a pulse in the time interval. Bit 0 is represented by its absence.

Pulse-Delay Binary Modulation (PDBM) or Binary Pulse-Position Modulation. The time interval is divided into two times slots. A pulse occupies one of these slots. Bit 1 is represented by the pulse that occupies the first slot. Bit 0 is represented by the pulse delayed to the second slot (Fig.6.2).

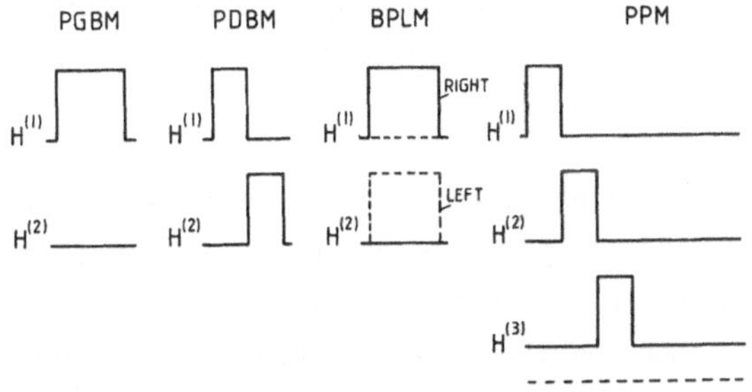

Fig.6.2. Digital modulation formats

Phase-Shift Keying (PSK). The phase of the wave is switched between two states; 0 and π corresponding to bits 1 and 0, respectively.

Pulse-Polarization Modulation (BPLM). The two states are: vertically polarized wave for bit 0, and horizontally polarized wave for bit 1. Right-handed and left-handed circularly polarized waves could equally well be used to denote bit 1 and 0, respectively.

M-ary Coding. In M-ary coding, M symbols are used. In each interval of time, one of these M symbols is transmitted and should be identified by the receiver. Examples are:

Multilevel-Pulse-Amplitude Modulation (PAM) in which case M levels of the pulse amplitude are used to denote the M symbols.

Pulse-Position Modulation (PPM), which uses the location of a pulse in one of M possible time intervals to denote the M symbols, as shown in Fig.6.2.

2) Optical communication systems are also classified according to whether the detection is *direct* or *heterodyne*. In heterodyne detection, the optical signal is mixed with a strong coherent reference beam before it is detected. This permits the recovery of phase and frequency information.

3) Optical receivers are classified according to the measurement they perform. *Photoelectron-counting receivers* measure the number of photoelectron counts in a sequence of time intervals and base all decisions (estimations or tests of hypotheses) on these counts. A more complex receiver measures the *instant of occurrence of photoelectrons*. A much simpler receiver allows the detected electrons to pass through an electric filter and measures the *photoelectric current*.

4) A final aspect of classifying optical communication systems is the photoelectron statistics corresponding to the signal and the background. This depends on the statistical nature of the fields, on the time constant of the post-detection filter (or the counting time in a photoelectron-counting receiver), and on the aperture area. The signal statistics depend on the characteristics of the propagation channel. For example, propagation in a turbulent atmosphere causes a drastic change of the statistics of the signal.

6.2 Estimation of Optical Signals. Direct Detection

In this section, we study the estimation of optical signals from observed photoelectron counts in fixed time intervals. Unlike many estimation problems in classical communication theory, the observation cannot, in general, be written as a sum of a signal and an additive signal-independent stationary noise. Because of the nature of the photodetection phenomenon, the uncertainty of the observation depends on the signal level. This significant feature must be borne in mind when conventional traditions of classical communication theory are applied to optical communication problems. For example, in many cases,that involve photodetection, the concept of signal-to-noise ratio becomes less useful, ambiguous, and sometimes even misleading as a criterion of performance [154, 155, 236, 439].

6.2.1 Estimation of the Intensity of a Stationary Optical Field

This subsection deals with measuring light intensity by use of photoelectron counters. In the past decade, high-speed photoelectron-counting techniques have come into wide use in many applications, such as recording light from stellar and quasi-stellar objects in astronomy, spectroscopy of faint objects, laser microprobe analyzers, and microfluorimetry, to name a few.

Suppose that we are allowed to take measurements in a total time interval $[0,T_0]$ during which the light intensity is assumed to be constant. A photodetector is used to intercept the field, and the number of photoelectrons $[n] = n_1, n_2, \ldots n_N$ emitted in N equal subintervals, each of width T, are recorded. If we obtain an estimate for the average rate of photoelectrons per second λ, then we can determine the light intensity in W cm^{-2} by multiplying this estimate by $h\nu/\eta A$, where A is the area of the aperture of the detector, η is its quantum efficiency, and $h\nu$ is the quantum of light energy. Given [n], what would be the optimum estimate for λ?

The simplest estimate is obviously the sample mean

$$\hat{\lambda} = \hat{n}/T \ , \quad \hat{n} = \sum_{i=1}^{N} n_i/N \ . \tag{6.1}$$

This is cleary an unbiased estimate ($\langle\hat{\lambda}\rangle = \lambda$). The maximum likelihood (ML) estimate can be obtained by determining the value of λ that maximizes P([n]). This depends on the statistical nature of the light, as is shown in the following examples. In many cases, it turns out that the ML estimate is itself the sample mean.

Coherent Light

In this case, $(n_1, n_2, \ldots n_N)$ are statistically independent Poisson-distributed variables with means λT. Here, the JPD

$$P([n]) = \prod_i (\lambda T)^{n_i} \exp(-\lambda T)/n_i! \tag{6.2}$$

is maximized by the sample mean, (6.1). The estimation error is easily shown to be

$$e_\lambda^2 = \mathrm{Var}(\hat{\lambda})/\lambda^2 = \frac{1}{N}\frac{1}{\lambda T} = \frac{1}{N}\frac{1}{\bar{n}} \ , \quad \bar{n} = \lambda T = \langle\hat{n}\rangle \ . \tag{6.3}$$

The normalized square error is therefore equal to the inverse of the average number of detected photoelectrons in the observation time $T_0 = NT$. This is the lowest possible error in an experiment that involves optical detection. It represents the uncertainty due to the nature of optical detection and is sometimes referred to as the shot-noise error. Other uncertainties usually add to this error. These will be discussed in subsequent sections.

Thermal Light

For simplicity, we assume that the samples are taken such that $(n_1, n_2, \ldots n_N)$ are statistically independent. We can then use the approximate formula for the photoelectron-counting statistics of thermal light (Sec. 5.2.2) and write

$$P([n]) = \prod_i \binom{n_i+\mathcal{N}-1}{n_i} (1+\lambda T/\mathcal{N})^{-\mathcal{N}}(1+\mathcal{N}/\lambda T)^{-n_i} \quad , \tag{6.4}$$

where $\mathcal{N}$ is the number of modes collected by the detector in the time interval T. We recall that this expression is exact when the counting time and detector area are either very small or very large compared to the coherence time and coherence area, respectively. The ML estimate of λ is the value that maximizes the right-hand side of (6.4). This is again the sample mean of (6.1).

The corresponding error can be obtained by use of (5.10),

$$\mathrm{Var}(\hat{\lambda}) = \frac{1}{T^2}\,\mathrm{Var}(\hat{n}) = \frac{1}{T^2}\,\frac{1}{N}\,\mathrm{Var}(n_i) = \frac{1}{T^2 N}(\bar{n}+\bar{n}^2/\mathcal{N}) \quad ,$$

from which

$$e_\lambda^2 = \frac{1}{N}\left(\frac{1}{\lambda T} + \frac{1}{\mathcal{N}}\right) \tag{6.5}$$

Comparison with (6.3) shows that an additional source of uncertainty is introduced because of the fluctuations in the field. If λT is very small or $\mathcal{N}$ is very large, this additional term is negligible. This occurs if the average number of counts per coherence term per coherence area is small.

An exact expression for e_λ^2, in the case of light with a Lorentzian spectrum (of width Γ) and a small detector, is [229]

$$e_\lambda^2 = \frac{1}{N}\left[\frac{1}{\lambda T} + \frac{1}{N} - \frac{1}{N^2}\frac{1}{N}\left(1 - e^{-2NN}\right)\right] , \qquad N = \gamma = \Gamma T \quad . \tag{6.6}$$

Obviously if the sample size N is sufficiently large, (6.5) is reproduced.

Coherent Light in Thermal Background

Here we consider a coherent light signal with an unknown intensity λ_s (counts per second) mixed with thermal background whose intensity λ_b is assumed to be known exactly from a separate long measurement. Based on the measured number of counts $n_1, n_2, \ldots n_N$, assumed independent, we seek an estimate for λ_s.

If the background light is wideband or if many of its modes are collected by the detector, then [n] approximately have Poisson distributions.

$$P([n]) = \prod_{i=1}^{N} (\lambda_s T + \lambda_b T)^{n_i} \exp(-\lambda_s T - \lambda_b T)/n_i! \quad .$$

The value of λ_s that maximizes the right-hand side is

$$\hat{\lambda}_s = \frac{1}{T}\frac{1}{N}\sum_i n_i - \lambda_b \quad . \tag{6.7}$$

This is the ML estimate. When the number of temporal and spatial modes is not very large we use the Laguerre statistics (cf. (5.165))

$$P([n]) = \prod_i \frac{(\lambda_b T/\mathcal{N})^{n_i}}{(1+\lambda_b T/\mathcal{N})^{n_i+\mathcal{N}}} \exp(-y)\, L_{n_i}^{\mathcal{N}-1}(-x)$$

$$y = \lambda_s T/(1+\lambda_b T/\mathcal{N}) \quad , \quad x = y/(\lambda_b T/\mathcal{N}) \quad , \tag{6.8}$$

where $\mathcal{N}$ is the number of modes in the mixture, as discussed in Sec. 5.2.4, page 221. Taking the logarithm and equating its derivative with respect to λ_s to zero, we obtain the nonlinear equation

$$\sum_i L_{n_i-1}^{\mathcal{N}}(-x)/L_{n_i}^{\mathcal{N}-1}(-x) = \lambda_b T/\mathcal{N} \quad . \tag{6.9}$$

A solution of (6.9) gives the ML estimate. This is of course rather complicated and the estimation accuracy is seemingly impossible to determine. In the special case when $\lambda_b T \ll \min[\lambda_s T, 1]$, $x \gg 1$ and we can approximate the Laguerre polynomials in (6.9) by the last terms of their expansion

$$L_n^{\alpha}(-x) = \sum_{m=0}^{n} \binom{n+\alpha}{n-m} \frac{x^m}{m!} \simeq \frac{x^n}{n!} \quad .$$

Equation (6.9) then yields

$$\hat{\lambda}_s = \frac{1}{NT} \sum_i n_i \quad .$$

If we retain the last two terms, (6.9) gives

$$\hat{\lambda}_s = \frac{1}{NT}\left[1-(1-\frac{1}{N})\frac{\lambda_b}{\hat{\lambda}_s}\right]\sum_i n_i - \frac{1}{NT}\frac{\lambda_b}{\hat{\lambda}_s N}\sum_i n_i^2 \quad . \tag{6.10}$$

We therefore reach the conclusion that the first and second moments of the sample n_i are necessary for the determination of $\hat{\lambda}_s$. If N is large, we reproduce the estimator in (6.7).

The error involved in this estimation is

$$e_{\lambda_s}^2 = \frac{1}{N}\left[\frac{1}{\lambda_s T}\left(1+\frac{\lambda_b}{\lambda_s}\right)+\frac{1}{N}\left(\frac{2\lambda_b}{\lambda_s}+\frac{\lambda_b^2}{\lambda_s^2}\right)\right] \quad . \tag{6.11}$$

The first and second terms are due to the uncertainties introduced by the detection process. The fourth term is due to fluctuations in the thermal background and the third term is a result of the mixing between the coherent signal and fluctuations in the thermal background.

If some a priori information about the intensity level λ_s is available, then we can also obtain a MAP estimator. For example, if λ_s has a Gaussian a priori probability distribution with mean $\bar{\lambda}_s$ and variance σ^2, we can substitute in the MAP equation (2.147) and obtain, in the case of large N (Poisson statistics),

$$\hat{\lambda}_s = \bar{\lambda}_s + \sigma^2 NT\left[\frac{1}{NT}\sum n_i - (\hat{\lambda}_s + \lambda_b)\right] \Big/ (\hat{\lambda}_s + \lambda_b) \quad .$$

This shows how the a priori information is corrected by the experimental measurement.

In the strong-signal limit,

$$\hat{\lambda}_s = \bar{\lambda}_s + \sigma^2 \sum_i \left[(n_i - \hat{\lambda}_s T)/\hat{\lambda}_s\right] .$$

This is the same result that we obtain when we use Laguerre statistics, (6.8), in the MAP equation and apply the strong-signal approximation.

MAP estimators with other a priori distributions are treated in [177].

Coherent Light in Thermal Background of Unknown Intensity

The intensity of a coherent light field is to be estimated in the presence of a thermal background.

Whenever the signal can be modulated or chopped, a separate observation of the background can be obtained. A reference signal from the modulator or chopper initiates two counting periods per modulation cycle; T when the signal is present, and T' when it is absent. An estimate of the signal λ_s can be obtained by subtraction, at the end of a predetermined number of modulation cycles N,

$$\hat{\lambda}_s = \frac{1}{NT} \sum_i n_i - \frac{1}{NT'} \sum_i n_i' \quad , \tag{6.12}$$

where n_i and n_i' are the number of photocounts during T and T', respectively.

This estimation is obviously unbiased. The normalized error is given by

$$e^2_{\lambda_s} = \frac{1}{N}\left[\frac{1}{\lambda_s T}\left(1 + \frac{\lambda_b}{\lambda_s}\right) + \frac{1}{\mathcal{N}}\left(\frac{2\lambda_b}{\lambda_s} + \frac{\lambda_b^2}{\lambda_s^2}\right) + \left(\frac{1}{\lambda_b T'} + \frac{1}{\mathcal{N}'}\right)\frac{\lambda_b^2}{\lambda_s^2}\right] \quad , \tag{6.13}$$

where $\mathcal{N}$ and $\mathcal{N}'$ are the number of modes of the thermal background in the intervals T and T', respectively. Note that the uncertainty in measuring λ_b is added. If $T, T' \gg \tau_c$, then $\mathcal{N}$ and $\mathcal{N}'$ are very large and (6.13) becomes

$$e^2_{\lambda_s} = \frac{1}{N}\left[\frac{1}{\lambda_s T}\left(1 + \frac{\lambda_b}{\lambda_s}\right) + \frac{1}{\lambda_s T'}\frac{\lambda_b}{\lambda_s}\right] . \tag{6.14}$$

For $\lambda_s = \lambda_b = 1$ count s^{-1} and $NT = NT' = 10^4$ s, the estimated normalized error is about 1.7 %. An optimum duty cycle $T/(T+T')$ for the estimation of λ_s can be obtained by minimizing the relative error e_{λ_s} with respect to T and T' under the condition that T+T' is constant. This leads to

$$T/T' = [\lambda_b/(\lambda_s+\lambda_b)]^{\frac{1}{2}} \quad . \tag{6.15}$$

Thermal Light in Thermal Background of Unknown Intensity

In this case, both the signal and the background are thermal. The background can be viewed separately. An example is a telescope viewing a star plus a small amount of night-sky background. Two apertures could be used, one viewing the star plus the background, and the other, only an equal area of adjacent sky background. A chopper switches back and forth between the two apertures. Another example is an experiment in infrared radiometry, in which synchronous detection at some chopping frequency can be used. During each cycle, the signal is observed for a time T and the background for a time T'. The estimation (6.12) has a normalized error

$$e^2_{\lambda_s} = \frac{1}{N}\left[\left(\frac{1}{\lambda_s T} + \frac{1}{N}\right) + \left(\frac{\lambda_b}{\lambda_s}\right)^2\left(\frac{1}{\lambda_b T} + \frac{1}{N_b} + \frac{1}{\lambda_b T'} + \frac{1}{N'_b}\right)\right] \quad , \tag{6.16}$$

where N and N_b are the number of modes in the signal and background in the time interval T, and N'_b is the number of modes in the background in the time interval T'.

Note that (6.16) is the same as (6.13) except for the additional error due to fluctuations of the signal. If we assume that the background radiation has a wide bandwidth and/or a very small coherence area, then N_b and N'_b are very large and (6.16) reduces to

$$e^2_{\lambda_s} = \frac{1}{N}\,(\lambda_s T+2\lambda_b T+\lambda_s^2 T^2/N)/(\lambda_s T)^2 \quad , \tag{6.17}$$

where we assumed, for simplicity, that T = T'.

Two limits are of interest:

In the *weak-signal limit* the error is dominated by the background

$$e^2_{\lambda_s} = \frac{2}{NT}\left(\lambda_b/\lambda_s^2\right) = \frac{2}{T_0}\,\lambda_b/\lambda_s^2 \quad . \tag{6.18}$$

The error depends on the total observation time $T_0 = NT$ and not on N and T separately. Any sampling scheme would yield the same accuracy.

The other *limit of strong signal* corresponds to

$$e_{\lambda_s}^2 = \frac{1}{NN} \quad , \tag{6.19}$$

which is equal to the inverse of the total number of modes observed in the time T_0. But N is the product of N_s and N_t, the number of spatial and temporal modes. Because $N_t \simeq 1$ if $T < \tau_c$, and $N_t \simeq T/\tau_c$ if $T > \tau_c$, we can write

$$e_{\lambda_s}^2 = \frac{\tau_c}{T_0} \frac{1}{N_s} \quad , \qquad T > \tau_c \quad ,$$

$$= \frac{T}{T_0} \frac{1}{N_s} \quad , \qquad T < \tau_c \quad . \tag{6.20}$$

This may seem to imply that, in the strong-signal limit, the accuracy can be unlimitedly improved by taking a shorter sampling time T. This is not so, because as T becomes too short, the assumption of statistically independent samples no longer holds.

The shortest time T that avoids correlation between the signal samples may be taken as the time that makes the separation between two signal samples equal to a coherence time, i.e., $T = \tau_c/2$. This corresponds to an error

$$e_{\lambda_s}^2 = \frac{\tau_c}{2T_0} \frac{1}{N_s} \quad , \qquad T = \tau_c/2 \quad . \tag{6.21}$$

Sometimes the coherence time is so short that it becomes impossible to take $T = \tau_c/2$. In this case, $T > \tau_c$ and the error becomes a factor of $\sqrt{2}$ larger

$$e_{\lambda_s}^2 = \frac{\tau_c}{T_0} \frac{1}{N_s} \quad , \qquad T > \tau_c \quad . \tag{6.22}$$

Minimum Detectable Signal

The minimum signal that is detectable in a total observation time of 1 s could be defined as the signal that corresponds to a 100 % relative error. Putting $e_{\lambda_s} = 1$ and $T_0 = 1/2$s in the weak-signal-limit expression, (6.18), we obtain

$$\lambda_s|_{min} = 2\sqrt{\lambda_b} \quad [\text{counts } s^{-1}] \quad .$$

This corresponds to an intensity of

$$I_s|_{min} = (h\nu/\eta A)2\sqrt{\lambda_b} \quad \text{W cm}^{-2} \quad . \tag{6.23}$$

If the background λ_b is limited by a dark current that corresponds to r_d counts cm^{-2}, then we can write $\lambda_b = r_d A_d$, where A_d is the area of the detector's cathode. Therefore, the minimum detectable intensity can be written in the form

$$I_s|_{min} = 2\sqrt{A_d}/AD^* \quad , \tag{6.24}$$

where $D^* = \eta/h\nu\sqrt{r_d}$ is the detectivity of the detector material [254].

A numerical example illustrates the order of magnitude of the quantities involved: If $A_d = 10^{-4} cm^2$, $A = 100\ cm^2$, $D^* = 10^{10}$, the minimum detectable intensity in 1 s is $2 \cdot 10^{-14}$ W cm^{-2}.

Separation of the Intensities of a Mixture of Two Optical Fields with Different Correlation-Time Scales

In the cases previously described, we assumed that the intensity of the background radiation is known or could be obtained from a separate measurement. If such a separate experiment is not physically possible, then the only alternative for estimating the signal intensity is to distinguish between the two fields according to their scale of coherence times. This requires a measurement of the photoelectron-counting correlation.

Because the signal and background are statistically independent, we can write

$$\bar{n} = \bar{n}_s + \bar{n}_b \tag{6.25}$$

and

$$\bar{n}^2 g_n(\tau) = \bar{n}_s^2 g_{I_s}(\tau) + \bar{n}_b^2 g_{I_b}(\tau) + 2\bar{n}_s\bar{n}_b \, \mathrm{Re}\left[g_{V_s}(\tau) g_{V_b}^*(\tau)\right] \quad , \tag{6.26}$$

where $g_{I_s}(\tau)$ and $g_{I_b}(\tau)$ are the normalized intensity correlation functions of the signal and background, respectively, $g_{V_s}(\tau)$ and $g_{V_b}(\tau)$ are the field correlation functions of the signal and background, respectively, and $g_n(\tau)$ is the photoelectron-counting correlation function. Eq. (6.26) is valid independently of the statistics of the signal and background, provided that the detector is small ($A \ll A_c$) and that the counting interval is short ($T \ll \tau_c$).

If the signal and background are thermal, then (6.26) becomes

$$\bar{n}^2[g_n(\tau)-1] = \bar{n}_s^2|g_{V_s}(\tau)|^2 + \bar{n}_b^2|g_{V_b}(\tau)|^2 + 2\,\bar{n}_s\bar{n}_b\,\mathrm{Re}\left[g_{V_s}(\tau)g_{V_b}^*(\tau) -1\right] \ . \qquad (6.27)$$

If the widths of the functions $|g_{V_s}(\tau)|$ and $|g_{V_b}(\tau)|$ are a priori known and are different, then we can presumably choose a time delay τ such that one of them, g_{V_b}, is equal to 1.0 and the other, g_{V_s}, is zero. In this ideal case, (6.27) becomes

$$\bar{n}^2[g_n(\tau) -1] = \bar{n}_s^2-2\bar{n}_b\bar{n}_s$$

which can be directly used together with (6.25) to determine $\bar{n}_s$ and $\bar{n}_b$.

If such a choice of τ is not possible, then a solution of (6.25) and (6.27) necessitates extra information on the functional forms of the correlation functions. DAVIDSON and IYER [120] who suggested this technique examined it in the more general situation when T is not necessarily smaller than τ_c and demonstrated experimentally the feasibility of the technique.

6.2.2 Estimation of an Optical Signal with a Time-Varying Intensity

Coherent Signal. Intensity Modulation (IM)

Let a coherent optical field with a time-varying intensity $\lambda(t)$ be observed with a photoelectron counter that can be gated to record the number of counts in fixed intervals of time. An estimate of $\lambda(t)$ as a function of time is sought.

This is the problem to be solved in order to demodulate an intensity-modulated coherent light. The problem also arises in other fields. For example, if an array of detectors is used to detect a time-varying image, then each detector has to estimate the intensity as a function of time at its location. If one detector is to scan a fixed image, then a two-dimensional generalization of the preceding problem serves for the estimation of the image. This has important applications in medicine [30, 360, 361].

The simplest approach to this problem is to observe the number of photoelectron counts n_i in the intervals $(t_i - T/2, t_i + T/2)$, $i = 1,2, \ldots$ and to determine an estimate of $\lambda(t_i)$ by using

$$\hat{\lambda}(t_i) = n_i/T \quad . \tag{6.28}$$

Because the signal varies during the interval T, this estimator is not unbiased

$$\langle\hat{\lambda}(t_i)\rangle = \frac{1}{T}\int_{t_i-T/2}^{t_i+T/2} \lambda(t')dt \neq \lambda(t_i) \quad .$$

The bias,

$$B = \langle\hat{\lambda}(t_i)\rangle - \lambda(t_i) = \frac{1}{T}\int_{t_i-T/2}^{t_i+T/2} [\lambda(t')-\lambda(t_i)]dt' \quad , \tag{6.29}$$

could be reduced by choosing a shorter sampling interval T. But a shorter T corresponds to a smaller average number of counts per interval and therefore to a higher relative variance of the estimator. A compromise between these two conflicting factors could be obtained by choosing a T that minimizes the mean square error

$$\epsilon^2_{\lambda(t)} = \langle[\hat{\lambda}(t)-\lambda(t)]^2\rangle/\lambda^2(t) = e^2_{\lambda(t)} + B^2/\lambda^2(t) \quad , \tag{6.30}$$

where

$$e^2_{\lambda(t)} = \mathrm{Var}[\hat{\lambda}(t)]/\lambda^2(t) = \frac{1}{T^2\lambda^2(t)}\int_{t_i-T/2}^{t_i+T/2} \lambda(t')dt' \quad . \tag{6.31}$$

PAPOULIS [361] found that if $\lambda(t)$ is sufficiently smooth in the interval T, then the optimum counting time is given by

$$T \simeq 2 \left\{ 9\lambda(t)/2 \left[\frac{\partial^2\lambda(t)}{\partial t^2}\right]^2 \right\}^{1/5} , \tag{6.32}$$

which corresponds to

$$\epsilon^2_{\lambda(t)} \simeq \frac{5}{4} \frac{1}{\lambda(t)T} . \tag{6.33}$$

This shows that the biasing effect has increased the error by about 25 %, compared to the error that results from constant intensity.

Note that the optimum sampling time is a function of t and therefore requires an adaptive system that follows variations of the signal and adjusts the sampling time accordingly.

A simpler approach is to sample the signal at its Nyquist rate, i.e., by taking

$$T = 1/\Gamma_s \tag{6.34}$$

where $\Gamma_s/2$ is the highest frequency in the signal $\lambda(t)$ (assumed to be band-limited). According to the criterion of minimum mean square error, this choice is not necessarily optimum and it corresponds to errors greater than those given by (6.33).

An alternative to estimating samples of the signal $\lambda(t)$ at times t_i is to express it as a sum of orthogonal functions and to use the observed $\{n_i\}$ to find an ML estimate of the coefficients of the expression. For this purpose, BIJAOUI [57] uses the Walsh-Hadamard transformation.

A generalized mathematical treatment of the problem is given by BROWN [70] and GRANDELL [185].

Coherent Signal in Thermal Background

Here we have a time-varying coherent signal $\lambda_s(t)$ in a stationary thermal background of known intensity λ_b. The simplest way of estimating $\lambda_s(t)$ from photoelectron counts is to sample at the Nyquist rate and use the recorded counts $n_1, n_2, \ldots$ to obtain the estimator

$$\hat{\lambda}_s(t_i) = n_i/T - \lambda_b .$$

If we assume that the signal is approximately constant during the interval T and neglect the bias term, we can use the results of Sec. 6.2.1, page 291 and write

$$e^2_{\lambda_s(t_i)} = \frac{1}{\lambda_s(t_i)T}\left[1 + \frac{\lambda_b}{\lambda_s(t_i)}\right] + \frac{1}{N}\left[\frac{2\lambda_b}{\lambda_s(t_i)} + \frac{\lambda_b^2}{\lambda_s^2(t_i)}\right] \quad . \tag{6.35}$$

If we assume that the samples $n_1, n_2, \ldots$ are statistically independent, then we can also use the more complicated ML estimate described in Sec. 6.2.1, page 291.

Note that if the bandwidth of the background is much wider than that of the signal, then sampling at the Nyquist rate would ensure the statistical independence of the samples. On the other hand, if a filter is used to limit the background to the bandwidth of the signal, we can assume that the filtered background becomes rectangular; then, its fluctuations at two times separated by $T = 1/\Gamma_s$ are uncorrelated (see 5.70)). This means that the samples $n_1, n_2, \ldots$ would be approximately uncorrelated.

Estimation of Parameters of an Intensity Distribution

Sometimes the intensity of the optical signal is a known function of time with one or several unknown parameters $[\theta]$,

$$\lambda_s(t) = \lambda_s(t,[\theta]) \quad , \quad [\theta] = \theta_1, \theta_2, \ldots \theta_S \quad .$$

From the measurement of the number of counts $[n] = n_1, n_2, \ldots n_N$ in N time intervals each of width T, how should one find an estimate for $[\theta]$?

The simplest approach is to use $[n]$ to find estimates for $\lambda(t_i)$ as is mentioned above. Knowing the expected errors $e_{\lambda(t_i)}$, a fitting technique can be used to determine $[\theta]$, as is discussed in Sec. 2.7.2, page 55 .

If the a priori probabilities of $[\theta]$ are known, then a MAP estimator can be used (Sec. 2.7.2). In the case of a coherent signal in thermal background, if we assume that the counts are statistically independent, the MAP equation (2.147) gives [257]

$$\sum_i \left(\frac{1}{\lambda_s T + \lambda_b T}\, n_i - 1\right)\frac{\partial \lambda_s}{\partial \theta_\ell} + \frac{\partial}{\partial \theta_\ell} \ln p([\theta]) = 0 \qquad \ell = 1, \ldots S \quad , \tag{6.36}$$

when N is very large, and

$$\frac{T}{1+\lambda_b T/N} \sum_i \left[\frac{1}{\lambda_b T/N} L_{n_i-1}^{N}(-x_i)/L_{n_i}^{N-1}(-x_i)-1\right] \frac{\partial\lambda_s}{\partial\theta_\ell} + \frac{\partial}{\partial\theta_\ell} \ln P([\theta]) = 0 \quad ,$$

$$x_i = \lambda_s(t_i,[\theta])T/[(1+\lambda_b T/N)\lambda_b T/N] \quad ,$$

in general.

A difficult search procedure should lead to the values of $[\theta]$ that satisfy (6.36). We illustrate this by one example. Let

$$\lambda_s(t,\theta) = a + b\cos(\omega t + \theta) \quad , \quad b < 1 \quad , \tag{6.37}$$

where a, ω, and b are known. Find a MAP estimator for θ using a Gaussian a priori distribution with zero mean and variance σ^2. Assume that N is very large.

By simple substitution in (6.36), we obtain

$$\frac{\hat{\theta}}{\sigma^2} = \frac{1}{T}\sum_i n_i \left[\frac{b\sin(\omega t_i+\hat{\theta})}{a+\lambda_b+b\cos(\omega t_i+\theta)}\right] - \sum_i b\sin(\omega t_i+\theta) \quad . \tag{6.38}$$

The sampling time T is assumed much smaller than a period $2\pi/\omega$. Therefore, if the number of samples N is sufficiently large, the second term in (6.38) will vanish. The parameter $\hat{\theta}$ can then be obtained by solving the nonlinear equation

$$\frac{\hat{\theta}}{\sigma^2} = \sum_i n_i h_i(\hat{\theta}) \quad , \quad h_i(\theta) = \frac{1}{T}\,\frac{b\sin(\omega t_i+\theta)}{a+\lambda_b+b\cos(\omega t_i+\theta)}$$

A solution can be obtained by iteration of the successive approximation,

$$\frac{\hat{\theta}^{(k)}}{\sigma^2} = \sum_i n_i h_i\left[\hat{\theta}^{(k-1)}\right] \quad , \qquad k = 1,2,\ \ldots \quad , \quad \hat{\theta}^{(0)} = 0 \quad .$$

This can be performed on a digital computer, or on an analog computer by using a correlator and a feedback circuit (phase-lock loop).

Equation (6.37) represents the intensity of an intensity-modulated PM-coded analog modulation system (IM/PM). It also represents the intensity of a phase-modulated signal (PM) detected by a heterodyne receiver, as will be discussed in the next section.

6.3 Estimation of Optical Signals. Heterodyne Detection

6.3.1 Estimation of a Coherent Time-Varying Optical Signal. AM, IM, FM, and PM Systems

Assume that a coherent optical signal $V_s(t) = V_s(t)\exp(j\omega_s t)$ is mixed with another constant coherent signal, which we call the *local oscillator* $V_0 = V_0\exp(j\omega_s t)$ (see Fig.6.3). The resultant intensity

$$I(t) = I_0+I_s(t) + \left\{V_0^* V_s(t)\exp[-j(\omega_0-\omega_s)t] + \text{c.c.}\right\}$$

is the sum of the two intensities plus an interference term, the beat between the two fields. This sum can be written in the form

$$I(t) = I_0+I_s(t)+2\, I_0^{\frac{1}{2}}\, I_s^{\frac{1}{2}}(t)\, \cos[\omega_i t+\theta_s(t)-\theta_0]$$

where $\omega_i = \omega_0 - \omega_s$ is the beating frequency and θ_0 and $\theta_s(t)$ are the phases of V_0 and $V_s(t)$, respectively.

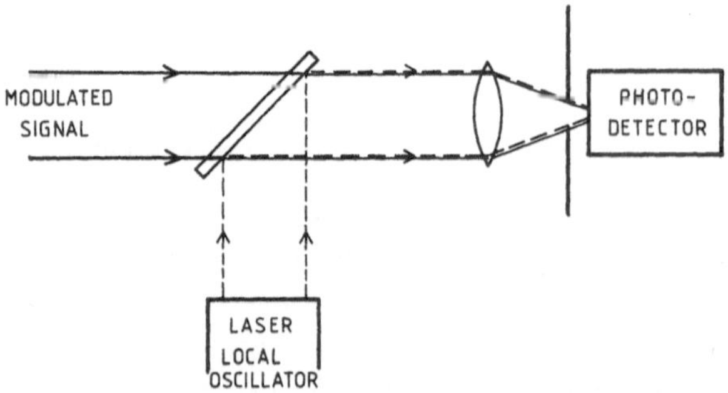

Fig.6.3. Heterodyne detection

If the detector area is sufficiently small, then the received photoelectrons follow an inhomogenous Poisson process with rate

$$\lambda(t) = \lambda_0+\lambda_s(t)+2\lambda_0^{\frac{1}{2}}\lambda_s^{\frac{1}{2}}(t)\cos[\omega_i t+\theta_s(t)-\theta_0] \quad , \tag{6.39}$$

where λ_0 and λ_s are proportional to I_0 and I_s, respectively. The function $\lambda(t)$ can be estimated from measured photoelectron counts in fixed short time intervals, as discussed in Sec. 6.2.2. We can therefore recover the magnitude of the signal $V_s(t)$, its phase, and its frequency. Addition of the reference field (the local oscillator) has enabled us to record the signal's phase and frequency, which would have been otherwise lost. This is one advantage of this technique of *heterodyning* or mixing with a coherent reference beam. The technique enables us to demodulate PM and FM as well as AM and IM signals. If $\omega_0 = \omega_s$, i.e., $\omega_i = 0$, the technique is called *homodyning*.

A second advantage (actually the main advantage) of heterodyning is that it enables us to reduce considerably the effect of background radiation. With thermal background radiation added to the mixture, (6.39) becomes

$$\lambda(t) = \lambda_0+\lambda_s(t)+\lambda_b+2\lambda_0^{\frac{1}{2}}\lambda_s^{\frac{1}{2}}(t)\cos[\omega_i t+\theta_s(t)-\theta_0] \quad . \tag{6.40}$$

Note that the thermal-background light does not produce additional interference terms with our original coherent fields.

Now if λ_0 is made much larger than λ_b and $\lambda_s(t)$, we can write

$$\lambda(t) \simeq \lambda_0+2\lambda_0^{\frac{1}{2}}\lambda_s^{\frac{1}{2}}(t)\cos[\omega_i t+\theta_s(t)-\theta_0] \quad . \tag{6.41}$$

The effect of background is neglected, whereas the signal is represented by its interference term with the strong local oscillator.

Now we turn to the accuracy of estimating optical signals using the heterodyne method. Consider the case of IM modulation and assume $\omega_i = 0$ (homodyne case). Eq. (6.41) becomes

$$\lambda(t) \simeq \lambda_0 + 2\lambda_0^{\frac{1}{2}}\lambda_s^{\frac{1}{2}}(t) \quad . \tag{6.42}$$

We measure the number of counts $n(t)$ in short time intervals of duration T and estimate λ by $\hat{\lambda}(t) = n(t)/T$. From $\hat{\lambda}(t)$, we use (6.42) to calculate $\hat{\lambda}_s(t)$. The estimator $\hat{\lambda}(t)$ has the uncertainty

$$\mathrm{Var}[\hat{\lambda}(t)] \simeq \lambda_0/T \quad ,$$

where the contribution of terms other than the strong local oscillator have been neglected.

This causes an uncertainty of $\hat{\lambda}_s(t)$ calculated from

$$\mathrm{Var}[\hat{\lambda}_s(t)] = \mathrm{Var}[\hat{\lambda}(t)] \left(\frac{\partial\lambda}{\partial\lambda_s}\right)^{-2} = \lambda_s/T \quad , \tag{6.43}$$

from which the normalized square error is

$$e^2_{\lambda_s(t)} = 1/\lambda_s T \quad . \tag{6.44}$$

Apparently, this is the same normalized error as that which would have been obtained if $\lambda_s(t)$ were to be measured by direct detection in the absence of background effects. The minimum signal detectable in 1 s corresponds to one photoelectron or $(h\nu/\eta A)$ W cm^{-2}. Although no improvement of the accuracy seems to have been achieved by heterodyning, detection at a higher-level signal gives the system more immunity against other sources of additional post-detection noise, such as electronic thermal noise, which is independent of the detected signal. This is a third advantage of heterodyning.

Note from (6.14) that if the background-photoelectron rate is much less than the signal-photoelectron rate ($\lambda_b \ll \lambda_s$), then the direct receiver approaches the same accuracy as the heterodyne receiver. In such a case, a direct detector is preferable, because it avoids the need for a stable local oscillator and for riogorous alignment procedure. Different aspects of the performance of heterodyne receivers have been the subject of many studies [87, 114, 142, 157, 181, 224, 291, 300, 311, 320, 341, 444, 446, 470, 472, 478]. The importance of heterodyning in photon-correlation spectroscopy is discussed further in Chap. 7.

Effect of the Detector Area. The Problem of Alignment

In deriving (6.40) and (6.42), we assumed a point detector. We show below how spatial integration of the mixed fields over the detector surface and their possible misalignment could be the limiting feature of the performance of a heterodyne receiver.

Let the field at a point r of the detector surface be written as the sum

$$V(\underline{r},t) = V_0(\underline{r},t)\exp(j\omega_0 t) + V_s(\underline{r},t)\exp(j\omega_s t) \quad ,$$

where the subscripts 0 ans s refer to the local oscillator and the signal, respectively. Therefore,

$$I(\underline{r},t) = I_0(\underline{r},t) + I_s(\underline{r},t) + [V_0^*(\underline{r},t)V_s(\underline{r},t)\exp(j\omega_i t) + \text{c.c.}] \quad .$$

Integrating over the detector surface, we obtain

$$\lambda(t) = \lambda_0(t) + \lambda_s(t) + 2\,\mathrm{Re}\left\{\exp(j\omega_i t)\int_A V_0^*(\underline{r},t)V_s(\underline{r},t)d\underline{r}\right\} \quad ,$$

where

$$\lambda_0(t) = \int_A I_0(\underline{r},t)dr \quad , \qquad \lambda_s(t) = \int_A I_s(\underline{r},t)d\underline{r} \quad .$$

We continue the analysis in the case $\omega_i = 0$ (homodyne detection). We assume that $V_0(r,t) = V_0(r)$ is independent of time, and $V_s(r,t) = V_s(r)S^{1/2}(t)$, where $S(t)$ is a real signal to be estimated. We choose $\int_\infty |V_s(r)|^2 dr = 1$, which means that the total power in the beam is $S(t)$ counts per s. This gives

$$\lambda(t) = \lambda_0 + \lambda_s(t) + 2\xi\lambda_0^{\frac{1}{2}}\lambda_s^{\frac{1}{2}}(t) \quad ,$$

$$\xi = \mathrm{Re}\left\{\int_A V_0^*(\underline{r})V_s(\underline{r})d\underline{r}\Big/\left[\int_A |V_0(\underline{r})|^2 d\underline{r} \cdot \int_A |V_s(\underline{r})|^2 d\underline{r}\right]^{\frac{1}{2}}\right\} \quad . \tag{6.45}$$

The factor ξ represents the effect of aperture integration. It attains its maximum value, 1, whenever $V_0(\underline{r}) \sim V_s(\underline{r})$, within a constant phase, over the detector surface. On the other hand, if $V_0(\underline{r})$ and $V_s(\underline{r})$ do not overlap sufficiently or if their phases mismatch, ξ could well be zero; the signal would then be lost. The normalized error of estimating $S(t)$ can be easily shown to be

$$e^2_{S(t)} = \frac{1}{S(t)T}\frac{1}{\Theta} \quad , \tag{6.46}$$

where the factor

$$\theta = \mathrm{Re}\left\{\left[\int_A V_0^*(\underline{r})V_s(\underline{r})d\underline{r}\right] / \left[\int_A I_0(\underline{r})d\underline{r}\right]^{\frac{1}{2}}\right\} \tag{6.47}$$

represents the contribution of the distributions of the mixed fields over the detector area to the estimation error. This factor has been calculated in several cases of interst [142, 165, 320].

Example 1. If the local-oscillator field is uniform and the signal field is an Airy function centered around the center of a circular detector of radius a, i.e.,

$$V_0(\underline{r}) = 1 \quad , \quad V_s(\underline{r}) = 2\,J_1(Kr)/Kr \quad ,$$

then

$$\theta = 2[1-J_0(Ka)]/Ka \quad .$$

As a function of a, the parameter θ increases from a value of zero at $a = 0$ to an optimum peak value of 0.85 at $Ka = 2.75$ (Fig.6.4a). If the detector is increased in size to match the Airy disc, θ drops to 0.73, thus increasing the error by a factor of 1.16 [142].

Example 2. The local oscillator produces a uniform plane wave perpendicular to the surface of the detector; the signal is a plane wave incident at a small angle θ and is assumed to be limited to an area A_s larger than the area of the detector, i.e.,

$$V_0(\underline{r}) = 1 \quad , \quad V_s(\underline{r}) = \frac{1}{\sqrt{A_s}} \exp\left[j\,\frac{\omega}{c}\,(x\theta + y\theta)\right] \quad .$$

The detector is a square of dimension a. From (6.47), the factor θ becomes

$$\theta = \frac{a}{\sqrt{A_s}}\,\mathrm{sinc}^2\left(\frac{\omega a}{2\pi c}\,\theta\right) \quad , \quad \mathrm{sinc}(x) = \sin(\pi x)/(\pi x) \quad .$$

As shown in Fig.6.4 b, it increases from a value of zero at $a = 0$ to a maximum value at an optimum detector size of approximately $a = 2c/\omega\theta$. It then drops again and reaches zero at $2\pi c/\omega\theta$.

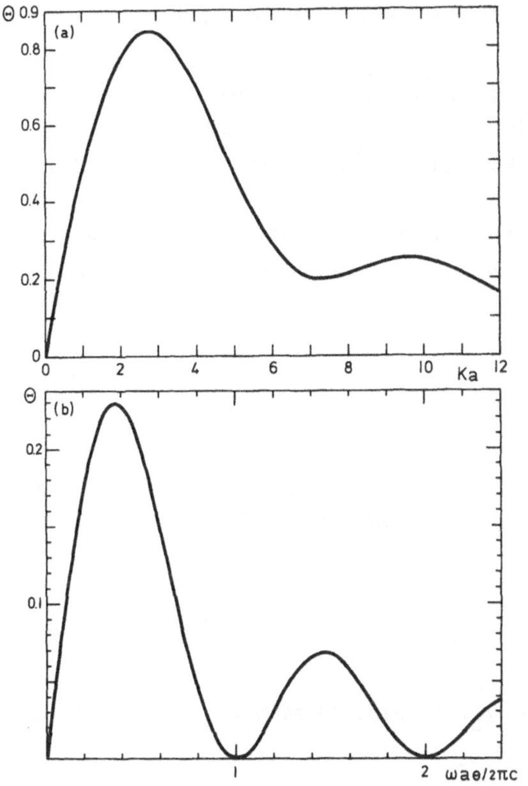

Fig.6.4a and b. The accuracy factor θ in signal estimation by a heterodyne receiver. a) and b) correspond to examples 1 and 2 in the text

For a fixed detector size a, the parameter θ drops sharply from its maximum value when the two beams are perfectly collimated ($\theta = 0$) to zero when $\theta = 2\pi c/\omega a$, i.e., when the angle θ is equal to the ratio between the wave length of light and the detector width. Indeed, this corresponds to a very small cone, which makes the matching between the local oscillator and the signal beam a critical problem. This is one disadvantage of heterodyning.

Any displacement between the centers of symmetry of the beams can be shown to lead to a considerable decrease of ξ and θ [165].

The above analysis enables us to return to our earlier assumption that the background light, which is spread over a large solid angle, does not mix with the light from the local oscillator. Actually, it does, but the parameter ξ for the mixture is very small. Including this small term would increase the estimation error, because of the uncertainty that results from the fluctuating background [199].

6.3.2 Homodyne Estimation of the Intensity of a Stationary Thermal Optical Signal in a Thermal Background

A thermal signal of constant average intensity λ_s is mixed with a strong coherent local oscillator whose intensity is λ_0. The average intensity of the thermal signal is to be estimated in the presence of a wide-band thermal background of average intensity λ_b.

The simplest approach is to repeat the procedure of Sec. 6.2.1, page 294. Measurement of the average photocount rate in the presence of the signal gives

$$\hat{\lambda} = \hat{n}/T \quad , \qquad \hat{n}_i = \frac{1}{N}\sum_i n_i \quad ,$$

which is an unbiased estimator of the sum $\lambda_0+\lambda_s+\lambda_b$. Another measurement of the average photocounts rate in the absence of the signal yields

$$\hat{\lambda}' = \hat{n}'/T \quad , \qquad \hat{n}_i' = \frac{1}{N}\sum_i n_i'$$

an unbiased estimate of the sum $\lambda_0+\lambda_b$. An anbiased estimate of λ_s is therefore

$$\hat{\lambda}_s = \hat{\lambda} - \hat{\lambda}' \quad . \tag{6.48}$$

The estimation error is

$$e^2_{\lambda_s} = \frac{1}{N}\left[\frac{1}{\lambda_s T}\left(1+\frac{2\lambda_0}{\lambda_s}+\frac{2\lambda_b}{\lambda_s}\right)+\frac{1}{N}\left(\frac{2\lambda_0}{\lambda_s}+1\right)\right] ,$$

where N is the number of modes of the signal.

This is apparently much greater than the corresponding error in the absence of a local oscillator (direct detection). Obviously, measurment of the difference between two small quantities by adding an equal constant noise-producing large quantity to each would result in a great loss of accuracy.

We have to find some other property that distinguishes the data n_i and n_i' from each other. How is the presence of the thermal signal manifest in the difference between the statistics of n_i and n_i' other than by very slightly increasing the mean? Clearly, the fluctuating thermal signal broadens the probability distribution of n_i, as compared to n_i', considerably. The simplest measure of this broadening is the variance. We have

$$\mathrm{Var}(n_i) = (\lambda_0+\lambda_s+\lambda_b)T + \frac{1}{N}\,(2\lambda_0\lambda_s+\lambda_s^2)T^2 \quad ,$$

and

$$\mathrm{Var}(n_i') = (\lambda_0+\lambda_b)T \quad ,$$

from which

$$\mathrm{Var}(n_i)-\mathrm{Var}(n_i') = \lambda_s T + \frac{1}{N}\,(2\lambda_0\lambda_s+\lambda_s^2)\,T^2 \simeq \frac{2}{N}\,\lambda_0\lambda_s T^2 \quad . \tag{6.49}$$

The signal λ_s is therefore proportional to the difference between the photoelectron-counting variances in its presence and in its absence. This is not a difference between two almost equal quantities, as in (6.48).

Let us now test the new unbiased estimator

$$\hat{\lambda}_s = (\hat{V}-\hat{V}')/(2\lambda_0 T^2/N) \quad , \tag{6.50}$$

where

$$\begin{aligned} \hat{V} &= \hat{G} - (\hat{n})^2 \quad , & \hat{G} &= \frac{1}{N}\sum_i n_i^2 \quad , \\ \hat{V}' &= \hat{G}' - (\hat{n}')^2 \quad , & \hat{G}' &= \frac{1}{N}\sum_i n_i'^2 \quad . \end{aligned} \tag{6.51}$$

This estimation error is given by

$$e_{\lambda_s}^2 = [\mathrm{Var}(\hat{V}) + \mathrm{Var}(\hat{V}')] \,/\, (2\lambda_0\lambda_s T^2/N)^2 \quad . \tag{6.52}$$

In order to determine the variances of the variance estimators $\hat{V}$ and $\hat{V}'$, we need the moments of n_i up to the 4th order. The photoelectron-counting moments for a mixture of a coherent and a thermal field were discussed in Sec. 5.2.4. Assuming that the samples are independent, and using the fact that $\lambda_0 \gg \lambda_s$ and λ_b, JAKEMAN et al. [236] obtained

$$e^2_{\lambda_s} = \frac{1}{N}(1 + 2x + 2x^2)/x^2 \quad , \qquad x = \lambda_s T/N \quad . \tag{6.53}$$

This should be compared to the corresponding expression in the case of direct detection, (6.17).

For small signals ($\lambda_s T/N << 1$), the error,

$$e^2_{\lambda_s} = \frac{1}{N}\frac{1}{(\lambda_s T/N)^2} \tag{6.54}$$

is a rapidly decreasing function of ($\lambda_s T/N$) that ultimately saturates in the strong-signal limit ($\lambda_s T/N >> 1$) to the constant level

$$e^2_{\lambda_s} = \frac{2}{N} \quad . \tag{6.55}$$

As in the case of direct detection (Sec. 6.2.1), the optimum choice of N, for a given λ_s, T_0, and τ_c, depends on the operating conditions.

In the *strong-signal limit*, (6.55) indicates that N should be as large as possible (or T as small as possible) without allowing the samples to be correlated. In synchronous detection, we can choose $T = \tau_c/2$ and interlace the on-off regions without losing much of the accuracy. Hence, $N = 2T_0/\tau_c$ samples are used and the error is given by

$$e^2_{\lambda_s} = \tau_c/T_0 \quad . \tag{6.56}$$

The error is therefore equal to the inverse square root of the total number of temporal modes observed in the time T_0.

The background choherence time may be too short to make the choice $T = \tau_c/2$ possible. In this case, the shortest counting time T_{min} has to be used; this corresponds to

$$e^2_{\lambda_s} = 2\, T_{min}/T_0 \quad . \tag{6.57}$$

Equations (6.56) and (6.57) should be compared to the corresponding equations, (6.21) and (6.22), in the case of direct detection. The comparison reveals that, in the strong-signal limit, direct detection is superior. Whereas the performance of a direct-detection system can be improved by increasing the detector area (increasing the number of spatial modes), the performance of a heterodyne receiver is limited by the area over which coherent mixing is achievable.

Now we discuss the opposite limit. In the *weak-signal limit,* (6.54) can be written in the form

$$e^2_{\lambda_s} = \frac{1}{T_0 T} \frac{N_s^2}{\lambda_s^2} \quad , \quad T < \tau_c \quad ,$$
$$= \frac{T}{T_0 \tau_c^2} \frac{N_s^2}{\lambda_s^2} \quad , \quad T > \tau_c \quad , \tag{6.58}$$

indicating that the optimum choice of T is $T = \tau_c$ in which case

$$e^2_{\lambda_s} = \frac{1}{T_0 \tau_c} \frac{N_s^2}{\lambda_s^2} \quad . \tag{6.59}$$

If $T_{min} > \tau_c$, we should take $T = T_{min}$ and obtain

$$e^2_{\lambda_s} = \frac{T_{min}}{T_0 \tau_c^2} \frac{N_s^2}{\lambda_s^2} \tag{6.60}$$

A comparison between (6.59) and (6.58), for heterodyne detection, and (6.18), for direct detection, is difficult because direct detection is limited by the background level. The minimum signal detectable in 1 s can be obtained from (6.59) and (6.58) by putting $T_0 = 1/2$ and $e_{\lambda_s} = 1$. This gives

$$I_s|_{min} = \frac{h\nu}{\eta A} N_s \sqrt{\frac{2}{\tau_c}} \quad , \quad T = \tau_c/2 \quad , \tag{6.61}$$

$$= \frac{h\nu}{\eta A} N_s \frac{\sqrt{2T_{min}}}{\tau_c} \quad , \quad T = T_{min} > \tau_c \quad . \tag{6.62}$$

Let $A = 100\ cm^2$, $N_s = 1$, $\nu = 5 \cdot 10^{14}$, $\eta = 0.5$, and $T_{min} = 10^{-7}$ s. If $\tau_c = 10^{-6}$ s, we can choose $T = 0.5 \cdot 10^{-6}$ s and use (6.61) to obtain $I_s|_{min} \approx 10^{-17}\ W\ cm^{-2}$.

If $\tau_c = 10^{-12}$, we are forced to choose $T_{min} = 10^{-7}$s, and (6.62) gives $I_{s|min} \approx 3 \cdot 10^{-12}$ W cm^{-2}.

Comparing these values with the corresponding value in the example of the direct-detection receiver given in Sec. 6.2.1, page 294, in which $I_{s|min} \approx 2 \cdot 10^{-14}$ W cm^{-2}, we conclude that the heterodyne receiver can, under certain conditions, surpass the direct receiver.

6.4 Detection of Optical Signals. Digital Communication Systems

In Sec. 6.1, some of the basic digital communication systems were introduced. This section examines the role played by photoelectron statistics in determining the optimum processors for these communication systems and their performances. The section is divided into five subsections, each dealing with certain statistics for the signal and the background. Section 6.4.1 assumes that both signal and background lead to Poisson photoelectron statistics. This applies to the case when the signal is either coherent or wide band (i.e., has a coherence time much shorter than the counting time) and when the background is wide band. It also corresponds to the case when the detector area is much larger than the coherence area of the field. Section 6.4.2 relaxes these assumptions to cover the situation when the counting time and the area of the detector are of the same order of magnitude as the coherence time and coherence area, respectively. This is the situation governed by Bose-Einstein or Laguerre statistics and their generalizations. It also covers the situation when the spectral profile of the detected background becomes important. In Sec. 6.4.3, we allow the signal to fluctuate, as in the case of optical-radar signals; in Sec. 6.4.4 we examine the effect of signal fluctuations due to atmospheric turbulence on the design and performance of the detectors. Heterodyne detection is examined in Sec. 6.4.5.

In each of these cases, the performance of the basic modulation formats is discussed. These formats reduce to problems of tests between M hypotheses, given the measured photoelectron counts in N intervals. The photoelectron probability distributions under the various hypotheses determine the decision strategy and the errors involved (cf. Sec. 2.7.1).

6.4.1 Detection of a Coherent Signal in a Wide-Band Thermal Background

When the background light has a large bandwidth, the counting statistics are approximately Poisson. Hypothesis $H^{(k)}$ then corresponds to a Poisson counting process n_i, i = 1,2, ... N, with means $\bar{n}_{si}^{(k)} + \bar{n}_b$, i = 1,2, ... N, i.e.,

$$P\left[[n]|H^{(k)}\right] = \prod_{i=1}^{N} \left[\bar{n}_{si}^{(k)} + \bar{n}_b\right]^{n_i} \exp\left[-\bar{n}_{si}^{(k)} - \bar{n}_b\right] / n_i! \quad . \qquad (6.63)$$

Therefore

$$\ln P\left[[n]|H^{(k)}\right] = -\sum_i \bar{n}_{si}^{(k)} + \sum_i n_i \ln\left[1+\bar{n}_{si}^{(k)}/\bar{n}_b\right] + \left[\sum_i n_i \ln \bar{n}_b - N\bar{n}_b - \ln(n_i!)\right] .$$

Because the third term is independent of the hypotheses, we can define the likelihood function

$$\phi^{(k)} = -\sum_i \bar{n}_{si}^{(k)} + \sum_i n_i h_{ik} \quad , \qquad h_{ik} = \ln\left[1+\bar{n}_{si}^{(k)}/\bar{n}_b\right] \quad . \qquad (6.64)$$

In many communication formats, the sum $\sum_i n_{si}^{(k)}$ is independent of k (total energy is independent of hypotheses), and we can use the likelihood function

$$\phi^{(k)} = \sum_i n_i h_{ik} \quad . \qquad (6.65)$$

According to the maximum-likelihood decision rule, hypothesis q is chosen if $\phi^{(q)}$ exceeds all other $\phi^{(k)}$. The numbers $\phi^{(k)}$ can be obtained by cross correlating the measured n_i with h_{ik} for each hypothesis. A simple comparison between the values $\phi^{(k)}$, k = 1, ... M leads to a decision.

Note that if $\bar{n}_{si}^{(k)} \ll \bar{n}_b$, $h_{ik} \sim \bar{n}_{si}^{(k)}$, and the processing becomes much easier.

In the preceding analysis, we assumed that the signal and background levels $\bar{n}_s$ and $\bar{n}_b$ are known. If this is not the case, it is necessary, to resort to more-complicated techniques of decision theory [63].

Now we consider some special formats of the signal $\bar{n}_{si}^{(k)}$, i = 1, ... N, k = 1, ... M that correspond to the basic modulation schemes discussed in Sec. 6.1.

Pulse-Gated Binary Modulation (PGBM).

Detection of a Radar Signal

This case corresponds to testing two hypotheses ($M = 2$) by counting in one time interval ($N = 1$). Hypothesis 1 (the absence of a radar signal or binary 0) corresponds to $\bar{n}_{si}^{(1)} = 0$; and hypothesis 2 (the presence of a radar signal or binary 1) corresponds to $\bar{n}_{si}^{(2)} = \bar{n}_s$.

From (6.64), we see that

$$\phi^{(1)} = 0 \quad ,$$

$$\phi^{(2)} = -\bar{n}_s + n \ln(1+\bar{n}_s/\bar{n}_b) \quad .$$

The decision rule $\phi^{(2)} \underset{1}{\overset{2}{\gtrless}} \phi^{(1)}$ corresponds to a comparison between the measured number of counts and a threshold level n_T,

$$n \underset{1}{\overset{2}{\gtrless}} n_T \ , \qquad n_T = \bar{n}_s/\ln(1+\bar{n}_s/\bar{n}_b) \quad . \tag{6.66}$$

If n exceeds n_T, choose $H^{(2)}$. If n_T exceeds n, choose $H^{(1)}$. If $n = n_T$, choose either $H^{(2)}$ or $H^{(1)}$, with equal probabilities.

For a fixed signal-to-background ration $\bar{n}_s/\bar{n}_b$, the threshold level increases linearly with the signal power. This distinguishes this system from the classical Gaussian communication channel, in which the threshold level is only a function of the signal-to-background ratio.

The performance of this threshold detector can be calculated by using

$$P(n|H^{(1)}) = \bar{n}_b^n \exp(-\bar{n}_b)/n! \quad ,$$

$$P(n|H^{(2)}) = (\bar{n}_s+\bar{n}_b)^n \exp(-\bar{n}_s-\bar{n}_b)/n! \quad ,$$

and writing the probability of errors (cf. Sec. 2.7.1)

$$Q_1 = \sum_{n=n_T+1}^{\infty} P\left[n|H^{(1)}\right] + \frac{1}{2} P\left[n_T|H^{(1)}\right] \quad , \tag{6.67}$$

and

$$Q_2 = \sum_{n=0}^{n_T-1} P\left[n|H^{(2)}\right] + \frac{1}{2} P\left[n_T|H^{(2)}\right] \quad . \tag{6.68}$$

For equilikely hypotheses, the average probability of error is given by

$$P_e = \frac{1}{2} Q_1 + \frac{1}{2} Q_2 \quad . \tag{6.69}$$

For a given $\bar{n}_s$ and $\bar{n}_b$, n_T is calculated from (6.66) and P_e is computed from (6.67,69). Results [108, 387] are displayed in Fig.6.5.

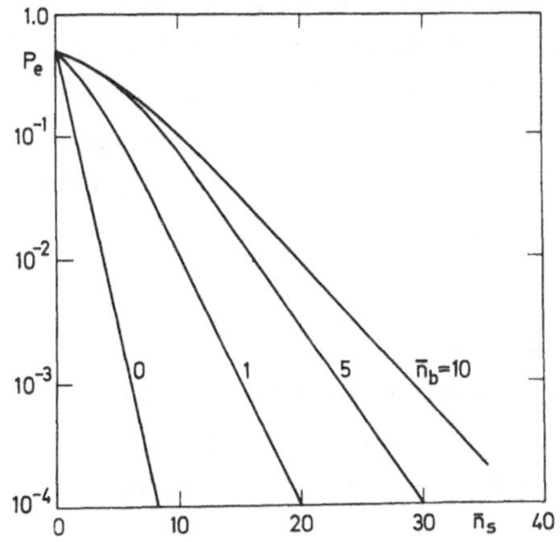

Fig.6.5. Dependence of the probability of error of a PGBM system on the average number of counts due to a coherent signal $\bar{n}_s$, and a wide-band thermal background $\bar{n}_b$ (after GAGLIARDI and KARP [165])

As expected, P_e decreases with an increase of the signal $\bar{n}_s$ or a decrease of the background $\bar{n}_b$. But, unlike the coherent Gaussian channel, the performance of our Poisson channel depends on both $\bar{n}_s$ and $\bar{n}_b$, rather than only on their ratio. From Fig.6.5, we see that in order to achieve an error probability less than 10^{-4}, a signal corresponding to more than 8 photoelectrons must be used, even in the absence of background. In the absence of background ($\bar{n}_b = 0$), $n_T = 0$ and

$$P_e = \frac{1}{2} \exp(-\bar{n}_s) \quad , \tag{6.70}$$

i.e., is equal to half the probability of obtaining no photoelectrons.

BUCHER [71] has found a useful upper bound for P_e

$$P_e \leq \exp[-\bar{n}_s E(\bar{n}_s/\bar{n}_b)] \quad , \tag{6.71}$$

where

$$E(x) = (y - \ln y - 1)/\ln(1 + x) \quad , \quad y = (1 + 1/x)\ln(1 + x) \, . \tag{6.72}$$

Note that E(x) increases with x and saturates at a value 1 for $x = \infty$. When the background is very small, the probability of error is less than $\exp(-\bar{n}_s)$, the probability that no photoelectrons are observed in the sampling interval.

Note that if the sampling time T is reduced while keeping $\bar{n}_s$ fixed, then $\bar{n}_b = \lambda_b T$ will decrease proportionally for a fixed number of background photoelectrons per second, λ_b. This improves the performance considerably at the expense of a larger information bandwidth and larger signal peak power. REIFFEN and SHERMAN [396], GOODMAN [178], and ABEND [2] showed that a very narrow pulse is the optimum choice. LACHS and QUARATO [284] studied the effect of pulse shape and duty cycle on the probability of error. They showed that the effect becomes less important as $\bar{n}_b$ decreases and as the duty cycle decreases.

Pulse-Delay Binary Modulation (PDBM). Binary PPM

In this format, we have two time slots (N = 2) and two hypotheses (M = 2). Hypothesis 1 corresponds to the presence of a signal pulse in slot 1 and nothing in slot 2. In hypothesis 2, the pulse is present in slot 2. Mathematically,

$$\bar{n}_{s1}^{(1)} = \bar{n}_s \quad , \quad \bar{n}_{s2}^{(1)} = 0 \quad ,$$

$$\bar{n}_{s1}^{(2)} = 0 \quad , \quad \bar{n}_{s2}^{(2)} = \bar{n}_s \; .$$

Because the total signal energy is independent of the hypotheses, we can use (6.65) and write

$$\phi^{(k)} = n_k \ln(1 + 2\,\bar{n}_s/\bar{n}_b) \quad , \quad k = 1,2,$$

where n_1 and n_2 are the number of counts in slot 1 and 2, respectively, and $\bar{n}_b/2$ is the number of background counts in one time slot. The decision rule $\phi^{(2)} \underset{1}{\overset{2}{\gtrless}} \phi^{(1)}$ is now equivalent to $n_2 \underset{1}{\overset{2}{\gtrless}} n_1$, or to

$$m \underset{1}{\overset{2}{\gtrless}} 0 \quad , \quad m = n_2 - n_1 \quad , \tag{6.73}$$

i.e., $H^{(2)}$ or $H^{(1)}$ is chosen depending on whether n_2 is greater or less than n_1. The case $n_1 = n_2$ is handled by an equilikely randomized decision between $H^{(1)}$ and $H^{(2)}$.

The performance of the system is then obtained from

$$Q_1 = \Pr[m > 0|H^{(1)}] + \frac{1}{2}\Pr[m = 0|H^{(1)}] \quad . \tag{6.74}$$

From symmetry, $Q_2 = Q_1$ and for equiprobable hypotheses,

$$P_e = Q_1 = Q_2 \quad . \tag{6.75}$$

It remains to determine the probability distribution of m, the difference between two Poisson random variables with means $\bar{n}_s + \bar{n}_b/2$ and $\bar{n}_b/2$, respectively.

By using standard methods of transformation of random variables, we get [153, 387]

$$P[m|H^{(1)}] = \exp(-\bar{n}_1-\bar{n}_2)(\bar{n}_2/\bar{n}_1)^{m/2} I_m(2\sqrt{\bar{n}_1\bar{n}_2}) \tag{6.76}$$

$$\bar{n}_1 = \bar{n}_s + \bar{n}_b/2 \quad , \quad \bar{n}_2 = \bar{n}_b/2 \quad .$$

This enables us [387] to compute P_e as a function of $\bar{n}_s$ and $\bar{n}_b$ (Fig.6.6).

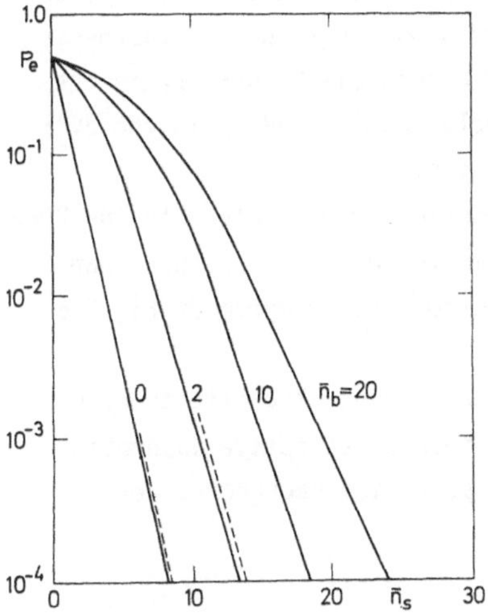

Fig.6.6. Dependence of the probability of error of a PDBM system on the average number of counts due to a coherent signal $\bar{n}_s$, and a wide-band thermal background $\bar{n}_b$. The dotted lines are obtained from (6.77) (after GAGLIARDI and KARP [165])

Comparison between Fig.6.6 and Fig.6.5 shows that the PDBM format is more efficient than the PGBM format. The signal power necessary to achieve the same P_e for the same signal-to-background ratio is about 2 dB less in PDBM than in PGBM at moderate error rates [297, 406].

A useful upper bound of P_e can be obtained [222] by replacing $I_m(.)$ in (6.76) with $I_0(.)$. Because $I_m(.) \leq I_0(.)$, it follows that

$$P_e \leq \left[\frac{1}{2} + \sqrt{\bar{n}_1}/(\sqrt{\bar{n}_2}-\sqrt{\bar{n}_1})\right] \exp(-\bar{n}_1-\bar{n}_2) I_0(2\sqrt{\bar{n}_1\bar{n}_2}) \quad . \tag{6.77}$$

The right side is an accurate approximation at P_e values less than 10^{-3} (see the dotted lines in Fig.6.6).

Another useful upper bound of P_e has been found by BUCHER [71]

$$P_e \leq \exp\left[-\bar{n}_s E(2\bar{n}_s/\bar{n}_b)\right] \quad , \qquad E(x) = \left[(1+x)^{1/2}-1\right]^2/x \quad . \tag{6.78}$$

Again for small background $E(\bar{n}_s/\bar{n}_b) = 1$ and P_e is bounded by the probability of no counts.

Binary Polarization Modulation (BPLM)

In a BPLM system, bits 0 and 1 are represented by right- and left-hand circularly polarized light pulses of time duration T. Two detectors are used. One is sensitive to right-hand polarized light; the other to left-hand polarized light. If n_1 and n_2 are the number of counts measured by the detectors, then in the presence of unpolarized background, n_1 and n_2 have Poisson distributions with means $\langle n_1\rangle = \bar{n}_s + \bar{n}_b/2$ and $\langle n_2\rangle = \bar{n}_b/2$ under hypothesis 1; and $\langle n_1\rangle = \bar{n}_b/2$ and $\langle n_2\rangle = \bar{n}_s + \bar{n}_b/2$ under hypothesis 2. Here, $\bar{n}_s$ is the average number of signal counts due to one polarization and $\bar{n}_b$ is the average number of background counts in both polarizations.

It is obvious that the situation is mathematically equivalent to the PDBM case except that n_1 and n_2 represent counts in two detectors rather than counts in two adjacent time intervals. The performance diagram of Fig.6.6 can therefore be used [386, 387].

Leakage could be regarded as part of the background. The effects of other nonideal behaviour of the system components, such as selective absorption or selective birefringence, and of partially polarized background were examined by PETERS and ARGUELLO [375].

Pulse-Position Modulation (PPM)

In this system, M time slots are observed. Hypothesis k corresponds to the case when slot k has a signal pulse with an average number of counts $\bar{n}_s$ while all others are empty (see Fig.6.2). Mathematically,

$$\bar{n}_{si}^{(k)} = \bar{n}_s \delta_{ik} \quad .$$

When M = 2, this reduces to the PDBM system.

From (6.65),

$$\phi^{(k)} = n_k \ln(1 + \bar{n}_s/\bar{n}_b) \quad ,$$

where $\bar{n}_b$ is the number of background counts per time slot. This is equivalent to

$$\phi^{(k)} = n_k \quad .$$

We should therefore use the decoding rule:

If $\quad n_k > n_j \quad \forall j \neq k \quad$ choose $H^{(k)}$

If $\quad n_k \geq n_j \quad \forall j \quad$ and $n_k = n_{j_1} = n_{j_2} = \ldots n_{j_r}$;

decide randomly between $H^{(k)}$, $H^{(j_1)}$, ... $H^{(j_r)}$, each with a probability $1/(1+r)$. Now we study the performance of this decoding system [164, 259].

Assuming that hypothesis $H^{(q)}$ is true, then the probability of choosing $H^{(q)}$ correctly is equal to

$$P_D = \Pr(n_q > n_j \forall_j \neq q) + \frac{1}{2} \sum_{j_1} \Pr(n_q = n_{j_1} > n_j \ \forall j \neq q, j_1)$$
$$+ \frac{1}{3} \sum_{j_1, j_2} \Pr(n_q = n_{j_1} = n_{j_2} > n_j \ \forall j \neq q, j_1, j_2) + \ldots + \frac{1}{M} \Pr(n_q = n_j \ \forall j) \quad . \tag{6.79}$$

Because counts in the different intervals are independent, this can be written in the form

$$P_D = \sum_{r=0}^{M-1} \frac{1}{r+1} \sum_{m=1}^{\infty} \Pr(n_q=m) \binom{M-1}{r} \prod_{s=j_1}^{s=j_r} \Pr(n_s=m) \prod_{\ell \neq q, j_1, \ldots j_r} \Pr(n_\ell < m)$$

$$+ \frac{1}{M} \prod_j \Pr(h_j = 0) \quad . \qquad (6.80)$$

Under hypothesis $H^{(q)}$,

$$P(n_j) = (\bar{n}_s \delta_{jq} + \bar{n}_b)^{n_j} \exp(-\bar{n}_s \delta_{jq} - \bar{n}_b)/n_j! \quad ,$$

from which

$$P_D = \frac{1}{M} e^{-M\bar{n}_s - \bar{n}_b}$$

$$+ \sum_{r=0}^{M-1} \frac{1}{r+1} \binom{M-1}{r} \sum_{m=1}^{\infty} \left(\bar{n}_b^m e^{-\bar{n}_b}/m!\right)^r \left(\sum_{\ell=0}^{m-1} \bar{n}_b^\ell e^{-\bar{n}_b}/\ell!\right)^{M-1-r} \quad . \qquad (6.81)$$

From symmetry, the probability of correct detection of $H^{(q)}$ is independent of q. Therefore it represents the average probability of correct detection. The average probability of error is then

$$P_e = 1 - P_D \quad , \qquad (6.82)$$

and is a function of $\bar{n}_s$, $\bar{n}_b$ and M.

Figure 6.7 [164, 165] shows the dependence of P_e on $\bar{n}_s$ and M for a fixed value of $\bar{n}_b$. The case M = 2 is the same as in Fig.6.6.

Again, P_e depends on $\bar{n}_s$ as well as on the ratio $\bar{n}_s/\bar{n}_b$. The probability of error P_e increases as M increase. This is not, however, a sufficient criterion for comparing systems with different M. After all, a system with larger M carries more information. An M-ary system communicates at a rate $R = \log_2 M/MT$ bits s^{-1} where T is the sampling time. If λ_b is the number of counts per second due to the background, then

$$\bar{n}_b = \lambda_b \log_2 M/MR \quad .$$

We should therefore compare systems with different M for a fixed value of λ_b/R and $\bar{n}_s$. The comparison shows [164] that for negligible background, the

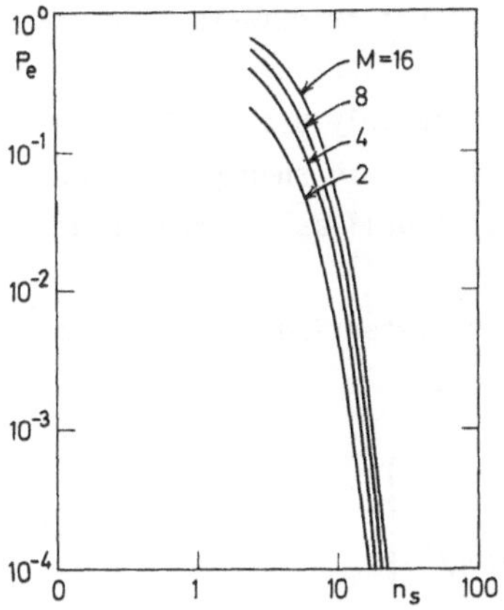

Fig.6.7. Error probability of a PPM system. The signal is coherent and the background is wide-band thermal, $\bar{n}_b=3$ (after GAGLIARDI and KARP [164])

system operation improves with decreasing M, and is best for M = 2. For large background, however, the converse is true and M = ∞ gives minimal probability of error. Detailed performance diagrams can be found in [165].

6.4.2 Detection of a Coherent Signal Corrupted by Thermal Background of Arbitrary Bandwidth. Arbitrary Area of Detector

When the background bandwidth becomes comparable to the information-processing bandwidth (i.e., the coherence time of the background τ_c becomes comparable to the sampling time T); and when the area of the detector is not much larger than the coherence area of the background field, the assumption of Poisson statistics does not hold. We should then use the formula (5.165) as discussed in Sec. 5.2.2 and write

$$P\left[[n]|H^{(k)}\right] = \prod_i \frac{(\bar{n}_b/N)^{n_i}}{(1+\bar{n}_b/N)^{n_i+N}} \exp\left[-\frac{\bar{n}_{si}^{(k)}}{1+\bar{n}_b/N}\right] L_{n_i}^{N-1}\left[-\frac{n_{si}^{(k)}}{(1+\bar{n}_b/N)\bar{n}_b/N}\right] , \tag{6.83}$$

where $N = N_t N_s$ represents the appropriate number of temporal and spatial modes. We have seen that if $T \ll \tau_c$ (narrow band background), $N_t = 1$. For a background with a rectangular spectrum such that $T \gg \tau_c$, $N_t = T/\tau_c$. Also $N_s = 1$, if $A \ll A_c$; and $N_s = A/A_c$, if $A \gg A_c$.

In the intermediate range of time-bandwidth product or detector area-coherence area ratio, it is necessary to resort to the exact statistics described in Sec. 5.2.2. Alternatively, we may approximately use (6.83) with N appropriately adjusted as discussed in Sec. 5.2.2. Care should be taken, however, that N is then a function of $\bar{n}_s$ and $\bar{n}_b$, as well as a function of the temporal and spatial spectral profiles.

In the following, we discuss the performance of detection systems based on (6.83). We first write the likelihood function as

$$\phi^{(k)} = \sum_i \bar{n}_{si}^{(k)}/(1+\bar{n}_b/N) + \sum_i \ln L_{n_i}^{N-1}\left[-\frac{\bar{n}_{si}^{(k)}}{(1+\bar{n}_b/N)\bar{n}_b/N}\right] \; . \tag{6.84}$$

If $\sum_i \bar{n}_{si}^{(k)}$ is independent of k, we can ignore the first term and write

$$\phi^{(k)} = \sum_i \ln L_{n_i}^{N-1}\left[-\frac{\bar{n}_{si}^{(k)}}{(1+\bar{n}_b/N)\bar{n}_b/N}\right] . \tag{6.85}$$

We can now consider some special cases.

Pulse-Gated Binary Modulation (PGBM).

Detection of a Radar Signal

Substituting $\bar{n}_{s1}^{(1)} = 0$, $\bar{n}_{s1}^{(2)} = \bar{n}_s$ in (6.83) and (6.84), we obtain

$$P[n|H^{(1)}] = \binom{n+N-1}{n} \frac{(\bar{n}_b/N)^n}{(1+\bar{n}_b/N)^{n+N}}$$

$$P[n|H^{(2)}] = \frac{(\bar{n}_b/N)^n}{(1+\bar{n}_b/N)^{n+N}} \exp\left(-\frac{\bar{n}_s}{1+\bar{n}_b/N}\right) L_n^{N-1}\left[-\frac{\bar{n}_s}{(1+\bar{n}_b/N)\bar{n}_b/N}\right] \quad ,$$

and

$$\phi^{(1)} = 0$$

$$\phi^{(2)} = -\bar{n}_s/(1+\bar{n}_b/N) + \ln L_n^{N-1}\left\{-\bar{n}_s/[(1+\bar{n}_b/N)\bar{n}_b/N]\right\} .$$

The decision rule $\phi^{(2)} \overset{2}{\underset{1}{\gtrless}} \phi^{(1)}$ is equivalent to

$$L_n^{N-1}\left\{-\bar{n}_s/[(1+\bar{n}_b/N)\bar{n}_b/N]\right\} \overset{2}{\underset{1}{\lessgtr}} \exp[-\bar{n}_s/(1+\bar{n}_b/N)] \quad . \tag{6.86}$$

Because of the monotonicity of Laguerre functions with respect to their indices, the decision rule (6.86) is equivalent to the threshold detector,

$$n \overset{2}{\underset{1}{\gtrless}} n_T \quad ,$$

where the threshold level n_T is obtained by finding the smallest integer for which

$$L_{n_T}^{N-1}\left\{-\bar{n}_s/[1+\bar{n}_b/N)\bar{n}_b/N]\right\} \geq \exp[-\bar{n}_s/(1+\bar{n}_b/N)] \quad . \tag{6.87}$$

The probability of error is then computed from

$$P_e = \frac{1}{2}Q_1 + \frac{1}{2}Q_2 \quad ,$$

$$Q_1 = \sum_{n=n_T+1}^{\infty} P[n|H^{(1)}] + \frac{1}{2}P[n_T|H^{(1)}] \quad , \tag{6.88}$$

$$Q_2 = \sum_{n=0}^{n_T-1} P[n|H^{(2)}] + \frac{1}{2}P[n_T|H^{(2)}] \quad . \tag{6.89}$$

For very large N, we should reproduce the results displayed in Fig.6.6, which correspond to Poisson statistics. The other extreme $N = 1$ is shown in Fig.6.8 [203, 281, 435].

By comparing Figs.6.5 and 6.8, we see that Poisson statistics ($N = \infty$) correspond to lower P_e than Laguerre statistics ($N = 1$), if all other parameters are fixed. Therefore, the assumption of Poisson statistics leads to more optimistic results. This is to be expected, because in the Poisson case, the only uncertainty is due to the nature of the phenomenon of photoelectron detection. In the Laguerre case, the uncertainty due to fluctuations of the background field is added. This is also manifest in the observation that for $\bar{n}_b \ll 1$, the difference between Laguerre and Poisson statistics becomes negligible. When $\bar{n}_b = 0$, both statistics give the same result, (6.70).

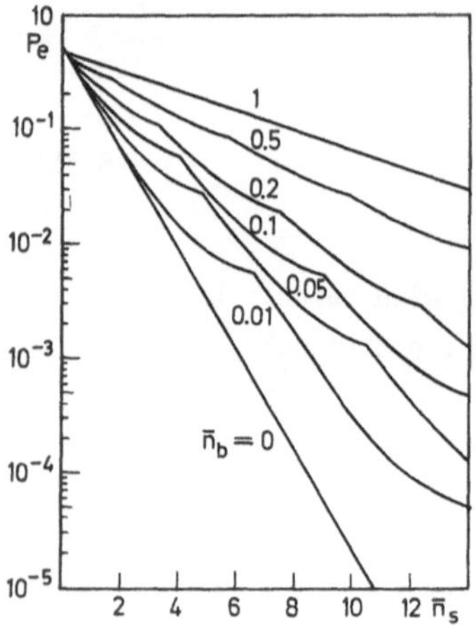

Fig.6.8. Error probability of a PGBM system. The signal is coherent and the background is narrow-band thermal (after Helstrom [203])

In the intermediate situation, $N = 2\text{-}10$, the background spectral profile plays an important role. In this case, it is necessary to resort to the exact photoelectron statistics discussed in detail in Sec. 5.2.2. Performance diagrams fro the rectangular and Lorentzian case were calculated by HELSTROM [206]. LACHS and RUGGIERI [286] give performance diagrams for the Lorentzian and Gaussian spectral profiles. Extensive computations are necessary for obtaining the threshold level and the probability of error.

Pulse-Position Modulation (PPM), Pulse-Delay Binary Modulation (PDBM), Pulse-Polarization Modulation (BPLM)

Putting $\bar{n}_{si}^k = \bar{n}_s \delta_{ik}$ in (6.85), we obtain

$$\phi^{(k)} = \ln L_{n_k}^{N-1} \left\{ -\bar{n}_s / [(1+\bar{n}_b/N)\bar{n}_b/N] \right\} \quad .$$

Again, the monotonicity of $L_{n_k}^{N-1}$ with respect to n_k enables us to put

$$\phi^{(k)} = n_k \quad .$$

This means that the ML test is equivalent to a simple comparison between the number of counts. This is the same as in the case of Poisson statistics. Using the decision rule described in Sec. 6.4.1, we can obtain the probability of error by using the appropriate probability distributions in (6.80). For M = 2 (PDBM or BPLM), this gives the expression [163]

$$P_e = 1-z^{2N}e^{-z\bar{n}_s}\left[\sum_{n=0}^{\infty} y^n L_n^{N-1}(-x) \sum_{m=0}^{\infty}\binom{N-1+m}{m} y^m - \frac{1}{2}\sum_{n=0}^{\infty}\binom{N-1+n}{n} y^{2n} L_n^{N-1}(-x)\right] , \tag{6.90}$$

$$z = (1+\bar{n}_b/N)^{-1} , \quad y = z\bar{n}_b/N , \quad x = z\bar{n}_s/(\bar{n}_b/N) . \tag{6.91}$$

For $N = 1$, this simplifies to [163]

$$P_e = \frac{1}{2}\exp[-\bar{n}_s/(1+2\bar{n}_b)] . \tag{6.92}$$

Figure 6.9 displays the dependence of P_e on $\bar{n}_s$, $\bar{n}_b$, and N. The case $N \gg 1$ corresponds to Poisson statistics and is plotted in the same figure in dotted lines. It is evident that an increase of N leads to an improvement of performance if all other parameters are fixed. The Poisson statistics correspond to the smallest error, as was the case for PGBM.

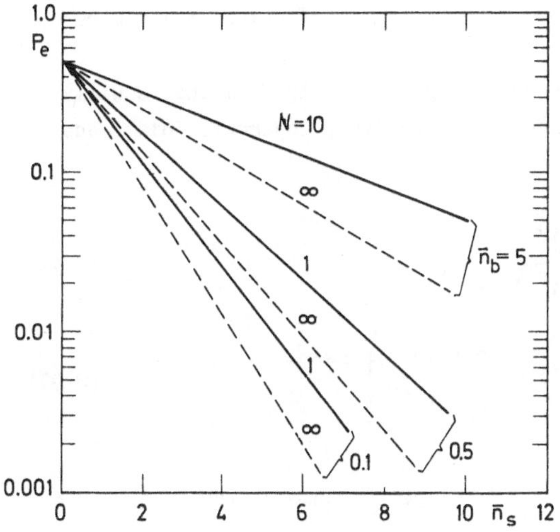

Fig.6.9. Error probability of a PDBM system. The signal is coherent; N modes of thermal background are collected by the detector (after GAGLIARDI [163])

Note that if $N = 1$, the counting times become so short compared to the background coherence time that our assumption of statistically independent counts n_1 and n_2 becomes doubtful. The effect of correlation between n_1 and n_2 was found to increase the probability of error [422].

When M is larger than 2, the expression of P_e becomes more complicated. In the case $N = 1$, LIU [299] found an upper bound on P_e.

6.4.3 Detection of a Fluctuating Signal Corrupted by Thermal Background

In all of the discussions of detection problems, so far, we assumed that the signal is coherent. This is indeed the case when modulated laser light is used to carry the signal. In some situations, however, the signal may fluctuate. For example, a laser-radar signal is coherent if it is produced by a return from a specular target. On the other hand, laser light reflected from a rough target is thermal [179]. In this section, we show how this could be incorporated in the study of the optimum detector. We limit the discussion to cases that lead to a binary test of hypotheses.

We assume that the signal is thermal, and that the receiver collects N modes. As discussed earlier, N accounts for the number of coherence areas collected by the detector (i.e., the ratio of the area of the detector and the area of the granularity of the return), the number of coherence times observed in the time window T, and also the number of polarization modes involved (cf. Sec. 5.2.3).

We also assume that the background is thermal, with N' modes. Usually N' is very large, making the background effectively coherent. This leads directly to the distributions

$$P[n|H^{(1)}] = \binom{n+N'-1}{n} \frac{(\bar{n}_b/N')^n}{(1+\bar{n}_b/N')^{n+N'}} \quad , \tag{6.93}$$

$$P[n|H^{(2)}] = P[n|H^{(1)}] \circledast \left[\binom{n+N-1}{n} \frac{(\bar{n}_s/N)^n}{(1+\bar{n}_s/N)^{n+N}}\right] \quad . \tag{6.94}$$

These distributions enable us to study the performance of the system as a function of $\bar{n}_s$, $\bar{n}_b$, N, and N'.

GOODMAN [179] investigated the detector performance in the case $N' = \infty$ that corresponds to a Poisson distribution in the absence of the signal. If $N = N' = \infty$, we have a Poisson distribution in the presence of the signal and we reproduce the error plots of Fig.6.5. GOODMAN found that a decrease of N results in an increase of the error. Again this is a direct consequence of the additional uncertainty introduced when the detector has sufficient resolution to follow the signal fluctuations. It is expected that N' would have a similar effect.

6.4.4 Detection in the Presence of Atmospheric Scintillation

When a coherent optical beam propagates through the clear atmosphere, it suffers temporal and spatial random fluctuations that could represent the most severe limitation to the performance of an optical communication system. These fluctuations are caused by small changes of the refractive index, due to temperature fluctuations that result from the turbulent mixing of the thermal layers of the atmosphere.

Fluctuations of the refractive index lead to spatial and temporal intensity fluctuations (known as scintillation or fading), distortion of the phase-front, loss of coherence, as well as spreading and wandering of the beam.

These effects are reviewed by STROHBEHN [463, 464], LAWRENCE and STROHBEHN [293], and BROOKNER [68, 69]. Several statistical models have been the subject of extensive theoretical and experimental investigations. At present, no model can be claimed to describe atmospheric propagation completely. The subject still has its paradoxes and confusions. However the log-normal model [469] is the most commonly used. In spite of discrepancies that have been observed in certain situations between the log-normal model and experiments (see, e.g., [119, 502]), it seems to be the most appropriate model for analysis of optical-communication problems. According to the log-normal model, the effect of atmospheric propagation on an optical field $V_0(\underline{x})$ is a multiplicative modulation

$$V(\underline{x}) = V_0(\underline{x})\exp\left[\frac{1}{2}\psi(\underline{x}) + j\phi(\underline{x})\right] \quad , \tag{6.95}$$

where $\psi(\underline{x})$ and $\phi(\underline{x})$ are real Gaussian stochastic processes. The intensity of the received field is then related to the intensity of the transmitted field by

$$I(\underline{x}) = I_0(\underline{x})\alpha(\underline{x}) \quad , \qquad \alpha(\underline{x}) = \exp[\psi(\underline{x})] \quad . \tag{6.96}$$

Accordingly, the intensity is modulated by a log-normal stochastic process $\alpha(x)$ (cf. Sec. 5.2.9).

The correlation time of the process $\psi(\underline{x})$ is often on the order of a millisecond or more. Therefore, for signal bandwidths of interest, the fading may be assumed contant over the detection time T. The correlation area of the process $\psi(\underline{x})$ is of the order of 1 cm^2 for visible light. Consequently, if the detector is smaller than 1 cm^2, we can assume that the integrated intensity is simply multiplied by a constant α that is log-normally modulated,

$$P(\alpha) = (\sqrt{2\pi}\sigma\alpha)^{-1}\exp\left[-(\ln\alpha + \tfrac{1}{2}\sigma^2)^2/2\sigma^2\right] \quad .$$

The log-variance σ^2 depends on many factors, such as the path length, the wavelength, the altitude, etc. The phase factor $\phi(\underline{x})$ does not affect the intensity fluctuations and is of no importance to direct-detection receivers. In heterodyne detection, the factor $\phi(\underline{x})$ must be considered.

The photoelectron statistics of several models of primary light beams affected by log-normal modulation have been examined in Sec. 5.2.9. It has also been argued that even in the presence of spatial integration over a large detector, the effect of fading would approximately be a multiplicative log-normal modulation with a smaller variance. Hence, we can use these results directly to examine the effect of atmospheric log-normal fading on the structure and performance of photoelectron-counting direct-detection receivers.

Pulse-Gated Binary Modulation (PGBM)
Detection of a Coherent Radar Signal

A number of models have been used for the photoelectron statistics of this problem:

I) TEICH and ROSENBERG [477], ROSENBERG and TEICH [406, 407], TITTERTEN and SPECK [480], and WEBB and MARINO [503] used the model,

(coherent signal × log-normal scintillation) + independent wide-band thermal background.

The model corresponds to Poisson statistics in the absence of the signal and Poisson statistics whose mean is averaged over a log-normal distribution in the presence of the signal

$$P[n|H^{(1)}] = \bar{n}_b^n \exp(-\bar{n}_b)/n! \quad , \tag{6.97}$$

$$P[n|H^{(2)}] = \left\langle (\bar{n}_s\alpha+\bar{n}_b)^n \exp(-\bar{n}_s\alpha-\bar{n}_b)/n! \right\rangle_\alpha \quad , \tag{6.98}$$

where α is a log-normally distributed RV whose mean is 1 and whose logarithm has variance σ^2.

II) FRIED and SCHMELTZER [156] used the same model but assumed that the background noise leads to Gaussian statistics. This should lead to the same results as in I) if the background level is sufficiently large. See also [331].

III) SOLIMENO et al. [455] studied the model

(coherent signal x log-normal scintillation) + interferring narrow-band thermal background.

This corresponds to Bose-Einstein statistics in the absence of the signal and Laguerre statistics averaged over a log-normal distribution in the presence of the signal

$$P[n|H^{(1)}] = \bar{n}_b^n/(1+\bar{n}_b)^{n+1} \quad , \tag{6.99}$$

$$P[n|H^{(2)}] = \left\langle \frac{\bar{n}_b^n}{(1+\bar{n}_b)^{n+1}} \exp\left(-\frac{\bar{n}_s\alpha}{1+\bar{n}_b}\right) L_n\left[-\frac{\bar{n}_s\alpha}{\bar{n}_b(1+\bar{n}_b)}\right] \right\rangle_\alpha \quad . \tag{6.100}$$

This model is applicable when the counting time is much shorter than the coherence time of the thermal background, and when the mixing efficiency is unity. Such a situation is unlikely unless picosecond pulses are used.

IV) The model:

(coherent signal x log-normal scintillation) + interferring thermal background with arbitrary bandwidth

was considered by LACHS and LAXPATI [282]. This corresponds to

$$P[n|H^{(1)}] = \underset{i}{\circledast}\, \bar{n}_{bi}^n/(1+\bar{n}_{bi})^{n+1} \quad , \tag{6.101}$$

$$P[n|H^{(2)}] = \left\langle \underset{i}{\circledast}\, \frac{\bar{n}_{bi}^n}{(1+\bar{n}_{bi})^{n+1}} \exp\left(-\frac{\bar{n}_s\alpha}{1+\bar{n}_{bi}}\right) L_n\left[\frac{-\bar{n}_s\alpha}{\bar{n}_{bi}(1+\bar{n}_{bi})}\right] \right\rangle_\alpha \quad , \tag{6.102}$$

where $\bar{n}_{bi}$, $i = 1,2, \ldots$ are the different modes of the thermal radiation (cf. Sec. 5.2.2) and $\circledast$ denotes multiconvolution. This model reproduces the models in I) and III) in the limits of large and small time-bandwidth products, respectively. The lengthy computations it requires are justified only in the intermediate range of time-bandwidth products.

LACHS and MINER [283] allowed the signal field to fluctuate and studied the effect of atmospheric turbulence on the performance of laser radar with a fluctuating target.

Because the model I) is the most likely to occur in present practical problems, we shall limit our discussions to it. We have already studied some of the basic properties of photoelectron-counting distribution due to log-normally faded light (Sec. 5.2.9). Using (6.98) in (5.253),(5.254), and (5.255), we obtain

$$P[n|H^{(2)}] = \frac{(M+\bar{n}_b)^n e^{-M-\bar{n}_b}}{n!} \frac{\exp\{-\frac{1}{2}\sigma^2 M[1-n/(M+\bar{n}_b)]\}}{\{1+\sigma^2 M[1-n/(M+\bar{n}_b)+nM^2/(M+\bar{n}_b)]\}^{\frac{1}{2}}} \quad , \tag{6.103}$$

where M satisfies

$$M = (\bar{n}_s+\bar{n}_b)\exp\left\{-\frac{1}{2}\sigma^2[1-2Mn/(M+\bar{n}_b) + 2M]\right\} . \tag{6.104}$$

The probability $P[n|H^{(1)}]$ is given by (6.97). The likelihood ratio is

$$\Lambda = \frac{\phi^{(2)}}{\phi^{(1)}} = \langle(1+\alpha\bar{n}_s/\bar{n}_b)^n \exp(-\bar{n}_s\alpha)\rangle_\alpha = f(n,\bar{n}_s,\bar{n}_b,\sigma) \quad , \tag{6.105}$$

where f is a monotonic increasing function of n. Consequently, the decision is equivalent to a threshold detector with an optimum threshold level given by

$$f(n_T,\bar{n}_s,\bar{n}_b,\sigma) = 1 \quad , \tag{6.106}$$

for equilikely hypotheses. For $\sigma = 0$, the optimum threshold is given by (6.66).

Now we can directly use (6.103) and compute the probability of error P_e, which will be a function of $\bar{n}_s$, $\bar{n}_b$, n_T, and the turburlence level σ.

Calculations show that the optimum threshold level decreases with an increase of the turbulence level σ. From (6.66), we already know that, for $\sigma = 0$, n_T increases with the signal-to-background ratio for a fixed background level. Computations from (6.106) show that, in the presence of turbulence ($\sigma > 0$), the increase of n_T with an increase of the signal-to-background ratio is less marked. For severe turbulence ($\sigma > 1.5$), n_T becomes insensitive to the signal-to-background ratio. As expected, the probability of error corresponding to optimum threshold is found to increase rapidly with an increase of σ. For moderate to high levels of σ, the increase of signal power that is necessary to achieve a probability of error equal to that obtained in the case of quiet atmosphere ($\sigma = 0$), for the same signal-to-background ratio can readily exceed 10 dB at moderate error rates (see Fig.6.10).

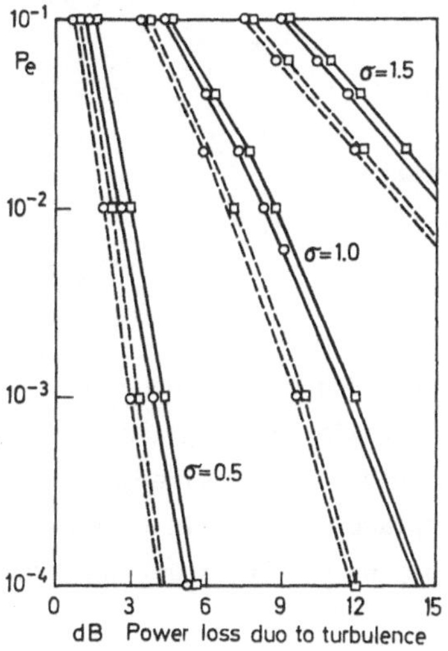

Fig.6.10. Increase of the signal $\bar{n}_s$ necessary to obtain a probability of error at a turbulence level σ, equal to that of the quiescent case ($\sigma = 0$): ———, PGBM, --------, PDBM; o , $\bar{n}_b = 1$; □, $\bar{n}_b = 4$ (after ROSENBERG and TEICH [406])

We know from Fig.6.5, which corresponds to $\sigma = 0$, that for a fixed $\bar{n}_b$, the error rate decreases with increase of the signal-to-background ratio. The larger σ becomes, the more insensitive the error rate becomes to change of the signal-to-background ratio.

The effect of using a suboptimum threshold level was studied by WEBB and MARINO [503]. The probability of error P_e increases sharply if n_T is different from its optimum value. In a practical situation, changing atmospheric conditions will deoptimize the system. An adaptive threshold detection may then prove useful.

Computations show that decrease of σ always results in improvement, even when the threshold is kept fixed. This suggests that we should choose the threshold that is optimum for the strongest expected scintillation.

Effect of Aperture Integration

The model described by (6.97) and (6.98) assumes that the effect of atmospheric scintillation is simply a log-normal modulation of the signal mean.

This is true only when the area of the detector is small compared to the area of coherence of the scintillation effect. Usually this is not the case. The area of the detector is, however, smaller than the area of coherence of the signal field. This allows us to average the scintillation effect over the aperture first and then use the average,

$$\bar{\alpha} = \frac{1}{A}\int_A \alpha(\underline{r})d\underline{r} \tag{6.107}$$

instead of α in (6.98).

As we have mentioned in Sec. 5.2.9, $\bar{\alpha}$ has variance,

$$\bar{\sigma} = \ln\left[(e^{\sigma^2}-1)/N_\sigma + 1\right] \quad , \tag{6.108}$$

where N_σ is the effective number of scintillation modes collected by the detector. It has also been argued that $\bar{\alpha}$ has an approximately log-normal distribution. If this approximation is used, all we need do to examine the effect of aperture scintillation averaging is to replace σ with $\bar{\sigma}$ in (6.103-106) and in Fig.6.10. Because $\bar{\sigma}$ is smaller than σ , we should expect improvement of performance due to aperture averaging.

Pulse-Position Modulation (PPM)

The performance of pulse-position modulation systems in the presence of atmospheric scintillation was studied by TEICH and ROSENBERG [477] and ROSENBERG and TEICH [406, 407] using the model:

(coherent signal × *log-normal scintillation)* + *wide band thermal background.*

The probability distributions are given by

$$P\left[[n]|H^{(k)}\right] = \left\langle \prod_{i=1}^{N}\left[\alpha\bar{n}_{si}^{(k)}+\bar{n}_b\right]^{n_i}\exp\left[-\alpha\bar{n}_{si}^{(k)}-\bar{n}_b\right]/n_i! \right\rangle_\alpha \quad , \tag{6.109}$$

where α is assumed constant over the interval MT. The orthogonal modulation format $\bar{n}_{si}^{(k)} = \bar{n}_s\delta_{ik}$ leads to

$$P\left[[n]|H^{(k)}\right] = \left\langle (\alpha\bar{n}_s+\bar{n}_b)^{n_k}\, \bar{n}_b^{\sum n_i - n_k} \exp(-\bar{n}_s\alpha - M\bar{n}_b)\right\rangle_\alpha \Big/ \prod_i (n_i!) \quad .$$

Likelihood functions can then be written in the form

$$\phi^{(k)} = <(1+\alpha\bar{n}_s/\bar{n}_b)^{n_k}\exp(-\alpha\bar{n}_s-\bar{n}_b)>_\alpha \quad . \tag{6.110}$$

This is obviously a monotonically increasing function of n_k (for any distribution of α). Hence, we can take

$$\phi^{(k)} = n_k \quad . \tag{6.111}$$

Again, we have the simple comparison test, just as in the absence of turbulence.

In order to evaluate the performance of the system, we have to compute the expectation value in (6.109) and use the usual equations for P_e. This is considerably simpler in the case M = 2 that corresponds to the PDBM or the BPLM systems. In such systems, P_e was calculated by ROSENBERG and TEICH [406] as a function of $\bar{n}_s$, $\bar{n}_b$, and the turbulence level σ. When $\sigma = 0$, the values of P_e, plotted in Fig.6.6, are reproduced. Results are best displayed in terms of the dB increase of $\bar{n}_s$ necessary to obtain a probability of error equal to that corresponding to $\sigma = 0$ at the same signal-to-background ratio. This is illustrated by the broken lines of Fig.6.10. For comparison, the corresponding errors for the PGBM system are also shown by solid lines.

We have already mentioned that the PDBM has an advantage of about 2 dB power gain over the PGBM in the absence of turbulence. Figure 6.10 shows that this gain can readily exceed 3 dB at moderate levels of turbulence and 5 dB near $\sigma = 1.5$. Thus, the binary orthogonal format employed in PDBM or BPDM systems appears to be more immune to the effect of atmospheric turbulence. Note that in the case of BPLM, the possible effect of atmospheric depolarization should be considered. But, as SALEH [413] noted, this does not appear to have an important contribution at optical frequencies.

Effect of Diversity on the Performance of a Radar-Signal Detector. Array Detectors

Here we assume that an array of detectors is used to detect the presence or absence of a radar signal. A quantitative assessment of the magnitude of the anticipated improvement of performance is very important.

Consider an array of D detectors that simultaneously measure the number of counts $[n] = n_1, n_2, \ldots n_D$ in a time interval [0,T]. The observation [n] is to be used to make a decision concerning the presence or absence of a coherent signal whose average is $\bar{n}_\ell$ at detector ℓ. Background radiation produces an average of $\bar{n}_{b\ell}$ counts at detector ℓ. The conditional-probability distributions are given by

$$P\left[[n]|H^{(1)}\right] = \prod_{\ell=1}^{D} \bar{n}_b^{n_\ell} \exp(-\bar{n}_b)/n_\ell! \quad , \tag{6.112}$$

$$P\left[[n]|H^{(2)}\right] = \left\langle \prod_{\ell=1}^{D} (\alpha_\ell \bar{n}_{s\ell} + \bar{n}_{b\ell})^{n_\ell} \exp(-\alpha_\ell \bar{n}_{s\ell} - \bar{n}_{b\ell})/n_\ell! \right\rangle_{[\alpha]} \quad , \tag{6.113}$$

where $[\alpha] = \alpha_1, \alpha_2, \ldots \alpha_D$ are random variables that represent the effect of fading at the different detectors.

We next write the likelihood ratio as

$$\Lambda = \left\langle \prod_{\ell=1}^{D} (1+\alpha_\ell \bar{n}_{s\ell}/\bar{n}_{b\ell})^{n_\ell} \exp(-\alpha_\ell \bar{n}_{s\ell}) \right\rangle_{[\alpha]} \quad , \tag{6.114}$$

The problem cannot be further pursued without specifying the statistics of $[\alpha]$.

The assumption that $\{\alpha_\ell\}$ are statistically independent, i.e., that the detectors intercept independent paths of the signal, simplifies the problem considerably. It enables us to write

$$\ln \Lambda = \sum_{\ell=1}^{D} \ln \langle (1+\alpha_\ell \bar{n}_{s\ell}/\bar{n}_{b\ell})^{n_\ell} \exp(-\alpha_\ell \bar{n}_{s\ell}) \rangle_{\alpha_\ell} = \sum_{\ell=1}^{D} \ln f_\ell(n_\ell) \quad , \tag{6.115}$$

where $f_\ell(n_\ell)$ is, in general, a nonlinear function of n_ℓ that depends on $\bar{n}_{s\ell}, \bar{n}_{b\ell}$, and the probability distribution of α_ℓ. The decision rule is therefore based on comparing the sum of a nonlinearly weighted number of counts in the detector. To give an example, assume that the detectors are identical, that they receive equal signal and background energies, and that $\{\alpha_\ell\}$ are identically distributed. Then f_ℓ is independent of ℓ and is given by

$$f(n) = \langle (1+\alpha\bar{n}_s/\bar{n}_b)^n \exp(-\bar{n}_s\alpha) \rangle_\alpha \quad , \tag{6.116}$$

as in (6.105).

In addition, if we assume the α is log-normally distributed, we can use (5.253-255) and obtain

$$f(n) = \left(1 + \frac{M}{\bar{n}_b}\right)^n e^{-M} \left\{1+\sigma^2 M \left[1 - \frac{n}{\bar{n}_b+M} + \frac{nM^2}{(\bar{n}_b+M)^2}\right]\right\}^{-1/2} \exp\left[-\frac{1}{2}\sigma^2\left(M - \frac{Mn}{\bar{n}_b+M}\right)\right], \quad (6.117)$$

where M is determined from (6.104).

Computation of the performance of the above system is not so simple. The decision rule,

$$\sum_{\ell=1}^{D} f(n_\ell) \underset{1}{\overset{2}{\gtrless}} \eta \quad , \quad (6.118)$$

divides the D-dimensional space of counts $n_1, n_2, \dots n_D$ into two decision regions, which have to be defined. The probability that the observation [n] lies in each of these regions, under each hypothesis has to be calculated; then the probability of error can be obtained. Results were obtained for D = 2 by ROSENBERG and TEICH [406]. Their conclusions can best be appreciated in terms of a comparison between two systems. The first is an array of 2 detectors that measure n_1 and n_2 and use their nonlinearly weighted sum to obtain a decision. The second system uses the direct sum $n = n_1 + n_2$ to obtain a decision. This second system is obviously much simpler, because it can be realized by using only one detector with double the area, i.e., an aperture-integrating detector. The performance of an aperture-integrating detector was discussed on page 332. Since n_1 and n_2 are assumed independent, we have to adjust the turbulence level for the effect of integrating over two independent modes, i.e., put $N_\sigma = 2$ in (6.108).

A comparison between the two systems leads to the conclusion that the array-detector gain will be significant only for severe turbulence and high background noise.

It remains to say a word about the effect of the correlation between the fading paths that we previously neglected. TEICH and ROSENBERG [476] computed the joint-probability distribution in (6.113) for correlated $\{\alpha_\ell\}$. ROSENBERG and TEICH [406] showed that the variation of the optimum threshold boundary, in the n_1-n_2 plane, with the correlation coefficient ρ, between α_1 and α_2, is minimal for low background level ($\bar{n}_b = 2$); for high background level ($\bar{n}_b = 8$) and larger σ, however, the effect is pronounced. The probability of error is sensitive to the degree of fading correlation especially at high background levels. In the limit of fully correlated, fading, the performance is identical to a single-detector receiver with twice the corresponding background. From Fig.6.10, we can see that there is more than an order-of-magnitude increase of P_e as ρ goes from 0 to 1 for moderate turbulence levels, and somewhat less than an order of magnitude at $\sigma = 1.5$.

6.4.5 Heterodyne Detection

The technique of heterodyning or mixing the signal field with a strong coherent field (local oscillator) was introduced in Sec. 6.3 as a method of combating background noise and of allowing phase- and frequency-modulation systems. Here we consider the performance of some digital systems based on this technique.

PGBM. Detection of a Radar Signal

If the signal is coherent, then both $P[n|H^{(1)}]$ and $P[n|H^{(2)}]$ are Poisson, with means approximately equal to $\bar{n}_0$ and $\bar{n}_0+2(\bar{n}_0\bar{n}_s)^{1/2}$, respectively ($\bar{n}_s$ and $\bar{n}_0$ are the average number of counts due to the signal and local oscillator, respectively).

The performance diagram of Fig.6.5 could then be used with an effective signal corresponding to $2(\bar{n}_0\bar{n}_s)^{1/2}$ counts and an effective background $\bar{n}_0$. The threshold level is given by

$$n_T = 2(\bar{n}_0\bar{n}_s)^{1/2}/\ln\,[1+2(\bar{n}_s/\bar{n}_0)^{1/2}] \approx \bar{n}_0 \quad . \tag{6.119}$$

Because $\bar{n}_0$ is usually large, the Poisson distributions could be replaced by Gaussian distributions and the performance diagrams of the usual Gaussian channel could be used.

By examining Fig.6.5, it is easy to see that when strong background is present, heterodyning could improve the performance.

Phase-Shift Keying (PSK)

It is possible to demodulate phase modulated light with a homodyne receiver. The detected intensity is

$$I \simeq I_0 + 2\,(I_0I_s)^{1/2}\cos(\theta) \quad .$$

If we choose the phase 0 as hypothesis 1, and the phase π as hypothesis 2, we obtain Poisson distributions for $P[n|H^{(1)}]$ and $P[n|H^{(2)}]$ with means

$\bar{n}_0+2(\bar{n}_0\bar{n}_s)^{1/2}$ and $\bar{n}_0-2(\bar{n}_0\bar{n}_s)^{1/2}$, respectively. If we regard the quantity $4\sqrt{\bar{n}_0\bar{n}_s}$ as the effective signal and $\bar{n}_0-2(\bar{n}_0\bar{n}_s)^{1/2}$ as the effective background, we can directly use the performance diagram of Fig.6.5.

Detection of a Fluctuating Radar Signal

In this problem, hypothesis 1 corresponds to a strong coherent local-oscillator field and negligible background field, whereas, hypothesis 2 corresponds to the sum of the strong coherent local osicllator and the thermal signal field, as well as a relatively negligible background field.

Under hypothesis 1, the number of counts n has a Poisson distribution with mean $\bar{n}_0$,

$$P[n|H^{(1)}] = \bar{n}_0^n \exp(-\bar{n}_0)/n! \quad .$$

Under hypothesis 2, n is Laguerre distributed with mean $\bar{n}_0 + \bar{n}_s$,

$$P[n|H^{(2)}] = \frac{(\bar{n}_s/N)^n}{(1+\bar{n}_s/N)^{n+N}} \exp\left(-\frac{\bar{n}_0}{1+\bar{n}_s/N}\right) L_n^{N-1}\left[-\frac{\bar{n}_0}{(1+\bar{n}_s/N)\bar{n}_s/N}\right] \quad ,$$

where N is the number of signal modes. If we pursue the problem algebraically to determine the ML test, as we did in all previous cases, we end with a complex nonlinear inequality that determines the division of the possible observed values of n into two regions, corresponding to the decisions $H^{(1)}$ and $H^{(2)}$. It turns out that the decision rule takes the form:

$$\begin{aligned}
&\text{If } n_{T1} < n < n_{T2}, && \text{choose } H^{(1)} \\
&\text{If } n > n_{T2} \text{ or } n < n_{T1}, && \text{choose } H^{(2)} \\
&\text{If } n = n_{T1} \text{ or } n = n_{T2}, && \text{choose } H^{(1)} \text{ with probability } 1/2 \quad .
\end{aligned} \tag{6.120}$$

This can be very easily illustrated by plotting $P[n|H^{(1)}]$ and $P[n|H^{(2)}]$ as in Fig.6.11. For hypotheses with equal a priori probability, n_{T1} and n_{T2} are determined by the intersection between these probability distributions. Note that the two distributions have approximately the same mean, but the one that corresponds to the presence of the thermal signal is considerably broader. This makes the distributions always intersect at two points.

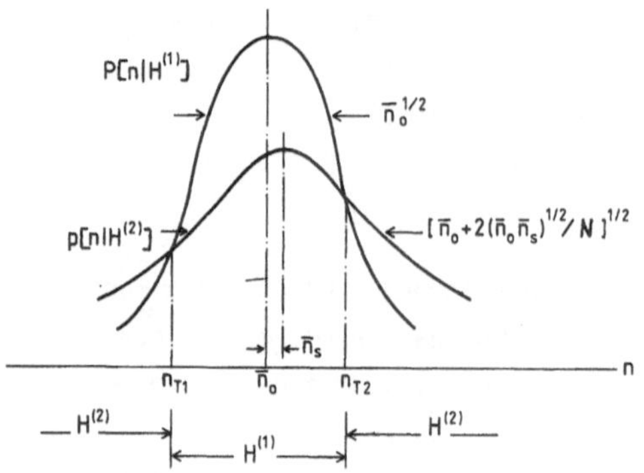

Fig.6.11. Photoelectron counting PD due to a coherent local oscillator $[H^{(1)}]$ and that due to a superimposed thermal signal $[H^{(2)}]$. Threshold levels n_{T_1} and n_{T_2} are determined by the intersection points

Since $\bar{n}_0 \gg \bar{n}_s$, we can approximately write the rule (6.120) in the form

$$|n - \bar{n}_0| \underset{1}{\overset{2}{\gtrless}} S \quad , \quad S = (n_{T2} - n_{T1})/2 \quad . \tag{6.121}$$

The parameter S determines a fluctuation threshold that is much more likely to be exceeded in the presence of the thermal signal. The probability of error can be determined from

$$P_e = \frac{1}{2} Q_1 + \frac{1}{2} Q_2$$

$$Q_1 = \sum_{n=n_{T1}+1}^{n_{T2}-1} P[n|H^{(1)}] + \frac{1}{2} P[n_{T1}|H^{(1)}] + \frac{1}{2} P[n_{T2}|H^{(1)}] \quad , \tag{6.122}$$

$$Q_2 = \sum_{n=0}^{n_{T1}-1} P[n|H^{(2)}] + \sum_{n=n_{T2}+1}^{\infty} P[n|H^{(2)}] + \frac{1}{2} P[n_{T1}|H^{(2)}] + \frac{1}{2} P[n_{T2}|H^{(2)}] \quad .$$

Detailed performance diagrams can be found in [236]. Note that as *N* increases, the broadening due to the presence of the signal becomes less, and the probability of error increases.

6.5 Inference Based on Instants of Occurrence of Photoelectrons

In Sec. 6.2.4, we examined the estimation and detection of optical signals using photoelectron-counting receivers. In this section, we assume that the receiver is capable of recording the instants of occurrence of the photoelectrons. With this additional available data, we should expect an improvement of performance. Obviously, the hardware of an instants-of-occurrence receiver is considerably more complex. We shall also see that the optimum data processing is significantly more complicated than in the case of photoelectron-counting receivers. Moreover, it is rather difficult to assess the performance of receivers based on instants of occurrence. This section points out these difficulties. It also examines the many situations in which photoelectron-counting receivers are optimum and no gain of performance is achieved by measuring the instants of occurrence.

6.5.1 Estimation Problems

Parameters of a Coherent Signal

Consider a coherent optical signal that corresponds to a density of photoelectron events $\lambda(t,\theta)$, where θ is a parameter to be estimated. Assume that the signal is observed during a time interval $[0,T]$ and that $\{t_j\} = (t_1, t_2, \dots t_n)$ are the instants of occurrence of the photoelectrons observed by the detector. Given $\{t_j\}$, what is the optimum estimate for θ?

We recall from Sec. 3.4.2 that the joint-probability density of $\{t_j\}$ is given by

$$P(\{t_j\}) = \exp\left[-\int_0^T \lambda(t,\theta)dt\right]\prod_{j=1}^{n}\lambda(t_j,\theta) \quad . \tag{6.123}$$

In order to obtain the ML estimate, we write the likelihood function $\phi = P(\{t_j\})$ and calculate the value of θ which makes it a maximum. Putting $\partial \ln\phi/\partial\theta = 0$ gives a nonlinear equation for the estimator $\hat{\theta}$ in terms of $\{t_j\}$.

$$\frac{\partial}{\partial\theta}\sum_{j=1}^{n}\ln\lambda(t_j,\theta) = \int_0^T \frac{\partial}{\partial\theta}\lambda(t,\theta)dt \quad . \tag{6.124}$$

Example 1. Let $\lambda(t) = \theta S(t)+b$, where $S(t)$ and b are known and θ is to be estimated. Eq. (6.124) then gives

$$\sum_{j=1}^{n} S(t_j)/[\hat{\theta}\, S(t_j)+b] = \int_0^T S(t)dt \quad . \tag{6.125}$$

This is an algebraic equation of power n. If $S(t) = S$ is independent of t, then

$$\hat{\theta} = (n/T - b)/S \quad ; \tag{6.126}$$

i.e., it is sufficient to know the total number of counts n in order to determine $\hat{\theta}$. Knowledge of the times at which the photoelectrons occur does not affect our estimate.
If we assume that $b \ll aS(t)\ \forall t \in [0,T]$, then we can approximate (6.125) by

$$\hat{\theta} = n/\int_0^T S(t)dt - (b/n)\sum_j 1/S(t_j) \quad ; \tag{6.127}$$

and if $b = 0$, we obtain

$$\hat{\theta} = n/\int_0^T S(t)dt \quad . \tag{6.128}$$

We conclude from this example that, in some situations, n is a sufficient statistic for the estimation of θ, but, in general, it is not.

Example 2. Let $\lambda(t) = S(t-\theta)+b$, where $S(t)$ is a known signal, b is a known background level, and θ is an unknown time delay. The signal has the form of a pulse that exists somewhere in the interval $[0,T]$. Its exact location θ is to be estimated. From (6.124), we can write

$$\frac{\partial}{\partial\theta}\sum_j \ln\,[S(t_j-\theta) + b] = 0 \quad . \tag{6.129}$$

This gives a nonlinear equation for θ in terms of $\{t_j\}$. A detailed study of this problem and its applications to PPM systems can be found in [32-34].

Parameters of a Thermal Signal

We recall from Sec. 5.3.1 that, for thermal light, the JPD of instants of occurrence in $[0,T]$ is given by

$$P(\{t_j\}) = Q(1)\langle n\rangle^n \sum_\pi \prod_{j=1}^{n} \eta(t_j,t_{\pi_j}) \quad , \tag{6.130}$$

where Q(1) is the probability of zero counts, $\langle n\rangle$ is the mean number of counts, and η is a function similar to the coherence function of the field (cf. (5.327)). We are interested in estimating the mean intensity of the field $\langle\lambda\rangle$. This can be obtained from an estimate of $\langle n\rangle$, by dividing by T. Since $\eta(t,t')$ depends on $\langle n\rangle$, an estimate $\hat{n}$ of $\langle n\rangle$ would depend on the times $\{t_j\}$ as well as on n.

In many cases, however, the function $\eta(t,t')$ is approximately proportional to the coherence function g(t,t'). This takes place (cf. Sec. 5.3.1) when $T >> \tau_c$, when $T > \tau_c$ and the spectrum is rectangular, or when the number of counts per coherence time is small. In these situations, $P(\{t_j\})$ factorizes into a product of a function of n and $\langle n\rangle$ and another of $\{t_j\}$. This means that the ML estimate of $\langle n\rangle$ depends on n and not on $\{t_j\}$. The total number of counts is then a sufficient statistic for the estimation of $\langle n\rangle$. When this approximation is not applicable, knowledge of $\{t_j\}$ could improve the estimate of $\langle n\rangle$.

Now assume that thermal light is modulated by a signal S(t),

$$\lambda(t) = \lambda_0(t)S(t) \quad , \quad \text{where, for simplicity,} \quad \langle\lambda_0(t)\rangle = 1 \quad .$$

Then,

$$P(\{t_j\}) = \left\langle \exp\left[-\int_0^T \lambda_0(t)S(t)dt\right] \prod_{j=1}^{n} \lambda_0(t_j)S(t_j) \right\rangle . \tag{6.131}$$

Because it is very difficult to determine this expectation exactly, we resort to approximation. If S(t) is much slower than the fluctuation in $\lambda_0(t)$, then we can write

$$\int_0^T \lambda_0(t)S(t)dt \simeq \langle\lambda_0(t)\rangle \int_0^T S(t)dt = \int_0^T S(t)dt \quad , \tag{6.132}$$

and therefore,

$$P(\{t_j\}) \simeq g^{(n)}(t_1,t_2, \ldots t_n)\exp\left[-\int_0^T S(t)dt\right] \prod_{j=1}^{n} S(t_j) \quad , \tag{6.133}$$

where $g^{(n)}(t_1, \dots t_n) = \left\langle \prod_{j=1}^{n} \lambda_0(t_j) \right\rangle$

is the intensity correlation function of the unmodulated thermal beam. Since $g^{(n)}$ is independent of S(t), we can write the likelihood function

$$\phi = \exp\left[-\int_0^T S(t)dt\right] \prod_{j=1}^{n} S(t_j) \quad . \tag{6.134}$$

This is exactly the likelihood function that corresponds to the coherent case, (6.124). Therefore, the analysis on page 339 applies here too.

To test the approximation in (6.132), we determine the root-mean-square normalized error. This is equal to

$$\frac{1}{\sqrt{N}}\left[\frac{1}{T}\int_0^T S^2(t)dt\right]^{1/2} \Bigg/ \left[\frac{1}{T}\int_0^T S(t)dt\right] \quad ,$$

where N is the number of thermal modes in [0,T]. If natural thermal light is used, then $N \gg 1$ and the approximation seems to be satisfactory [416].

When N is not very large, the approximation becomes less accurate and an exact evaluation of (6.131) must be provided.

Successful data processing based on measurement of the time of arrival of the Mth photoelectron was attempted [117, 389]. In general, these measurements contain less information than in the times of occurrence of all photoelectrons. But they may perform better than counters in fixed time intervals [389].

Estimation of Time-Varying Random Parameters of an Optical Signal

In many situations, the parameter to be estimated is itself a random function of time with a known statistical law. The density of events can be written as $\lambda[\theta(t),t]$ where $\theta(t)$ is a stochastic process. From the measured instants of occurrence $\{t_j\}$ we seek an estimate of the realization $\theta(t)$ whose occurrence led to the observation $\{t_j\}$.

We consider an example. Coherent light of intensity λ_0 (counts per second) is scattered from a moving rough surface. The scattered light $\lambda(t) = \lambda_0\theta(t)$ is modulated with a thermal signal $\theta(t)$. The instants of occurrence $\{t_j\}$ are measured in a time interval [0,T]. We are interested in estimating $\theta(t)$, $t \in [0,T]$ as a function of time. The estimate portrays the effect of the random variations in the rough surface. Since $\lambda(t)$ is thermal, $P(\{t_j\})$ is given by (6.130). The simplest estimation is the MAP estimator. It can be determined from (2.147) [304]

$$\hat{\lambda}(t = t_{n+1}) = \left\langle \exp\left[-\int_0^T \lambda(t')dt'\right] \prod_{j=1}^{n+1} \lambda(t_j) \right\rangle \Big/ \left\langle \exp\left[-\int_0^T \lambda(t')dt'\right] \prod_{j=1}^{n} \lambda(t_j) \right\rangle$$

$$= \left[\sum_{\pi_{n+1}} \prod_{j=1}^{n+1} \eta(t_j,t_{\pi j})\right] \Big/ \left[\sum_{\pi_n} \prod_{j=1}^{n} \eta(t_j,t_{\pi j})\right] \quad . \tag{6.135}$$

The width of the function $\eta(t,t')$ is the memory time of the photoelectrons τ_η. If $\tau_\eta < T$, then many of the terms of sums in (6.135) would be negligible. Only those points that lie within a memory time from t would have contribution to the sums. This simplifies the expression considerably. For example, if the density of photoelectrons is so low that no photoelectrons occur in the vicinity of t, then (6.135) reduces to

$$\hat{\lambda}(t) = \langle n \rangle\, \eta(t,t) \quad . \tag{6.136}$$

If only one photoelectron occurs at a time t_ℓ in the neighbourhood of t, then

$$\hat{\lambda}(t) = \langle n \rangle\, [\eta(t,t) + |\eta(t,t_\ell)|^2/\eta(t_\ell,t_\ell)] \quad , \tag{6.137}$$

and so on.

When $\theta(t)$ follows other probability laws, the problem complicates significantly. Suppose that the process $\theta(t)$ is a priori known to be described by a stochastic differential equation (cf. Sec. 2.4.5), e.g.,

$$\frac{d\theta}{dt} = f(\theta,t) + g(\theta,t)w(t) \quad . \tag{6.138}$$

The instants of occurrence $\{t_j\}_{0,T}$, in [0,T], are measured, and $\theta(t)$ is to be estimated at all times $t \in [0,T]$. We first consider a causal estimate, i.e., an estimate of $\theta(t)$ based on past observations $\{t_j\}_{0,t}$. If we obtain

$P[\theta(t)|\{t_j\}_{0,t}]$, the posterior probability distribution of $\theta(t)$ given $\{t_j\}_{0,t}$, then we can directly determine the MAP and MMSE estimators of $\theta(t)$ (cf. Sec. 2.7.2). SNYDER [451, 453] showed that if $\theta(t)$ satisfies (6.138), then the posterior probability density $P = P[\theta(t)|\{t_j\}_{0,t}]$ satisfies the differential equation

$$dP = L(P)dt + P.[\lambda(\theta,t)-\hat{\lambda}(t)]\,[\hat{\lambda}(t)]^{-1}[dN_0(t)-\hat{\lambda}(t)dt] \quad , \qquad (6.139)$$

where $L(.)$ is the forward Kolmogrov differential operator defined in (2.78), $N_0(t)$ is the discrete representation of the point process $\{t_j\}$ described in Fig.3.1, and

$$\hat{\lambda}(t) = \langle\lambda(\theta,t)|\{t_j\}_{0,t}\rangle = \int \lambda(\theta,t)P(\theta|\{t_j\}_{0,t})d\theta \qquad (6.140)$$

is the MMSE estimator of the intensity.

The combination of (6.139) and (6.140) gives an integro-differential equation for $P(\theta|\{t_j\}_{0,t})$.

An estimate for $\theta(t)$ based on measurements in the interval $[0,T]$, $0 \le t < T$, can be obtained [452, 453] by using the relation,

$$dP(\theta|\{t_j\}_{0,T}) = L[P(\theta|\{t_j\}_{0,t})]\exp[\psi_{t,T}(\theta)]dt \quad , \qquad (6.141)$$

where

$$\psi_{t,T}(\theta) = -\int_t^T [\langle\lambda(0,t')|\{t_j\}_{t,t'}\rangle - \langle\lambda(\theta,t')|\{t_j\}_{0,t'}\rangle]dt'$$
$$+\int_t^T \ln\,[\langle\lambda(\theta,t')|\{t_j\}_{t,t'}\rangle/\langle\lambda(\theta,t')|\{t_j\}_{0,t'}\rangle]dN_0(t') \quad . \qquad (6.142)$$

Equations (6.139-142) appear to be analytically intractable. Although numerical solutions are possible, they require complicated numerical algorithms. Substantial progress has been made in finding approximations to these equations. Details of these techniques and their applications can be found in [145, 221, 451, 453, 454]. Applications to experimental data were accomplished by DAVIDSON et al. [118].

6.5.2 Detection Problems

Coherent Signals

Consider the problem of M-ary test of hypotheses of coherent signals in constant background. Hypothesis $H^{(k)}$ corresponds to a photoelectron density

$$\lambda^{(k)}(t) = \lambda_s^{(k)}(t) + \lambda_b \quad .$$

Decision is based on the measured instants of occurrence $\{t_j\}$, whose JPD under hypothesis $H^{(k)}$ is

$$P[\{t_j\}|H^{(k)}] = \exp\left\{- \int_0^T [\lambda_s^{(k)}(t)+\lambda_b]dt\right\} \prod_{j=1}^{n} [\lambda_s^{(k)}(t_j)+\lambda_b] \quad . \tag{6.143}$$

The likelihood functions are

$$\phi^{(k)} = - \int_0^T \lambda_s^{(k)}(t)dt + \sum_{j=1}^{n} \ln [\lambda_s^{(k)}(t_j) + \lambda_b] \quad . \tag{6.144}$$

For hypotheses with equal signal energies

$$\phi^{(k)} = \sum_{j=1}^{n} \ln[\lambda_s^{(k)}(t_j) + \lambda_b] \quad . \tag{6.145}$$

For binary test of hypothesis, $\lambda_s^{(1)}(t) = 0$, $\lambda_s^{(2)}(t) = \lambda_s(t)$, and (6.144) leads to the threshold test

$$\sum_{j=1}^{n} \ln [1 + \lambda_s(t_j)/\lambda_b] \underset{1}{\overset{2}{\gtrless}} n_T \quad , \qquad n_T = \int_0^T \lambda_s(t)dt \quad . \tag{6.146}$$

We therefore conclude that logarithmic weighting is generally required. If the signal-to-background ratio is very small, this becomes approximately a linear weighting of $\lambda_s(t_j)$. Note, however, that if the signals $\lambda_s^{(k)}(t)$ are independent of time during the observation time [0,T], then the total number of counts n is a sufficient statistic. In a PDBM, it can be easily shown that the number of counts n_1 and n_2 in [0,T/2] and [T/2,T] are sufficient statistics. Detailed studies of the performance of PPM systems can be found in [32].

Thermal Signals

If the signals are thermal, then

$$P[\{t_j\}|H^{(k)}] = \left\langle \exp\left\{-\int_0^T [\lambda_s^{(k)}(t)+\lambda_b]dt\right\} \prod_{j=1}^{n} [\lambda_s^{(k)}(t_j)+\lambda_b] \right\rangle . \qquad (6.147)$$

The likelihood functions can be written as (MACCHI and PICINBONO [304])

$$\phi^{(k)} = \left\langle \exp\left[-\int_0^T \lambda_s^{(k)}(t)dt\right] \prod_{j=1}^{n} [1+\lambda_s^{(k)}(t_j)/\lambda_b] \right\rangle , \qquad (6.148)$$

or

$$\phi^{(k)} = \sum_{p=1}^{n} \lambda_b^{-p} \sum_{i_1<i_2<\dots<i_p}^{n} \left\langle \exp\left[-\int_0^T \lambda_s^{(k)}(t)dt\right] \prod_{i=i_1}^{i_p} \lambda_s^{(k)}(t_i) \right\rangle . \qquad (6.149)$$

The expectation in (6.149) can be obtained by using (6.130),

$$\phi^{(k)} = Q^{(k)}(1) \sum_{p=1}^{n} \lambda_b^{-p} \sum_{i_1<i_2<\dots<i_p}^{n} \sum_{\pi_{i_p}} \prod_{i=i_1}^{i_p} \eta(t_i,t_{\pi_i}) \quad . \qquad (6.150)$$

The test is rather complicated. It requires analytic expressions for Q(1) and $\eta(t,t')$ for each hypothesis and the computation of the lengthy permutations and summations in (6.150). In the limit of small signal-to-background, we can retain the first term of (6.150) only. Eq. (6.150) simplifies to

$$\phi^{(k)} = Q^{(k)}(1) \sum_{j=1}^{n} \eta^{(k)}(t_j,t_j) \quad .$$

Note that if the mean signal intensities are independent of time, then $\eta(t_j,t_j)$ is independent of t_j and n becomes a sufficient statistic. If the signal levels are high, or if they are time dependent, then $\{t_j\}$ must be included in the computation of the likelihood functions. Assessment of the performance of the system in all these cases is yet to be accomplished.

6.6 Concluding Remarks

In this chapter, we have assumed certain structures of optical receivers and examined their performance under different conditions of signal and background statistics. By no means have we exhausted the possible receiver structures, nor have we attempted to find the most-optimum ones. HELSTROM has pioneered fundamental research on the optimum detection of signals carried by an optical field. Employing a quantum description of the optical field, he reformulated the basic detection and estimation problems in quantum language and obtained what is known as quantum detection and estimation theory. We refer the reader to the review articles [200, 201, 204, 207, 209] and to HELSTROM'S book [208] for further reading on this subject.

Another aspect of optical communication that has not been considered in this chapter is the important problem of the flow of information in optical channels. This problem has been studied by many authors; the reader is urged to pursue this subject independently in the references [34, 44, 64, 141, 161, 162, 182, 183, 214-216, 251, 315, 411, 506]. The material of this chapter and of Part II of this book should facilitate such a study.

7. Applications to Spectroscopy

This chapter is concerned with determining the spectrum of an optical field by observing the photoelectrons detected by a photodetector in a finite period of time.

The simplest and most obvious method of obtaining spectral information from an optical field is to use an optical filter that scans the spectrum by allowing a narrow band of frequencies to pass and to measure its intensity. This may be achieved by using a prism that deflects different frequencies in different directions or by using a dispersive material that absorbs or reflects all frequencies except a selected narrow band. Interferometers can be used very efficiently to select a narrow band of frequencies by allowing the wave that has the desired frequency to interfere constructively with its own reflections while all other waves interfere destructively. The Fabry-Perot étalon and the diffraction grating are good examples of such interferometers.

Another method of measuring the spectrum is to make use of the fact that the power spectral distribution is the Fourier transform of the field's correlation function (cf. Sec. 4.1.2). Measurement of the field's correlation function is then equivalent to the determination of its spectrum. There are several interferometric methods of measuring the field's correlation function. For example, in the Michelson interferometer described in Sec. 4.1.3 and Fig.4.6, the relative delay between two interferring beams is changed by introducing an optical-path-length difference $d = c\tau$ and the intensity of the resultant interference is measured as a function of d.

The preceding two methods depend on measuring light intensities. This can be achieved by photoelectron-counting techniques; the accuracy of the measurement can be assessed by the methods presented in Chap. 6.

The third method, which is the subject of this chapter, is based on the possibility of inferring the spectrum from observation of fluctuations of the photoelectron events without need for a scanning bank of filters or an interference arrangement.

No doubt, the reader remembers from Chap.5 that the spectral profile plays an important role in determining the statistics of the photoelectron process:

counting distributions, time-interval distributions, moments, correlations, etc. Some of these statistical distributions give direct access to the field's correlation function (and consequently its power spectrum). These will be examined in detail in the present chapter.

An upper limit of bandwidths that can be measured by the photoelectron-statistics method is imposed by the highest speed of the necessary electronic instrumentation. Commercially available fast electronic circuits can measure variations in the range of 0.01 μs. This sets the widest bandwidth observable by photoelectron-counting techniques at 100 MHz. No lower bound exists on the width of spectra observed by this method. In principle, bandwidths as narrow as 1 Hz can be observed. Thus, the photoelectron-counting technique covers the vast range of linewidths from 1 Hz to 10^8 Hz, i.e., extends the resolution indefinitely.

The photoelectron-counting technique complements other spectroscopic techniques in covering the whole range of linewidths. In order to illustrate this, refer to Fig.7.1, which is a plot of the possible bandwidths $\Delta\nu$, their corresponding resolutions $\nu_0/\Delta\nu$ (where ν_0 is a typical optical frequency, $5 \cdot 10^{14}$), their corresponding coherence times $\tau_c = 1/\Delta\nu$, and their corresponding path lengths $d = c\tau_c$. Interferometers are limited to a range of geometrically reasonable delay d, above which they become impractical. A Fabry-Perot étalon that introduces a multipath delay is limited to d = 70 m delay, which corresponds to a resolution of 10^8 (i.e.), covers bandwidths greater than 5 MHz).

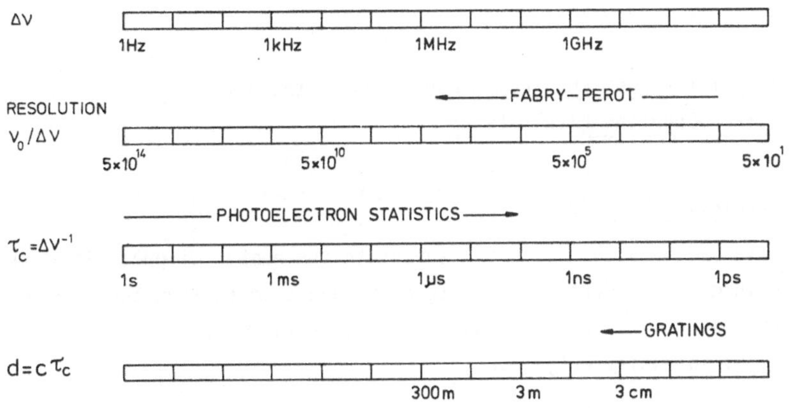

Fig.7.1. Operating domains of high-resolution spectrometers

This makes the method of photoelectron-statistics spectroscopy a very powerful tool for observing slower phenomena that could not have been considered before.

The use of photoelectron statistics as a spectroscopic tool has always been associated with laser light. This is simply because the natural broadening in conventional light sources is so great that a resolution greater than that which is available in conventional spectrometers is rarely required. The laser light itself has a very narrow bandwidth. But if it is scattered from a target, it can be slightly broadened and a measurement of the broadening is of great use in acquiring information about the target itself. Laser-scattering spectroscopy has been applied to the study of several phenomena including Brownian motion, theory of liquids, polymer solutions, phase transitions, liquid crystals, turbulence, anemometry, and laser radar. Several textbooks and review articles cover this subject in depth [3, 52, 95, 107, 109, 384, 431]. This book is not concerned with phenomena that lead to spectral broadening. Instead, it is about methods of measuring narrow spectral profiles, using photoelectron statistics.

This chapter emphasizes the method of autocorrelation of photoelectron-counting fluctuations because of its importance and wide use (Sec. 7.1). The subsequent section, Sec. 7.2, discusses the estimation of spectral parameters from measurements of probability distributions of photoelectron counts and time statistics of photoelectron events. Measurement of the lifetime of pulses of light is treated in Sec. 7.3. In Sec. 7.4, the problem of estimating the spatial spectrum (or the spatial coherence function) is discussed.

7.1 Estimation of the Spectrum from Measurement of the Autocorrelation of Photoelectron-Counting Fluctuations

7.1.1 Autocorrelation Function

We have seen in Sec. 5.3 that if $n(t)$ represents the number of photoelectron counts in an interval of duration T centered around time t, then the correlation function of $n(t)$ is proportional to the correlation function of the integrated light intensity $W(t)$, i.e.,

$$\langle n(t)n(t+\tau)\rangle = \langle W(t)W(t+\tau)\rangle \quad , \quad \tau \neq 0$$

or

$$G_n(\tau) = G_W(\tau) \triangleq G^{(2)}(\tau) \quad , \tag{7.1}$$

where W is in units of number of counts. The correlation function of the integrated intensity $G_W(\tau)$ is related to the correlation function of the intensity $G_I(\tau)$ by the double integrals of (3.77) and $G_I(\tau)$ itself is related to the optical-field correlation function $G^{(1)}(\tau) \triangleq G_V(\tau)$ by relations depending on the statistical nature of the field. The basic quantities involved are illustrated by the symbols

$$V \longrightarrow I \longrightarrow W \longrightarrow n$$

$$G_V \longrightarrow G_I \longrightarrow G_W = G_n$$

$$G^{(1)} \longrightarrow G^{(2)} \quad .$$

We are seeking to estimate $G^{(1)}$ from the measurement of $G^{(2)}$. From the results of Sec. 5.3, we compile below some of the relations between $G^{(1)}$ and $G^{(2)}$ (or their normalized versions $g^{(2)}$ and $g^{(1)}$) for several optical fields of practical importance.

Thermal Light

I) short sampling time ($T \ll \tau_c$), small detector area ($A \ll A_c$)

$$g^{(2)}(\tau) = 1 + |g^{(1)}(\tau)|^2 \quad . \tag{7.2}$$

II) short sampling time ($T \ll \tau_c$), arbitrary detector area

$$g^{(2)}(\tau) = 1 + f(A)|g^{(1)}(\tau)|^2 \quad , \tag{7.3}$$

where $f(A) = N_s^{-1}$ is the inverse of the number of spatial modes collected by the detector (cf. (5.101) and Fig. 5.12).

III) For an arbitrary T/τ_c, the relation between $g^{(2)}(\tau)$ and $g^{(1)}(\tau)$ depends on the spectral profiles, as illustrated by the examples:

Spectrum	$g^{(1)}(\tau)$	$g^{(2)}(\tau) - 1$	
Lorentzian	$\exp(-\Gamma\lvert\tau\rvert)$	$\mathrm{sh}^2\gamma\exp(-2\Gamma\lvert\tau\rvert)\ ,\quad \gamma = \Gamma T$	(7.4)
Gaussian	$\exp(-\Gamma^2\tau^2)$	$\frac{\exp(-\gamma^2)}{\gamma^2}\sinh(2\gamma\Gamma\tau)\exp(-\Gamma^2\tau^2)$ $+\frac{1}{\gamma^2}[\xi(\Gamma\tau+\gamma)+\xi(\Gamma\tau-\gamma)-2\xi(\Gamma\tau)]\ ,$ $\xi(x) = x\Phi(x)$	(7.5)
Sum of Lorentzians	$\sum_i \alpha_i\exp(-\Gamma_i\lvert\tau\rvert)$	$\sum_{i,j}\alpha_i\alpha_j\mathrm{sh}^2\gamma_{ij}\ \exp[-(\Gamma_i+\Gamma_j)\lvert\tau\rvert]\ .$ $\gamma_{ij} = (\Gamma_i+\Gamma_j)T/2$	(7.6)
Brillouin	$\alpha_1\exp(-\Gamma_1\tau) +$ $2\alpha_2\exp(-\beta\tau)\cos\Delta\omega\tau$	$\sum_{i,j=1}^{3}\alpha_i\alpha_j\mathrm{sh}^2\gamma_{ij}\ \exp[-(\Gamma_i+\Gamma_j)\lvert\tau\rvert]\ ,$ $\gamma_{ij} = (\Gamma_i+\Gamma_j)T/2,\ \alpha_3 = \alpha_2\ ,\ \Gamma_{\substack{2\\3}} = \beta\pm j\Delta\omega\ .$	(7.7)

These equations show that the absolute value of $g^{(1)}(\tau)$ can be determined from the measured $g^{(2)}(\tau)$. If the spectrum is symmetric, then $g^{(1)}(\tau)$ is real and can be completely determined from $g^{(2)}(\tau)$.

Mixture of Thermal and Coherent Light

I) For $A \ll A_c$, $T \ll \tau_c$ and $T\Delta\omega \ll 1$ ($\Delta\omega$ is the difference between the central frequencies),

$$g^{(2)}(\tau) = 1 + c\lvert g^{(1)}(\tau)\rvert^2 + b\ \mathrm{Re}\{g^{(1)}(\tau)\exp(j\Delta\omega\tau)\}\ , \tag{7.8}$$

$$c = \bar{n}_{th}^2/\bar{n}^2\ ,\quad b = 2\bar{n}_c\bar{n}_{th}/\bar{n}^2\ , \tag{7.9}$$

where $\bar{n}_c$ and $\bar{n}_{th}$ are the mean number of counts due to the coherent and thermal parts. If $g^{(1)}(\tau)$ is real (symmetric spectrum), we have

$$g^{(2)}(\tau) = 1 + c\lvert g^{(1)}(\tau)\rvert^2 + b\lvert g^{(1)}(\tau)\rvert\cos(\Delta\omega\tau)\ , \tag{7.10}$$

and if $\Delta\omega = 0$, then

$$g^{(2)}(\tau) = 1 + c|g^{(1)}(\tau)|^2 + b|g^{(1)}(\tau)| \quad . \tag{7.11}$$

Notice that if $\bar{n}_c$ is much larger than $\bar{n}_{th}$, then $b \gg c$ and we have

$$g^{(2)}(\tau) \simeq 1 + b\,\mathrm{Re}\{g^{(1)}(\tau)\exp(j\Delta\omega\tau)\} \quad , \tag{7.12}$$

$$\simeq 1 + b\,\mathrm{Re}\{g^{(1)}(\tau)\} \quad , \quad \Delta\omega = 0 \quad , \tag{7.13}$$

$$\simeq 1 + b|g^{(1)}(\tau)|\cos(\Delta\omega\tau) \quad , \text{ (symmetric spectrum)} \quad . \tag{7.14}$$

II) The thermal part has a Lorentzian spectrum, arbitrary T and A, $\Delta\omega = 0$:

$$g^{(2)}(\tau) = 1 + c\exp(-2\Gamma|\tau|) + b\exp(-\Gamma|\tau|) \tag{7.15}$$

$$c = (\bar{n}_{th}/\bar{n})^2 \mathrm{sh}^2\gamma f(A) \quad , \quad b = (2\bar{n}_{th}\bar{n}_c/\bar{n}^2)\mathrm{sh}^2\left(\frac{\gamma}{2}\right) f_D(A) \quad ,$$

where $f_D(A)$ is given by (5.181) and is illustrated by Fig.5.16. Eq. (7.13) demonstrates that the real part of $g^{(1)}$ of a thermal field can be determined by mixing the field with a strong coherent field (hetrodyning) and measuring the $g^{(2)}$ of the resultant field. If $g^{(2)}$ of the thermal field alone is measured, then $|g^{(1)}|$ can be determined via (7.2). From $|g^{(1)}|$ and Re $\{g^{(1)}\}$, the complex degree of coherence $g^{(1)}$ can be calculated, from which the profile of an asymmetric spectrum can be obtained. This is one advantage of heterodyning.

<u>Mixture of Coherent Light and Light with Unspecified Statistics</u>

If $T \ll \tau_c$, $A \ll A_c$, $\Delta\omega T \ll 1$, then

$$g^{(2)}(\tau) = 1 + c\left[g_s^{(2)}(\tau) - 1\right] + b|g_s^{(1)}(\tau)|\cos(\Delta\omega\tau) \quad , \tag{7.16}$$

where

$$c = \bar{n}_s^2/\bar{n}^2 \quad , \quad b = 2\bar{n}_c\bar{n}_s/\bar{n}^2 \tag{7.17}$$

and s denotes the part of light that has unspecified statistics. If $\bar{n}_c >> \bar{n}_s$, then

$$g^{(2)}(\tau) \simeq 1 + b|g_s^{(1)}(\tau)|\cos(\Delta\omega\tau) \quad , \tag{7.18}$$

i.e., by using a heterodyning technique the spectrum of a not necessarily thermal light field can be determined. This is another advantage of heterodyning.

Sum of N Independent Components of Light

For light that is composed of N statistically independent identically distributed components each having a constant amplitude and a uniformly distributed fluctuating phase, and for $T << \tau_c$ and $A << A_c$, the results of Sec. 4.2.8 give

$$g^{(2)}(\tau) = 1 + \left(1 - \frac{1}{N}\right)|g^{(1)}(\tau)|^2 \quad . \tag{7.19}$$

Coherent Light Modulated by Gaussian Noise

From (4.116), we have

$$g^{(2)}(\tau) = 1 + 2|g^{(1)}(\tau)|^2 \quad . \tag{7.20}$$

The last two cases are interesting because they demonstrate that the linear relation between $g^{(2)}$ and $|g^{(1)}|^2$ is not limited to thermal fields but can be valid for other fields as well.

In summary, the correlation function $g^{(1)}(\tau)$ of the field, and hence the spectrum, can be easily determined from the photoelectron-counting autocorrelation function $g^{(2)}(\tau)$ in many cases of interest.

7.1.2 Measurement of the Autocorrelation Function

Now we discuss methods of measuring the autocorrelation function $G^{(2)}(\tau)$. The total time interval $[0,T_0]$ during which the optical field can be observed is divided into N_0 equal intervals each of which is of duration T, $T_0 = N_0T$. By counting the photoelectrons detected in each interval, we obtain N_0 numbers $n(t_1)$, $n(t_2)$, ... $n(t_{N_0})$ which we call the sample and from which we hope to compute an estimate of the correlation $G^{(2)}(\tau)$.

From the sample of counts, we can construct the sample correlation function defined by

$$\hat{G}^{(2)}(\tau_\ell) = \frac{1}{N}\sum_{j=1}^{N} n(t_j)n(t_{j+\ell}) \quad , \quad \ell = 1,2, \ldots L \quad . \tag{7.21}$$

$$\tau_\ell = \ell T \quad , \quad N = N_0 - L \quad , \quad L << N_0 \quad ,$$

which should be a good estimate for the actual correlation $G^{(2)}(\tau_\ell)$ at the time delays T,2T, ... LT.

Because $n(t_1), \ldots n(t_N)$ are random variables, their sample correlation $\hat{G}^{(2)}$ is also a random variable whose expectation value is

$$\langle\hat{G}^{(2)}(\tau_\ell)\rangle = \frac{1}{N}\sum_{j=1}^{N}\langle n(t_j)n(t_{j+\ell})\rangle = G^{(2)}(\tau_\ell) \quad . \tag{7.22}$$

This means that $\hat{G}^{(2)}$ is an unbiased estimate of $G^{(2)}$; if N is sufficiently large, it should be quite adequate. The accuracy of such an estimate is discussed in Sec. 7.1.8.

Software Correlator

The sum in (7.21) can be computed by interfacing the photoelectron counter to a computer. The computer receives the numbers $\{n(t_j)\}$ in digital form, computes (7.21) using the proper software, and displays the results on line.

A basic limitation of such a system is that it is very slow. During one sampling time, the number of counts should be transferred to and stored in the memory of the computer, its contribution to the sum (7.21) computed for all channels and the new sum stored. This requires a long period of time (not less than 10 ms).

The situation may be improved if we store the number of counts of only a fraction of the samples in the computer memory and then interrupt the counting until the computer computes the sums of (7.21) and stores them in their corresponding channels. Then counting can be resumed to cover another fraction of the samples and so on until the desired N samples are covered. Such a system can handle sampling times down to 10 μs but it has a lower information-gathering capacity (signal-utilization efficiency) because of the idle time during which counting is stopped. Examples of the previously described software correlators are described by HALLET et al. [190] and VAN RESANDT [495].

Hardware Correlator

In order to obtain higher speeds, a hardware correlator is needed, designed to calculate directly the sums in (7.21) and to transfer the resultant L numbers after the end of the computation to the display or to a computer. An electronic instrument that is to compute $G^{(2)}(\tau_\ell)$ from the sample of counts should perform the operations: counting-delay-multiplication-addition, as illustrated in Fig.7.2. One scheme that can accomplish this (Fig.7.3) is to use a memory that is capable of storing L numbers, the number of counts of the last L samples.

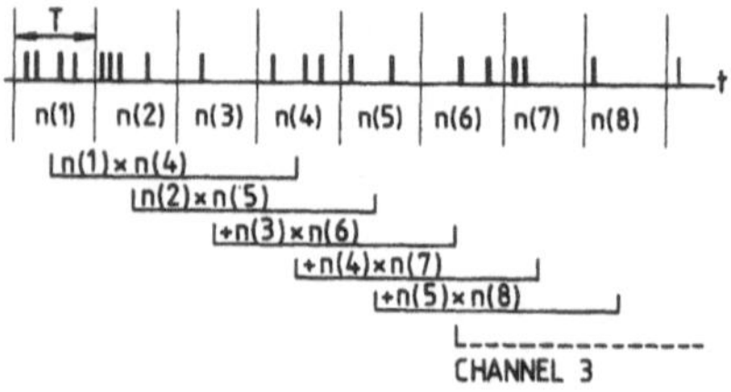

Fig.7.2. Computation of the photoelectron counting correlation function

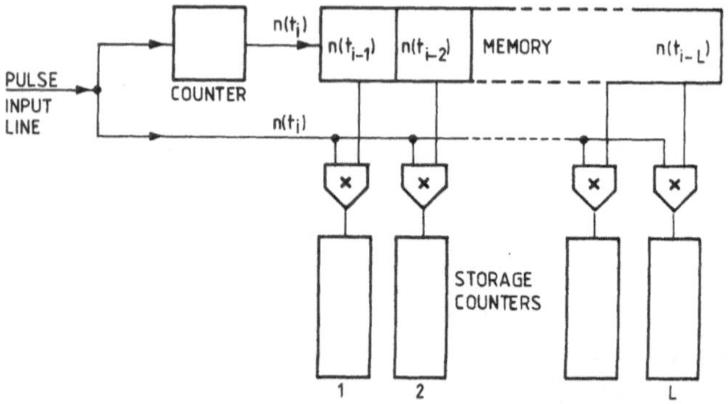

Fig.7.3. Outline of a photoelectron-counting correlator

At a time t_i, the batch of numbers $n(t_{i-1})$, $n(t_{i-2})$, ... $n(t_{i-L})$ is available in the memory. Therefore the products $n(t_i)n(t_{i-1})$, $n(t_i)n(t_{i-2})$, ... $n(t_i)n(t_{i-L})$ can be computed simultaneously using L multipliers. The resultant L numbers are then sent to L adders. This should be completed within one sampling time. At the next time interval t_{i+1}, the next batch of numbers $n(t_i)$,$n(t_{i-1})$, ... $n(t_{i-L+1})$ is stored in the memory and the multipliers produce $n(t_{i+1})n(t_i)$, $n(t_{i+1})n(t_{i-1})$, ... $n(t_{i+1})n(t_{i+1-L})$ which are transferred to the adders and added to the previous set. After N time intervals the adders contain

$$\sum_{j=0}^{N-1} n(t_{i+j})n(t_{i+j-1}) \ , \ \sum_{j=1}^{N} n(t_{i+j})n(t_{i+j-2}) \ , \ \ldots \ \sum_{j=1}^{N} n(t_{i+j})n(t_{i+j-L}) \ ,$$

which are proportional to

$$\hat{G}^{(2)}(\tau_\ell) \ , \qquad = 1,2, \ \ldots \ , L \ .$$

An example of the above scheme is the correlator constructed by ASH and FORD [24]. It uses a 3-bit shift register as a memory. The slowest link in such an instrument is the process of multiplication, which limits its speed considerably.

Sequential-Processing Correlator

A faster correlator may be obtained if we allow a time lapse between the batches longer than one sampling time. If M sampling times are allowed, then we obtain in channel ℓ the sum

$$\sum_{j=1}^{N} n(t_{jM})n(t_{jM+\ell}) \quad .$$

Thus, if one batch contains $n(t_{i-1})$, $n(t_{i-2})$, ... $n(t_{i-L})$, the next contains $n(t_{i+M-1})$, $n(t_{i+M-2})$, ... $n(t_{i+M-L})$ and so on, as illustrated in Fig.7.4.

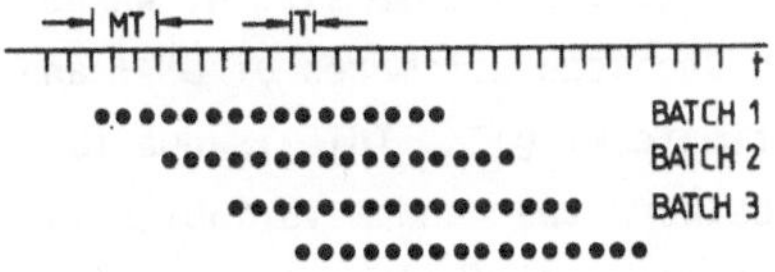

Fig.7.4. Sampling scheme for computation of the correlation function

The total time necessary for measuring N batches is approximately MNT, i.e., the duration of the experiment is M times as long. In order to determine the dependence of the signal-utilization efficiency of such a system on M, the time necessary to estimate the correlation function to a given accuracy has to be calculated.

Using methods of calculating the statistical accuracy that will be discussed in Sec. 7.1.8, OLIVER [355] showed that in the shot-noise limit (small degeneracy parameter, $\bar{n}_d \ll 1$) the experimental time necessary to achieve the same accuracy is M times longer than in the case of a real-time processor (M = 1). This is to be expected, because in this limit the samples are independent. In the large-counting rate limit, $\bar{n}_d > 100$, the required processing time is only a factor γM longer ($\gamma = T/\tau_c$), i.e., the loss of signal utilization due to sequential processing is normally much smaller. In this case we can conclude that real-time processing is essential only for low counting rates.

7.1.3 Normalized Autocorrelation Function

Whereas the autocorrelation function $G^{(2)}$ carries information on the spectral profile and the intensity of the detected light, the normalized autocorrelation $g^{(2)}$ depends on only the spectral profile and is therefore more suitable for the purpose of spectral estimation.

An estimate of $g^{(2)}$ can be obtained by normalizing our estimate of $G^{(2)}$ by use of an estimate of the average count rate. Thus,

$$\hat{g}^{(2)}(\tau) = \hat{G}^{(2)}(\tau)/\hat{n}^2 \quad , \tag{7.23}$$

where $G^{(2)}$ is given by (7.21) and the sample mean $\hat{n}$ can be obtained from

$$\hat{n} = \frac{1}{N}\sum_{j=1}^{N} n(t_j) \quad . \tag{7.24}$$

We have already shown in Chap. 6 that the sample mean is an unbiased estimate of $\bar{n}$, whose variance is inversely proportional to the sample size N. We observe, however, that although $\hat{G}^{(2)}$ and $\hat{n}$ are unbiased estimates of $G^{(2)}$ and $\bar{n}$, respectively, $\hat{g}^{(2)}$ is not an unbiased estimate of $g^{(2)}$. This is due to the fact that the expectation of the ratio between two random variables is not, in general, equal to the ratio between their expectations. Because, for large N, $\hat{G}^{(2)}$ and $\hat{n}$ are approximately equal to $G^{(2)}$ and $\bar{n}$, respectively, we should expect that $\hat{g}^{(2)}$ approximates $g^{(2)}$. The accuracy to be expected is discussed in Sec. 7.1.8.

Sometimes drifts may occur in the mean count rate for reasons unconnected with the fundamental statistical properties of the optical field under study. If such drifts occur during the observation time, the spectral profile estimated by (7.23) would be distorted.

It is, in general, advantageous to account for such drifts by dividing the allowed observation time into a number of subintervals and finding the average of estimates of the normalized correlation functions based on data in each subinterval. We would then be interested in computing

$$\hat{g}^{(2)}(\tau_\ell) = \frac{1}{M}\sum_{m=0}^{M-1}\left[\frac{1}{S}\sum_{j=mS+1}^{mS+S} n(t_j)n(t_{j+\ell})\right] \Bigg/ \left[\frac{1}{S}\sum_{j=mS+1}^{mS+S} n(t_j)\right]^2 \quad ,$$

where S is the number of sampling times in each subinterval and M = N/S is the number of subintervals. An optimum choice of the number of subintervals depends on the time constant of the drift of the mean count rate. However, the number of sampling times in a subinterval should not be made very small because this may lead to severe distortions due to the biasing effect of the normalization [355].

7.1.4 Clipped Correlation

As is illustrated in Fig. 7.2, an electronic instrument that is to compute $\hat{G}^{(2)}$ from the sample of counts $\{n(t_i)\}$ should perform the operations: counting-delay-multiplication-addition. The operation of multiplication is the most time- and storage-space-consuming of these operations, especially if the numbers $\{n(t_i)\}$ are large. The technique of clipping (or one-bit quantization) has been introduced to dispense with the multiplication operation [227, 242, 383].

A clipping level K is chosen and the numbers $\{n(t_i)\}$ are associated with another set of "one-bit" numbers $\{n_k(t_i)\}$ defined by

$$\begin{aligned} n_k(t) &= 1 \qquad n(t) > k \\ &= 0 \qquad n(t) \le k \quad . \end{aligned} \tag{7.25}$$

The singly clipped correlation function is then obtained by cross correlating $\{n_k(t_i)\}$ with $\{n(t_i)\}$,

$$G_k^{(2)}(\tau) = \langle n_k(t)n(t+\tau)\rangle \quad . \tag{7.26}$$

As will be shown later in this section, this function is directly related to the field-correlation function $g^{(1)}(\tau)$ in many case of practical importance; therefore it carries the same information as the full correlation function $G^{(2)}(\tau)$. Moreover, as will be shown in Sec. 7.1.8, spectral parameters can be estimated from $G_k^{(2)}(\tau)$, with an accuracy not much less than the accuracy of estimations based on $G^{(2)}(\tau)$. One special case is worth mentioning. If the average count rate is very small, then the numbers $\{n(t_i)\}$ take the values one or zero even without clipping. Therefore, clipping at a level zero should not affect these numbers; hence $G_0^{(2)}(\tau) = G^{(2)}(\tau)$ for $\bar{n} \ll 1$.

The clipped correlation function can be estimated from the sample of counts by computing

$$\hat{G}_k^{(2)}(\tau) = \frac{1}{N} \sum_{j=1}^{N} n_k(t_j) n(t_{j+\ell}) \quad . \tag{7.27}$$

Because n_k takes the values one or zero, the computation is then reduced to the operations: counting-comparison-delay-gated addition. No multiplication is required. See Fig.7.5. An instrument based on these operations is discussed at the end of this subsection.

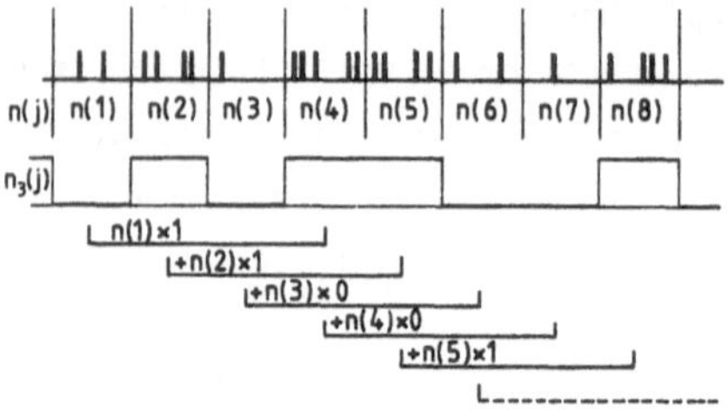

Fig.7.5. Computation of the clipped correlation function. K=3, τ=3T

The *doubly clipped correlation* function is similarly defined by

$$G_{kk'}^{(2)}(\tau) = \langle n_k(t) n_{k'}(t+\tau) \rangle \tag{7.28}$$

and can be estimated from

$$\hat{G}_{kk'}^{(2)}(\tau_\ell) = \frac{1}{N} \sum_{j=1}^{N} n_k(t_j) n_{k'}(t_{j+\ell}) \quad . \tag{7.29}$$

This is equivalent to replacing the sample of counts by a series of binary digits (zeros and ones) and performing their correlation.

The clipped correlation functions can be normalized:

$$g_k^{(2)}(\tau) = G_k^{(2)}(\tau)/\bar{n}_k\bar{n} \quad , \qquad \bar{n}_k = \langle n_k \rangle \quad , \qquad \bar{n} = \langle n \rangle \quad , \tag{7.30}$$

$$g_{kk'}^{(2)}(\tau) = G_{kk'}^{(2)}(\tau)/\bar{n}_k\bar{n}_{k'} \quad . \tag{7.31}$$

These normalized functions can be computed from

$$\hat{g}_k^{(2)}(\tau_\ell) = \hat{G}_k^{(2)}(\tau_\ell)/\hat{n}_k\hat{n} \tag{7.32}$$

and

$$\hat{g}_{kk'}^{(2)}(\tau_\ell) = \hat{G}_{kk'}^{(2)}(\tau_\ell)/\hat{n}_k\hat{n}_{k'} \quad , \tag{7.33}$$

where

$$\hat{n}_k = \frac{1}{N}\sum_{j=1}^{N} n_k(t_i) \tag{7.34}$$

is an estimate of the average number of clipped counts.

It remains now to show how the spectrum (or the field's correlation function) is related to $G_k^{(2)}$ and $G_{kk'}^{(2)}$. In general, a simple relation exists only for some special (but important) cases. In order to derive the relation between $G_k^{(2)}$ or $G_{kk'}^{(2)}$ and $G^{(1)}$, we need the following equations:

By definition

$$G_k^{(2)}(\tau) = \langle n_k(0)n(\tau)\rangle = \sum_{n=k+1}^{\infty}\ \sum_{m=0}^{\infty} mP(n,m) \quad . \tag{7.35}$$

Substituting for P(n,m) in terms of the mgf (3.81) and rearranging the summations, we get

$$G_k^{(2)}(\tau) = \sum_{n=k+1}^{\infty} \frac{(-1)^n}{n!} \frac{\partial^n}{\partial s_1^n} \sum_{m=1}^{\infty} m \frac{(-1)^m}{m!} \frac{\partial^m}{\partial s_2^m} Q(s_1,s_2)\bigg|_{s_1 = s_2 = 1}$$

Changing the index of summations to m' = m-1 and using the MacLaurin expansion,

$$Q(s_1,0) = -\sum_{m=0}^{\infty} \frac{(-1)^m}{m!} \frac{\partial^m}{\partial s_2^m} Q(s_1,s_2)\Big|_{s_2 = 1} ,$$

we can finally write

$$G_k^{(2)}(\tau) = -\sum_{n=k+1}^{\infty} \frac{(-1)^n}{n!} \frac{\partial^n}{\partial s_1^n} \frac{\partial}{\partial s_2} Q(s_1,s_2)\Big|_{s_1 = 1,\ s_2 = 0}$$

$$= \bar{n} + \sum_{n=0}^{k} \frac{(-1)^n}{n!} \frac{\partial^n}{\partial s_1^n} \frac{\partial}{\partial s_2} Q(s_1,s_2)\Big|_{s_1 = 1,\ s_2 = 0} , \qquad (7.36)$$

and hence the recurrence relation

$$G_k^{(2)}(\tau) = G_{k-1}^{(2)} + \frac{(-1)^k}{k!} \frac{\partial^k}{\partial s_1^k} \frac{\partial}{\partial s_2} Q(s_1,s_2)\Big|_{s_1 = 1,\ s_2 = 0} . \qquad (7.37)$$

Similarly, the doubly clipped correlation function is related to the mgf by

$$G_{kk'}^{(2)}(\tau) = \sum_{n=k+1}^{\infty} \sum_{m=k'+1}^{\infty} P(n,m) = \sum_{n=k+1}^{\infty} \sum_{m=k'+1}^{\infty} \frac{(-1)^{n+m}}{n!m!} \frac{\partial^n}{\partial s_1^n} \frac{\partial^m}{\partial s_2^m} Q(s_1,s_2)\Big|_{s_1 = s_2 = 1} . \qquad (7.38)$$

In particular,

$$G_{00}^{(2)}(\tau) = \sum_{n=1}^{\infty} \sum_{m=1}^{\infty} P(n,m) = 1 - 2P(0) + P(0,0) = 1 - 2Q(1) + Q(1,1). \qquad (7.39)$$

Also, the average number of clipped counts is given by

$$\bar{n}_k = \sum_{n=k+1}^{\infty} P(n) = \bar{n} - \sum_{n=0}^{k} \frac{(-1)^n}{n!} \frac{\partial^n}{\partial s^n} Q(s)\Big|_{s = 1} . \qquad (7.40)$$

Equipped with (7.36-40), we shall determine $g_k^{(2)}$ and $g_{kk'}^{(2)}$ for several special cases of practical importance.

Thermal Light

I) $T \ll \tau_c$, $A \ll A_c$. In this case $Q(s)$ and $Q(s_1,s_2)$ are given by (5.5) and (5.272). By substitution in (7.36) and (7.40) we get

$$\bar{n}_k = \left(\frac{\bar{n}}{1+\bar{n}}\right)^{k+1} \tag{7.41}$$

and

$$g_k^{(2)}(\tau) = 1 + c_k|g^{(1)}(\tau)|^2 \quad , \qquad c_k = \frac{1+k}{1+\bar{n}} \quad . \tag{7.42}$$

Comparing this result with the full autocorrelation case, (7.2), we see that in this particular case, the effect of clipping is simply a multiplicative factor that can be made equal to one if the clipping level k is chosen such that $k = \bar{n}$. Clipping does not distort the time dependence of the intensity correlation function. Only the relative importance of the signal to the background changes. Notice that as the clipping level k increases, c_k increases and the signal is amplified. Yet the uncertainty of the signal is expected to increase because, for large k, most of the information is discarded. This will be discussed further in Sec. 7.1.8. The doubly clipped autocorrelation function, on the other hand, is not simply related to $g^{(1)}$, e.g., for an arbitrary k and k' = 0, we obtain (from (7.38) and (7.41))

$$g_{k0}^{(2)}(\tau) = 1 + \frac{1}{\bar{n}} - \frac{1}{\bar{n}}\left[\frac{1 - \frac{\bar{n}}{1+\bar{n}}|g^{(1)}(\tau)|^2}{1 - \left(\frac{\bar{n}}{1+\bar{n}}\right)^2|g^{(1)}(\tau)|^2}\right]^{k+1} \quad , \tag{7.43}$$

which for k = 0 becomes

$$g_{00}^{(2)}(\tau) = \frac{1 + \frac{1-\bar{n}}{1+\bar{n}}|g^{(1)}(\tau)|^2}{1 - \left(\frac{\bar{n}}{1+\bar{n}}\right)^2|g^{(1)}(\tau)|^2} \quad . \tag{7.44}$$

For arbitrary k and k', more complicated formulas are obtained [229]. These are of little practical use, If $\bar{n} \ll 1$, (7.44) becomes

$$g_{00}^{(2)}(\tau) \approx 1 + |g^{(1)}(\tau)|^2 = g^{(2)}(\tau) \quad ,$$

i.e., gives the full correlation function as expected.

II) $A \ll A_c$, T/τ_c Arbitrary, Lorentzian Spectrum $|g^{(1)}(\tau)| = \exp(-\Gamma|\tau|)$. In this case, $Q(s)$ and $Q(s_1,s_2)$ are given by (5.54) and (5.294). Substitution in (7.36) and (7.40) gives

$$g_k^{(2)}(\tau) = 1 + c_k\, |g^{(1)}(\tau)|^2 \quad ,$$

where c_k is a constant

$$c_k = \mathrm{sh}\gamma \sum_{n=k+1}^{\infty} \left(-\frac{d}{ds}\right)^n \left[Q(s)y(s)\right]\Big|_{s=1} \Big/ \sum_{m=k+1}^{\infty} \left(-\frac{d}{ds}\right)^m Q(s)\Big|_{s=1} \qquad (7.45)$$

and $Q(s)$ and $y(s)$ are given by (5.54) and (5.294).

Again, single clipping results in multiplication of the time-dependent part of the autocorrelation function by a constant, a surprising but encouraging result.

III) $T \ll \tau_c$, A/A_c arbitrary. Here we resort to the expressions of $Q(s)$ and $Q(s_1,s_2)$ given by (5.83) and (5.311), which are written in terms of μ_j, the eigenvalues of the spatial coherence function of the field. Using these in our (7.36) and (7.40), we get [262]

$$g_k^{(2)}(\tau) = 1 + c_k|g^{(1)}(\tau)|^2 \quad , \qquad (7.46)$$

where c_k are related to the coefficients μ_j, e.g.,

$$c_0 = \frac{\sum_j (\bar{n}\mu_j)^2/(1+\bar{n}\mu_j)}{\bar{n}\left[\prod_j (1+\bar{n}\mu_j)-1\right]} \quad . \qquad (7.47)$$

For larger k, lengthier expressions are obtained.

An analysis of the effect of sampling time and detector area can also be found in [234, 272].

IV) Arbitrary A/A_c, Arbitrary T/τ_c, Arbitrary Spectrum (Approximate Formula). For arbitrary A/A_c and T/τ_c, as well as arbitrary spectral profile, we can use the approximate formulas (5.91) and (5.321-323). These give

$$g_k^{(2)}(\tau) = 1 + c_k\,[g^{(2)}(\tau) - 1] \quad , \tag{7.48}$$

where

$$\begin{aligned} c_k &= \frac{\bar{n}}{\bar{n}_k}\binom{N+k}{k}\frac{N^{N+1}\,\bar{n}^k}{(N+\bar{n})^{N+k+1}} \\ \bar{n}_k &= 1 - \sum_{m=0}^{k}\binom{N+m-1}{m}\frac{N^N\,\bar{n}^m}{(N+\bar{n})^{N+m}} \end{aligned} \quad . \tag{7.49}$$

This shows that, in this approximation, the full autocorrelation function can be completely recovered from the single-clipped autocorrelation function. The approximate formula (7.48) and the exact formula in the case of Lorentzian light (7.45), were compared by SALEH and HENDRIX [424], who demonstrated that the accuracy of the approximate formula is reasonable. By examining the graphs of $g_k^{(2)}(\tau)$ for light with Brillouin spectrum, which were computed by BLAKE and BARAKAT [58] using numerical methods, we see that the formula (7.48) is also reasonable.

Mixture of Coherent and Thermal Light

The photoelectron-counting statistics of mixtures of coherent and thermal light were discussed in Sec. 5.2.4 and 5.3.2. Eqs. (7.8-15) show that the coherence function (spectrum) of the thermal field can be easily recovered from the photoelectron-counting autocorrelation function. Here we examine whether this also applies to the clipped autocorrelation function.

I) $T \ll \tau_c$, $A \ll A_c$, $\Delta\omega T \ll 1$, symmetric Spectrum. Substituting for $Q(s)$ and $Q(s_1,s_2)$ from (5.127) and (4.103) we get [228]

$$g_k^{(2)}(\tau) = 1 + c_k|g^{(1)}(\tau)|^2 + b_k|g^{(1)}(\tau)|\cos(\Delta\omega\tau) \quad , \tag{7.50}$$

where

$$c_k = \frac{\bar{n}_{th}^2}{\bar{n}\bar{n}_k}\,\frac{\partial}{\partial\bar{n}_c}\left(\bar{n}_c\,\frac{\partial}{\partial\bar{n}_c}\,\bar{n}_k\right) \tag{7.51}$$

$$b_k = \frac{2\bar{n}_{th}\bar{n}_c}{\bar{n}\bar{n}_k} \frac{\partial}{\partial \bar{n}_c} \bar{n}_k \tag{7.52}$$

and

$$\bar{n}_k = 1 - \exp\left(\frac{-\bar{n}_c}{1+\bar{n}_{th}}\right) \sum_{n=0}^{k} \frac{\bar{n}_{th}^{\,n}}{(1+\bar{n}_{th})^{n+1}} L_n \left[\frac{-\bar{n}_c}{\bar{n}_{th}(1+\bar{n}_{th})}\right]$$

$$= 1 - \exp\left(\frac{-\bar{n}_c}{1+\bar{n}_{th}}\right)\left\{e_k\left[\frac{\bar{n}_c}{(1+\bar{n}_{th})^2}\right] - \left(\frac{\bar{n}_{th}}{1+\bar{n}_{th}}\right)^k e_k\left[\frac{\bar{n}_c}{\bar{n}_{th}(1+\bar{n}_{th})}\right]\right\}, \tag{7.53}$$

where

$$e_k(x) = \sum_{n=0}^{k} x^n/n! \quad .$$

It is interesting to find that the singly clipped correlation function (7.50) has the same form as the full correlation function except for the constants c_k and b_k. Because the ratio c_k/b_k is no longer equal to $\bar{n}_{th}/2\bar{n}_c$, we may wonder if it is still possible to make the third term in (7.50) dominate by choosing $\bar{n}_c$ much larger than $\bar{n}_{th}$. Calculations [228] show that in the limit $\bar{n}_{th} \ll 1$, (7.51) and (7.52) give once more

$$c_k/b_k = \bar{n}_{th}/2\bar{n}_c \quad .$$

Thus, for all values of k, it is possible to make the third term dominate.

In the case $\bar{n}_{th} \ll 1$, $\bar{n}_{th} \ll \bar{n}_c$, (7.50-53) give

$$g_k^{(2)}(\tau) \approx 1 + b_k|g^{(1)}(\tau)| \cos\Delta\omega\tau \ , \quad b_k = 2\bar{n}_{th}\bar{n}_c^k e^{-\bar{n}_c}/\Gamma(1+k,\bar{n}_c) \tag{7.54}$$

$$\bar{n}_k \approx \Gamma(1+k,\bar{n}_c)/k! \quad , \tag{7.55}$$

where $\Gamma(a,b)$ is the incomplete gamma function.

If, on the other hand, $\bar{n}_{th} \gtrsim 1$, the possibility that $b_k \gg c_k$ depends on the used clipping level k. For example, for k = 0

$$c_0/b_0 \approx \frac{\bar{n}_{th}}{2\bar{n}_c}\left(1 - \frac{\bar{n}_c}{1 + \bar{n}_{th}}\right)$$

and the choice $\bar{n}_c = 1 + \bar{n}_{th}$ satisfies our requirement.

For $K >> 1$

$$c_k/b_k \approx \frac{\bar{n}_{th}}{2\bar{n}_c}\left[1 - \frac{\bar{n}_c\bar{n}_{th}}{(1 + \bar{n}_{th})^2}\right] \quad .$$

If we choose $\bar{n}_c$ such that

$$\frac{(1 + \bar{n}_{th})^2}{\bar{n}_{th}} > \bar{n}_c > (1 + \bar{n}_{th}) \quad ,$$

we see that c_0/b_0 is negative, whereas $c_k/b_k \,|_{k >> 1}$ is positive. Therefore, a clipping level k must exist that makes c_k/b_k very small. In an actual experiment, such manipulations may be rather difficult.

Sum of N Independent Components of Light

This model has been studied in Sec. 4.2.8. The single-fold mgf, $Q(s)$, is given by (4.160). When it is substituted in (7.40), we get for $k = 0$

$$\bar{n}_0 \approx \frac{\bar{n}}{1 + \bar{n}}\left[1 + \frac{\bar{n}}{2(1+\bar{n})^2}\frac{1}{N}\right] \quad ,$$

where terms of order N^{-2} and higher have been neglected.

In order to determine $g_k^{(2)}(\tau)$, we need the two-fold mgf, $Q(s_1,s_2)$. Because this is not available, we limit ourselves to the lowest-order case $k = 0$ and rewrite (7.36) in the form

$$G_0^{(2)}(\tau) = \bar{n} - \langle e^{-W_1}W_2\rangle \quad .$$

For point detectors and short counting times, this can be written in terms of the intensity moments $\langle I_1^m I_2\rangle$, $m = 0,1,\ldots$ CHEN and TARTAGLIA [88] found a general expression for these moments, by using an extension of the analysis used in Sec. 4.2.8 to obtain $\langle I_1 I_2\rangle$. Their calculations lead to

$$g_0^{(2)}(\tau) = 1 + c_0|g^{(1)}(\tau)|^2$$

$$c_0 \approx \frac{1}{1+\bar{n}}\left[1 - \frac{\bar{n}^2+\bar{n}+2}{2(1+\bar{n})^2}\,\frac{1}{N}\right] ,$$

where terms of order N^{-2} and higher have been neglected. TARTAGLIA and CHEN [467] have considered the effect of the detector area on the above formulas.

Coherent Light Modulated by Gaussian Noise

In this model, we can use (4.119, 120) and (7.36) and obtain [43]

$$g_0^{(2)}(\tau) = 1 + c_0|g^{(1)}(\tau)|^2 ,$$

$$c_0 = \frac{2\bar{n}}{1+2\bar{n}}\,\frac{1}{[(1+2\bar{n})^{\frac{1}{2}}-1]}$$

For $k > 0$, the relation between $g_k^{(2)}(\tau)$ and $g^{(1)}(\tau)$ becomes much more complicated.

Clipped Digital Correlator

An L-channel clipped correlator is a sequential digital machine that computes the sum $\sum_i n_k(t_i)n(t_{i-\ell})$, $\ell = 1,2, \ldots L$. The machine has two input lines (Fig.7.6), the direct input line that contains the photoelectron pulses as they come, and the clipped-input line that feeds a shift register with the one-bit numbers $n_k(t_i)$ at the end of each interval.

During the time interval t_i, the shift register contains the clipped counts of the previous L time intervals $n_k(t_{i-\ell})$, $\ell = 1,2, \ldots L$. If $n_k(t_{i-\ell}) = 1$, the pulses at the direct input line are allowed by the ℓth AND gate to be counted by the ℓth counters and the number $n(t_i) = n(t_i)\,n_k(t_{i-\ell})$ is added to their contents. This is done simultaneously for the L channels. At the end of the interval t_i, the number $n_k(t_i)$ is sent to the shift register, which will now contain $n_k(t_{i+1-\ell})$, $\ell = 1,2, \ldots L$. The numbers $n_k(t_{i+1})n(t_{i+1-\ell})$, $\ell = 1,2, \ldots L$ are added to the counters and so on. The desired sums are finally obtained.

The clipped counts $n_k(t_i)$ can be obtained from the counts $n(t_i)$ by using a clipping gate. This can be a comparator circuit in which $n(t_i)$ is compared to the clipping level k. Alternatively [355] a scaler with a preset overflow k can be used to count $n(t_i)$ (Fig.7.6b). If k is exceeded,

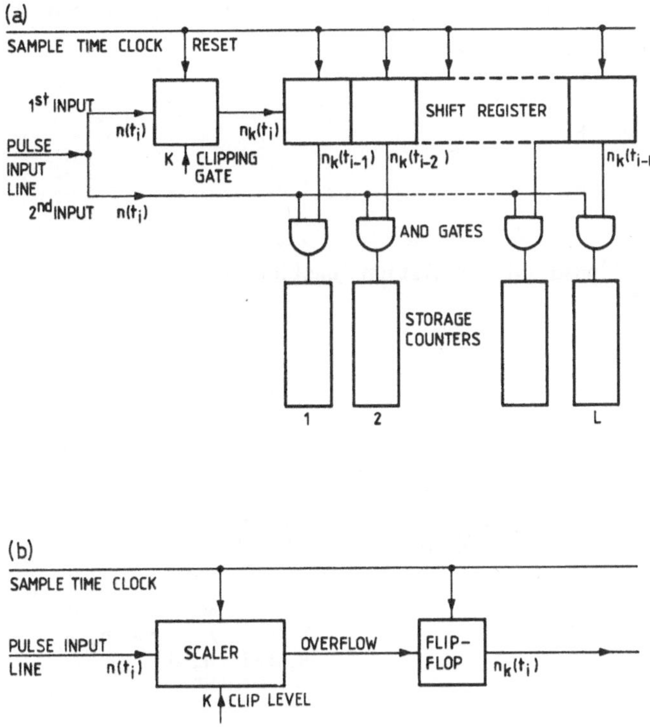

Fig.7.6a and b. Single-clipping correlator; a) correlator, b) clipping gate

the overflow pulse sets a flip-flop whose output is $n_k(t_i)$. At the end of the sampling time, the counter and the flip-flop are reset. A detailed description of a clipped digital correlator can be found in [90, 100]. Double clipping can be realized by passing the two input lines through the clipping gate.

A singly clipped correlator suffers from the problem of missing photoelectron pulses that may come at the leading or trailing edges of the sample interval. This causes an uncertainty of the sample-time duration.

In the double-clipping scheme, this uncertainty is avoided. This makes double clipping more accurate, particularly at high speeds [148].

7.1.5 Complementary Clipping

Complementary clipping is very similar to normal clipping except that the bit 1 is chosen when the number of counts in an interval is less than or equal to the clipping level. The bit zero is chosen if it exceeds the clipping level. This is exactly the opposite of the scheme of clipping discussed in the previous section [89]. If we define

$$n_{ck}(t) = 1 \qquad n \leq k$$
$$\qquad = 0 \qquad n > k$$

and the complementary clipped autocorrelation function by

$$G_{ck}^{(2)}(\tau) = \langle n_{ck}(0)n(\tau)\rangle \quad ,$$

then

$$G_{ck}^{(2)}(\tau) = \sum_{n_1=0}^{k} \sum_{n_2=0}^{\infty} n_2 P(n_1,n_2) = \sum_{n_1=0}^{\infty} \sum_{n_2=0}^{\infty} n_2 P(n_1,n_2) - \sum_{n_1=k+1}^{\infty} \sum_{n_2=0}^{\infty} n_2 P(n_1,n_2)$$
$$= \langle n\rangle - \langle n_k(0)n(\tau)\rangle \quad ,$$

i.e.,

$$G_{ck}^{(2)}(\tau) = \bar{n} - G_k^{(2)}(\tau) \quad . \tag{7.56}$$

This means that the complementary-clipped-correlation function carries the same information as the clipped correlation function. The advantage of using one method over the other can be shown by considering the numerical size of $G_k^{(2)}(\tau)$ and $G_{ck}^{(2)}(\tau)$. For simplicity, let us take the case of thermal light ($T \ll \tau_c$, $A \ll A_c$) and $\tau = 0$; this gives, (7.42),

$$G_k^{(2)}(0) = \bar{n}\bar{n}_k\left(1 + \frac{1+k}{1+\bar{n}}\right) = \left(\frac{\bar{n}}{1+\bar{n}}\right)^{k+2} (2 + \bar{n} + k) \quad ;$$

whereas,

$$G_{ck}^{(2)}(0) = \bar{n} - \left(\frac{\bar{n}}{1+\bar{n}}\right)^{k+2} (2 + \bar{n} + k) \quad .$$

In the limit $\bar{n} \gg 1$, $\quad G_k^{(2)}(0) \approx \bar{n} + k$,

whereas, $\quad G_{ck}^{(2)}(0) \simeq k$,

i.e., the number of counts to be stored is much smaller in the case of complementary clipping. Complementary clipping thus alleviates the problem of overflow. This is especially useful in heterodyne measurements in which intense coherent light is used.

In the other limit, $\bar{n} \ll 1$, $\quad G_k^{(2)}(0) = \bar{n}^{k+2}(2+k)$,

$$G_{ck}^{(2)}(0) = \bar{n} - \bar{n}^{k+2}(2+k),$$

i.e., $\quad G_k^{(2)} \ll G_{ck}^{(2)}$,

which makes it more advantageous to use the normal clipping system.

7.1.6 Randomly Clipped Autocorrelation

In the previous section, we have seen that for thermal light (stationary Gaussian light) it is possible to recover the coherence function of the field from the clipped correlation function. For non-Gaussian light, this is generally no longer possible (unless a heterodyne technique is used). Random clipping is a method designed to handle non-Gaussian light [238].

Let the clipping level k vary randomly from one sample to another with a probability distribution q(k). If this variation is statistically independent of the fluctuations in the photoelectron counts themselves, then the average clipped autocorrelation function is given by

$$\langle G_k^{(2)}(\tau)\rangle_k = \sum_{k=0}^{\infty} q(k) G_k^{(2)}(\tau) \quad . \tag{7.57}$$

Substituting from (7.35) and rearranging the summations, we get

$$\langle G_k^{(2)}(\tau)\rangle_k = \sum_{m=1}^{\infty} \sum_{n=1}^{\infty} \left[\sum_{k=0}^{n} q(k)\right] mP(n,m) \quad .$$

If we assume that k has a uniform distribution over the range 0 to s, i.e.,

$$\begin{aligned} q(k) &= \frac{1}{s} \ , && k < s \\ &= 0 \ , && k \geq s \ , \end{aligned}$$

then

$$\langle G_k^{(2)}(\tau)\rangle_k = \frac{1}{s}\sum_{m=1}^{\infty}\sum_{n=1}^{s-1} nmP(n,m) + \sum_{m=1}^{\infty}\sum_{n=s}^{\infty} mP(n,m) \quad . \tag{7.58}$$

Now if s is chosen large enough that

$$P(n,m) \simeq 0 \quad \text{for } n \geq s \quad ,$$

we can write

$$\langle G_k^{(2)}(\tau)\rangle_k \simeq \frac{1}{s}\sum_{m=1}^{\infty}\sum_{n=1}^{\infty} nmP(n,m) = \frac{1}{s}\, G^{(2)}(\tau) \quad . \tag{7.59}$$

We thus see that, with uniformly distributed random clipping, the averaged clipped autocorrelation is proportional to the full autocorrelation for optical fields with any statistics.

Calculations show [238] that, for thermal light, if $s > 10\,\bar{n}$, the error in (7.59) is less than 1 %. Of course, the value of s necessary to justify this approximation depends on the statistical model.

In practice, uniform random clipping can be achieved by using an electronic random-number generator to determine the clipping level. A simpler method is to use the so-called *ramp clipping* [468]. The clipping level is varied periodically in a sawtooth waveform. If the phase of this function is independent of the sampling-counting process, then a uniformly random clipping level will be produced (Fig.7.7).

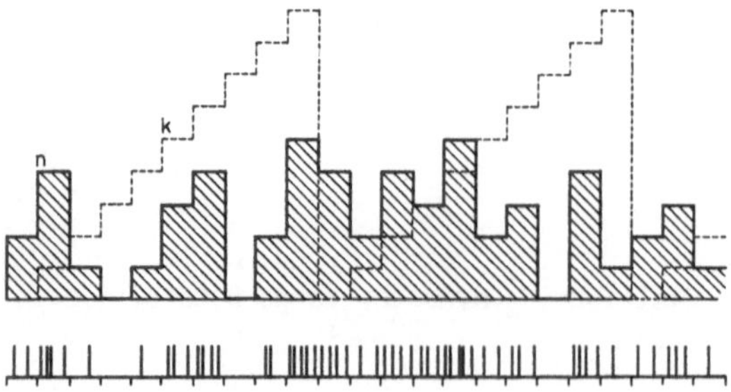

Fig.7.7. Ramp clipping

Another technique for achieving uniform random clipping is the technique of scaling.

7.1.7 Scaled Autocorrelation

Scaling is a simple technique that approximates uniform random clipping [238, 274, 390, 432]. Instead of counting all of the photoelectron events, we count one event and skip s events, where s is a scaling factor (see Fig.7.8). Let

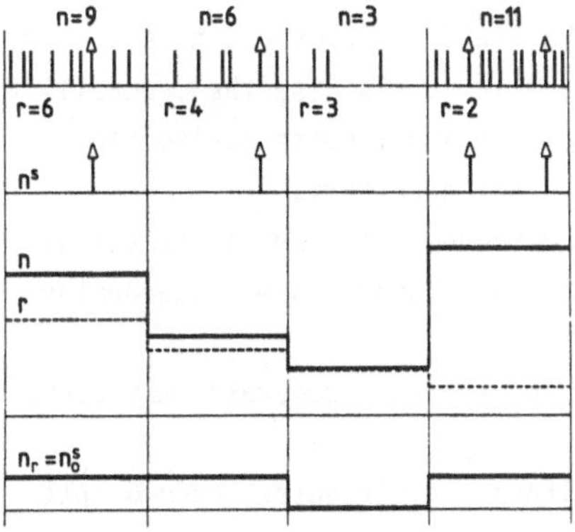

Fig.7.8. Scaling, s = 6

$n^s(t)$ be the number of scaled events in the interval centered around time t and let $n_0^s(t)$ be the number of clipped-at-zero scaled events,

$$
\begin{aligned}
n_0^s(t) &= 1 \qquad n^s(t) > 0 \\
&= 0 \qquad n^s(t) \leq 0 \quad .
\end{aligned}
$$

If r represents the number of events preceding the first scaled pulse in the interval t, then

$$n_0^s(t) = n_r(t) \quad , \tag{7.60}$$

where

$$
\begin{aligned}
n_r(t) &= 1 \qquad \text{if } n > r \\
&= 0 \qquad \text{if } n \leq r \quad .
\end{aligned}
$$

This can be easily seen by noting that $n > r$ is equivalent to $n^s > 0$.

Thus, the number of zero-clipped scaled events equals the number of r-clipped events. Because r is random, we see that scaling plus clipping at zero is equivalent to random clipping. If we assume that r has a uniform distribution between 0 and s-1 we get

$$\langle n_0^s(0)n(t)\rangle \simeq \frac{1}{s}\langle n(0)n(t)\rangle \quad , \tag{7.61}$$

i.e., the zero-clipped scaled correlation is proportional to the full correlation regardless of the statistics of the field and provided that the scaling level is sufficiently high. Note that in such a case the probability of obtaining more than one scaled event per sampling time is negligible; in such a case, clipping at zero is actually not necessary.

Although scaling is equivalent to uniform random clipping, it is actually much simpler to construct a scaling circuit than a random-number generator or a ramp generator.
Synchronization between the sampling clock and the ramp generator may occur, with a distorted output as a result [239].

Now we test the assumption that r is uniformly distributed. Assume that the counting begins at time $t = 0$ and that r counts preceed the first scaled pulse in the interval surrounding $t = (m+1)T$. This means that the first m intervals contain a total number of counts $n = \ell s-r$, where ℓ is any positive integer. If the total duration of m intervals is much larger than the coherence time of the field ($mT \gg \tau_c$) then n must have a Poisson distribution (see Sec. 5.2.2) whose rate is $\bar{N} = m\langle n\rangle$. This means that

$$q(r) = \sum_{\ell=1}^{\infty} \frac{\bar{N}^{\ell s-r} e^{-\bar{N}}}{(\ell s-r)!} \quad . \tag{7.62}$$

For $s > 1$, this can be rearranged to give

$$q(r) = \frac{1}{s}\left\{1 + e^{-\bar{N}} \sum_{k=1}^{s-1} \exp\left[\bar{N}\cos\left(\frac{2\pi k}{s}\right)\right] \cos\left[\frac{2\pi kr}{s} + \bar{N}\sin\left(\frac{2\pi k}{s}\right)\right]\right\} \quad . \tag{7.63}$$

For large $\bar{N}$, the second term decreases rapidly and we get $q(r) \simeq 1/s$. Calculations [238] show that for $s \leq 4$ and $\bar{N} > 5$, q(r) is uniform to better than 1 %. For larger s, this degree of uniformity can be obtained only for

$\bar{N} = m\langle n\rangle > 5s^2/2\pi^2$. For example, if $\bar{n} = 1$ and $s = 10$, uniformity is obtained after about 25 counting intervals.

7.1.8 Statistical Accuracy of Estimating the Autocorrelation Function

In this section, the errors in estimating the autocorrelation function in its full, normalized, and clipped forms are calculated. These errors arise when we attempt to calculate ensemble averages over the counting process from observations taken in a finite period of time, the allowed observation time. The effect of the total number of observed intervals, the counting rate, the sampling time, and the detector area on these errors are discussed. The problem of the statistical accuracy of measured correlation functions was discussed by many authors [59, 127, 197, 223, 228, 245, 246, 261, 272, 273, 383, 423].

Full Autocorrelation Function

In Sec. 7.1.1 we have seen that a good estimate of the autocorrelation function $G^{(2)}(\tau)$ is the sample autocorrelation function,

$$\hat{G}^{(2)}(\tau_\ell) = \frac{1}{N} \sum_{j=1}^{N} n(t_j)n(t_{j+\ell}) \quad . \tag{7.64}$$

We have also seen that because of the finiteness of the sample size, $\hat{G}^{(2)}$ is a random variable whose expectation value is the true autocorrelation $G^{(2)}$. The simplest measure of the accuracy of estimation is the variance of $\hat{G}^{(2)}$. This can be obtained from (7.64) by

$$\mathrm{Var}\,[\hat{G}^{(2)}(\tau_\ell)] = \frac{1}{N^2} \sum_{j=1}^{N} \sum_{i=1}^{N} \langle n(t_j)n(t_{j+\ell})n(t_i)n(t_{i+\ell})\rangle - [G^{(2)}(\tau_\ell)]^2 \;.$$

By making use of the stationarity of the field and the identity

$$\frac{1}{N}\sum_{j=1}^{N} \sum_{i=1}^{N} f_{j-i} = \sum_{m=-(N-1)}^{(N-1)} \left(1-\frac{|m|}{N}\right) f_m \;,$$

$$= f_0 + 2\sum_{m=1}^{N-1} \left(1-\frac{m}{N}\right) f_m \;, \quad (\text{if } f_m = f_{-m}) \;, \tag{7.65}$$

we can write

$$\mathrm{Var}[\hat{G}^{(2)}(\tau_\ell)] = \frac{2}{N}\sum_{m=1}^{N-1}\left(1-\frac{m}{N}\right)\langle n(t_{m+\ell})n(t_m)n(t_\ell)n(0)\rangle$$

$$+\frac{1}{N}\langle n^2(t_\ell)n^2(0)\rangle - |G^{(2)}(\tau_\ell)|^2 \quad . \tag{7.66}$$

The expectations in (7.66) can be written in terms of the moments of the integrated intensity of the detected field by using the rules of doubly stochastic Poisson processes (3.64, 65, 70). This gives

$$\mathrm{Var}[\hat{G}^{(2)}(\tau_\ell)] = \frac{2}{N}\sum_{m=1}^{N-1}\left(1-\frac{m}{N}\right)\langle W(t_{m+\ell})W(t_m)W(t_\ell)W(0)\rangle - |G^{(2)}(\tau_\ell)|^2$$

$$+\frac{2}{N}\left(1-\frac{\ell}{N}\right)\langle W(t_{2\ell})W(t_\ell)W(0)\rangle$$

$$+\frac{1}{N}\langle W^2(t_\ell)W^2(0)\rangle + \frac{1}{N}\langle W^2(t_\ell)W(0)\rangle \tag{7.67}$$

$$+\frac{1}{N}\langle W(t_\ell)W^2(0)\rangle + \frac{1}{N}\langle W(t_\ell)W(0)\rangle \quad .$$

The multifold moments of the integrated intensity can be obtained from the mgf by using the general rule (2.22) or by relating them to the moments of the light intensity, using the multiple integrals of (3.77). The computations are lengthy but straightforward. Below, we compile expressions for $\mathrm{Var}[\hat{G}^{(2)}]$ in some special cases of practical importance.

I) Shot-Noise Limit. We start with the limiting case of very-low counting rate. It is interesting to see that, in this case, the last term of (7.67) dominates and we get

$$\mathrm{Var}[\hat{G}^{(2)}(\tau_\ell)] \simeq \frac{1}{N}\,G^{(2)}(\tau_\ell) \quad , \qquad \bar{n} \ll 1 \quad , \tag{7.68}$$

irrespective of the nature of the optical field. This means that the relative error is inversely proportional to the measured signal $G^{(2)}(\tau)$, which signifies that the error of measurement reaches its maximum at the points where the function $G^{(2)}(\tau_\ell)$ reaches its minimum. This is of course typical of shot-noise errors.

II) Completely Coherent Light. In this case, the integrated intensity W does not fluctuate and (7.67) gives

$$\mathrm{Var}[\hat{G}^{(2)}(\tau_\ell)] = \frac{\bar{n}^4}{N}\left(\frac{4}{\bar{n}} + \frac{1}{\bar{n}^2}\right) . \tag{7.69}$$

III) Thermal Light with Arbitrary Spectrum, Small Detector, Short Sampling Time. In this case, the use of (7.67) together with (5.28) and (5.301) gives [423]

$$\mathrm{Var}[\hat{G}^{(2)}(\tau_\ell)] = \frac{\bar{n}^4}{N_c}\left[a_\ell + \frac{b_\ell}{\bar{n}_d} + \frac{g_\ell^{(2)}}{\bar{n}\bar{n}_d}\right] , \tag{7.70}$$

where $N_c = \gamma N$ is the number of observed coherence times,

$$a_\ell = 4 + z_0 + z_{2\ell} + 2|g_\ell|^2 + 2\,\mathrm{Re}\{g_\ell^* y_\ell + u_\ell\} , \tag{7.71}$$

$$b_\ell = 4\,(1 + 2|g_\ell|^2) , \tag{7.72}$$

and

$$g_\ell^{(2)} = 1 + |g_\ell|^2 . \tag{7.73}$$

Here,

$$\begin{aligned} y_\ell &= \tau_c^{-1} \int_{-\infty}^{\infty} g(t) g^*(t-\tau_\ell)dt \\ z_\ell &= \tau_c^{-1} \int_{-\infty}^{\infty} |g(t)|^2 |g(t-\tau_\ell)|^2 dt \\ u_\ell &= 2\tau_c^{-1}\,\mathrm{Re}\left\{\int_{-\infty}^{\infty} g^*(t) g^*(t) g(t-\tau_\ell) g(t+\tau_\ell)dt\right\} , \end{aligned} \tag{7.74}$$

where

$$g_\ell = g(t_\ell) = g^{(1)}(t_\ell) ,$$

and $\bar{n}_d=\bar{n}\tau_c/T$ is the average number of counts per coherence time. In the limit of very-small degeneracy parameter ($\bar{n}_d \ll 1$ and $\bar{n} \ll 1$), the shot-noise limit, the last term in (7.70) dominates and (7.68) is reproduced.

On the other hand, if $\bar{n} \gtrsim 1$ (and hence $\bar{n}_d \gg 1$), the shot noise is of much less importance and the first term of (7.70) dominates. When the spectrum is Lorentzian,($|g_\ell| = \exp(-\Gamma|\tau_\ell|)$), the coefficients a_ℓ and b_ℓ simplify to

$$a_\ell = 5 + 13|g_\ell|^2 - 3|g_\ell|^4 \quad , \tag{7.75}$$

$$b_\ell = 4 + 12|g_\ell|^2 + 6|g_\ell|^4 \quad . \tag{7.76}$$

IV) Thermal Light, Lorentzian Spectrum, Arbitrary T/τ_c: In this case, the use of (5.55) and (5.301-306) gives for $\mathrm{Var}[\hat{G}(\tau_\ell)]$ the same expression, (7.70), with the constants a_ℓ and b_ℓ now taking the form [246]

$$\begin{aligned} a_\ell = {} & \gamma\left(1 + \frac{1}{\gamma} - \frac{\mathrm{sh}\gamma}{\gamma}\, e^{-\gamma}\right)^2 + \gamma\left[\left(4 + \frac{2}{\gamma}\right) \mathrm{sh}\gamma \; -2\cosh\gamma\right]^2 |g_\ell|^2 \\ & -\gamma(1 - \mathrm{sh}^2\gamma|g_\ell|^2)(1 + 3\,\mathrm{sh}^2\gamma|g_\ell|^2) \\ & +\left(\frac{1-e^{-2\gamma}}{2\gamma}\right)\left\{2 + (5-e^{2\gamma})|g_\ell|^2 + \mathrm{sh}^2\gamma(e^{2\gamma}+1)^{-1}[1-(3 + 6e^{2\gamma})|g_\ell|^4]\right\} \\ & + 2[1 + \mathrm{sh}^2\gamma|g_\ell|^2(4+3|g_\ell|^2)] + 2\gamma(\ell-1)\mathrm{sh}^2\gamma|g_\ell|^2(4+3\mathrm{sh}^2\gamma|g_\ell|^2) \quad , \end{aligned} \tag{7.77}$$

$$b_\ell = 2\left(2+ \frac{1}{\gamma} - \frac{\mathrm{sh}\gamma}{\gamma}\, e^{-\gamma}\right)+2\left[\mathrm{sh}\gamma\left(6+ \frac{2}{\gamma}\right) 2\cosh\gamma\right]\mathrm{sh}\gamma|g_\ell|^2 + 6\mathrm{sh}^2\gamma|g_\ell|^4 \quad , \tag{7.78}$$

where now

$$g_\ell^{(2)} = 1 + \mathrm{sh}^2\gamma|g_\ell|^2 \quad . \tag{7.79}$$

Again, in the limit $\bar{n}_d \ll 1$, (7.70) and (7.77-79) give (7.68); in the limit $\gamma \ll 1$, the coefficients in (7.75, 76) are reproduced. The dependence of the normalized error $e = \{\mathrm{Var}[\hat{G}^{(2)}(\tau)]\}^{\frac{1}{2}}/[G^{(2)}(\tau)-\bar{n}^2]$ on the parameters γ, $\bar{n}$, and τ_ℓ is illustrated in Fig.7.9. As might be expected from the structure of (7.70), this error drops sharply with increase of $\bar{n}_d$ for small $\bar{n}_d$ and then saturates to a constant value for large $\bar{n}_d$. No gain of accuracy is obtained by having $\bar{n}$ exceed a certain level.

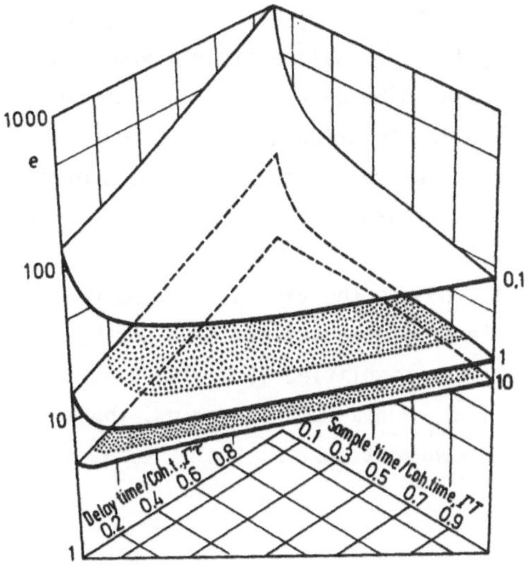

Fig.7.9. Percentage error in the measured correlation function of thermal light with Lorentzian spectrum, $e = 100\{\mathrm{Var}[\hat{G}^{(2)}(\tau)]\}^{1/2}/(G^{(2)}(\tau) - \bar{n}^2)$, as function of delay time and sampling time for the shown values of $\bar{n}_d$, the average number of photoelectrons per coherence time. $N_c = N\Gamma T = 10^4$ observed coherence times, the ordinate scales as $N_c^{-1/2}$ (after JAKEMAN et al. [246])

Figure 7.9 shows that the greatest percentage error occurs with large time delays (when the signal is small). Moreover, the strongest variation of the percentage error with delay time occurs at small values of γ.

The same figure illustrates that an increase of γ causes the percentage error to decrease. This is more pronounced at the shot-noise limit, for which

$$e_\ell^2 = \frac{\gamma}{N_c}\,[(\mathrm{sh}\gamma|g_\ell|)^{-4} + (\mathrm{sh}\gamma|g_\ell|)^{-2}] \quad . \tag{7.80}$$

The percentage error is proportional to $N_c^{-1/2}$. In Fig.7.9, the value $N_c = 10^4$ has been used. For other values of N_c, the error can be obtained by using the proper scaling. In order to assess the order of magnitude of the errors involved in such measurements, we mention that for an average 1 photoelectron per coherence time ($\bar{n}_d = 1$), and when 10^4 coherence times ($N_c = 10^4$) are observed, the error varies from around 20 % for small delay times to around 100 % for large delay times when γ is small. If the coherence time is 0.1 ms, we have to count for more than 100 s to obtain reasonable accuracy (2-10 %).

V) Mixture of Coherent Light and Thermal Light $T \ll \tau_c$, $\bar{n}_{th} \ll \bar{n}_c$, $\bar{n}_{th} \ll 1$.

As discussed in Sec. 7.1.1, coherent light may be deliberately mixed with the thermal light whose spectral properties are to be determined. An improvement of the accuracy of estimating the correlation function of the thermal part is to be expected, because the information is carried by a cross product of the signal by the less-noisy coherent component, rather than by the square of the signal.

In principle, (7.67) could be used to compute the variance of the resultant autocorrelation function, but such lengthy computations have not been performed. The problem can, however, be more simplified if we limit ourselves to the useful special case of very small $\bar{n}_{th}$ and a much larger $\bar{n}_c$. The main contribution to $\mathrm{Var}[\hat{G}_\ell^{(2)}]$ comes from the coherent part of the mixture; according to (7.69) it is equal to

$$\mathrm{Var}[\hat{G}_\ell^{(2)}] \simeq \frac{1}{N}\,(4\bar{n}_c^3 + \bar{n}_c^2) \quad . \tag{7.81}$$

As in (7.14), $G_\ell^{(2)} \simeq \bar{n}^2 + 2\bar{n}_{th}\bar{n}_c|g_\ell|\cos(\Delta\omega\tau)$. The relative error $e_\ell = \{\mathrm{Var}[\hat{G}_\ell^{(2)}]\}^{1/2}/[G_\ell^{(2)} - \bar{n}^2]$ is given by

$$e_\ell = \frac{1}{\sqrt{N}}\,\frac{(4\bar{n}_c + 1)^{1/2}}{2\bar{n}_{th}|g_\ell|\cos(\Delta\omega\tau_\ell)} \quad .$$

Comparing this to the situation without heterodyning (but with $\bar{n}_{th} \ll 1$) where

$$G_\ell^{(2)} \simeq \bar{n}_{th}^2(1 + |g_\ell|^2)$$

and

$$\mathrm{Var}[\hat{G}_\ell^{(2)}] \simeq \frac{\bar{n}_{th}^2}{N}\,(1 + |g_\ell|^2) \quad ,$$

from which

$$e_\ell = \frac{1}{\sqrt{N}} \frac{(1+|g_\ell|^2)^{1/2}}{\bar{n}_{th}|g_\ell|^2} \quad ,$$

we see that heterodyning decreases the signal by a factor $(2\bar{n}_{th}/\bar{n}_c|g_\ell|)^2$, whereas it decreases the standard deviation by a factor $(\bar{n}_{th}/\bar{n}_c)(1+|g_\ell|^2)^{1/2}/(4\bar{n}_c+1)^{1/2}$. For small $\bar{n}_c$, this means a reduction of the relative error by division by $2\sqrt{2}$ when τ is small and by a larger quantity for larger τ. This is an additional advantage of heterodyning. For $\bar{n}_c > 1$, however, the accuracy may be made worse by heterodyning.

Normalized Full Correlation

As is mentioned in Sec. 7.1.3, an estimate of the normalized full autocorrelation function can be obtained by use of

$$\hat{g}^{(2)} = \hat{G}^{(2)}/(\hat{n})^2 , \tag{7.82}$$

where $\hat{G}^{(2)}$ and $\hat{n}$ are defined by (7.21) and (7.24). Because the expectation value of a ratio is not equal to the ratio of the expectation values, we immediately see that $\langle\hat{g}^{(2)}\rangle \neq g^{(2)}$, i.e., our estimate is not unbiased. The bias and the variance of $\hat{g}^{(2)}$ are the simplest measures of its usefulness.

Exact expressions for these are very difficult to determine because of the difficulty of calculating the expectation value of a ratio between two random variables with unknown statistics. Therefore, we resort to approximation by expanding $\hat{n}$ in (7.82) around its mean value $\bar{n}$

$$\hat{g}^{(2)} = \frac{\hat{G}^{(2)}/\bar{n}^2}{\left(1+\frac{\hat{n}-\bar{n}}{\bar{n}}\right)^2} \approx \frac{\hat{G}^{(2)}}{\bar{n}^2}\left[1-2\left(\frac{\hat{n}}{\bar{n}}-1\right)+3\left(\frac{\hat{n}}{\bar{n}}-1\right)^2\right] \quad . \tag{7.83}$$

Also, expanding $\hat{G}^{(2)}$ around its mean $G^{(2)}$, we get

$$\hat{g}^{(2)} = g^{(2)} + \delta g \qquad \delta g = -2g^{(2)}\cdot\left(\frac{\hat{n}}{\bar{n}}-1\right) + 3g^{(2)}\cdot\left(\frac{\hat{n}}{\bar{n}}-1\right)^2 - 2\left(\frac{\hat{G}-G}{\bar{n}^2}\right)\left(\frac{\hat{n}}{\bar{n}}-1\right) \quad ,$$

where higher-order terms have been neglected. With this approximate relation and the definitions of $\hat{G}$ and $\hat{n}$ above we can compute this bias and the variance of $\hat{g}^{(2)}$. The bias is given by

$$\langle\delta g\rangle = 3\, g^{(2)} \cdot \frac{\mathrm{Var}(\hat{n})}{\bar{n}^2} - \frac{2}{\bar{n}^3} \langle \hat{G}\cdot(\hat{n}-\bar{n})\rangle$$

and the variance is given by

$$\mathrm{Var}[\hat{g}^{(2)}] \simeq \langle\delta g^2\rangle - \langle\delta g\rangle^2 \tag{7.84}$$

$$\mathrm{Var}[\hat{g}^{(2)}] = \frac{1}{\bar{n}^4}\, \mathrm{Var}(\hat{G}) + \Delta \quad , \tag{7.85}$$

$$\Delta = 4g^{(2)} \cdot \left\{ g^{(2)} \cdot \left[\frac{\mathrm{Var}(\hat{n})}{\bar{n}^2} + 1 \right] - \frac{1}{\bar{n}^3} \langle \hat{G}\hat{n} \rangle \right\} \quad . \tag{7.86}$$

The term $\langle\delta g\rangle^2$ has been neglected because it is much smaller than other terms (i.e., the bias is considerably smaller than the root-mean-square error).

The first term in (7.85) has already been determined (see pages 375-380). Also, the first term of (7.86) has already been determined (see page 290 of Sec. 6.2.1) in connection with errors in estimating the mean count rate. The third term can be obtained from

$$\langle \hat{G}\hat{n} \rangle = \frac{1}{N^2} \sum_j \sum_i n(t_{j+\ell}) n(t_j) n(t_i)$$

$$= \frac{1}{N} \sum_{m=1}^{N-1} \left(1 - \frac{m}{N}\right) [\langle n(t_{m+\ell}) n(t_m) n(0)\rangle + \langle n(t_\ell) n(0) n(t_m)\rangle] + \frac{1}{N} \langle n(t_\ell) n^2(0)\rangle$$

$$= \frac{1}{N} \sum_{m=1}^{N-1} \left(1 - \frac{m}{N}\right) [\langle W(t_{m+\ell}) W(t_m) W(0)\rangle + \langle W(t_\ell) W(0) W(t_m)\rangle]$$

$$+ \frac{1}{N} \left(1 - \frac{\ell}{N}\right) \langle W(t_\ell) W(0)\rangle + \frac{1}{N} [\langle W(t_\ell) W^2(0)\rangle + \langle W(t_\ell) W(0)\rangle] \quad . \tag{7.87}$$

The expectations in (7.87) can be determined from the moment-generating functions by use of (2.22). For a given field with a known mgf of orders 1, 2, 3, and 4, it is a matter of straightforward but lengthy computations to determine the variance of $\hat{g}^{(2)}$.

Let us now examine some special cases.

I) Coherent Light. In this case, $\langle W(t_1)W(t_2)W(t_3)W(t_4)\rangle = \bar{n}^4$. This leads to $\Delta = -4/N\bar{n}$. From (7.69) and (7.85), we get

$$\mathrm{Var}[\hat{g}_\ell^{(2)}] = \frac{1}{N}\frac{1}{\bar{n}^2} \quad . \tag{7.88}$$

It is interesting to see that in this case

$$\mathrm{Var}[\hat{g}_\ell^{(2)}] \,/\, [g_\ell^{(2)}]^2 < \mathrm{Var}[\hat{G}_\ell^{(2)}]/[G_\ell^{(2)}]^2 \quad ,$$

i.e., the relative accuracy has been enhanced by normalization!

II) Shot-Noise Limit. By examining (7.85-87) and (6.5), it is easy to show that, in the shot-noise limit, the contribution of the term Δ in (7.85) is negligible and we have again

$$\mathrm{Var}[\hat{g}_\ell^{(2)}] \simeq \frac{1}{N}\frac{1}{\bar{n}^2}\, g_\ell^{(2)} \quad . \tag{7.89}$$

Note that the error of $\hat{g}_\ell^{(2)}$ increases rapidly as the count rate decreases. The relative error is given by

$$e_\ell = \mathrm{Var}^{\frac{1}{2}}[g_\ell^{(2)}]/[g_\ell^{(2)}-1] = \frac{1}{\sqrt{N}}\frac{1}{\bar{n}}\left\{[g_\ell^{(2)}-1]^{-1} + [g_\ell^{(2)}-1]^{-2}\right\}^{\frac{1}{2}} \quad , \tag{7.90}$$

i.e., increases as $[g_\ell^{(2)}-1]$ decreases.

Equation (7.90) can be used to study the effect of the area of the detector. In the limit, $T \ll \tau_c$, $g_\ell^{(2)}-1 = f(A)|g_\ell^{(1)}|^2$, i.e., decreases with increase of the detector area. Because $\bar{n}$ is proportional to A, we see from (7.90) that two opposing effects result from an increase of A: an increase of $\bar{n}$ that reduces the error, and a decrease of the signal $[g^{(2)}-1]$ that increases the error. For small detector area $[f(A) \sim 1]$, the error e_ℓ decreases as A^{-1}. For large A/A_c, $f(A) \sim A^{-1}$ and the error tends to a constant value. The graph (Fig.7.10) of e_ℓ against A/A_c (obtained from (7.90) and the formula (5.101) of f(A) for $|g_\ell(\tau)| = 1$) shows that e_ℓ drops sharply with A/A_c and saturates for A larger than about two areas of coherence. Little is gained by increasing the detector area beyond this value. Note, however, that by increasing the area, the average count rate $\bar{n}$ becomes so large

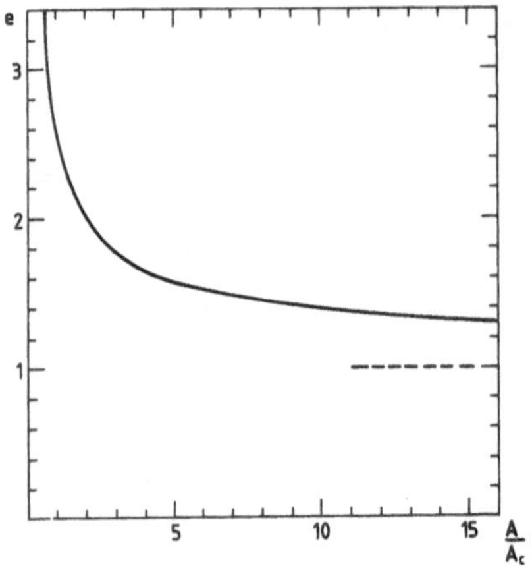

Fig.7.10. Dependence of the percentage error in the measured normalized correlation function of thermal light, $e = 100\{\mathrm{Var}[g^{(2)}(\tau)]\}^{1/2}/[g^{(2)}(\tau)-1]$, on the ratio A/A_c, for $|g^{(1)}(\tau)| = 1$ in the shot-noise limit. The sampling time is assumed short. One photoelectron per coherence area is detected. $N = 10^4$

that the shot-limit approximation no longer holds, and the contribution of the intensity fluctuation to the error must be considered. This contribution would eventually make the error in Fig.7.10 increase again. An optimum area then exists at which the error is a minimum.

III) Thermal Light with Arbitrary Spectrum, $T \ll \tau_c$, Small Detector Area. In this case, the variance of $\hat{g}_\ell^{(2)}$ is given by (7.85), where $\mathrm{Var}[\hat{G}_\ell^{(2)}]$ is given by (7.70-74), the excess error due to normalization is given by [423]

$$\Delta = \frac{1}{N_c}\left(d_\ell + \frac{1}{\bar{n}_d} f_\ell\right)$$

$$d_\ell = 4\,[g_\ell^{(2)}]^2 + 8\,g_\ell^{(2)}\,\mathrm{Re}\{g_\ell^* y_\ell\}$$

$$f_\ell = 4\,[g_\ell^{(2)}]^2 \tag{7.91}$$

and y_ℓ is given by (7.74).

The variance of $g^{(2)}$ is then given by the general formula,

$$\mathrm{Var}[\hat{g}^{(2)}] = \frac{1}{N_c}\left[(a_\ell + d_\ell) + \frac{1}{\bar{n}_d}(b_\ell + f_\ell) + \frac{1}{\bar{n}\bar{n}_d}\,g_\ell^{(2)}\right] \;. \tag{7.92}$$

In the shot-noise limit, the contribution of the term Δ is negligible and (7.89) is reproduced. For Lorentzian light, substitution in (7.92) gives

$$a_\ell + d_\ell = \frac{1}{2}\,[1+8e^{-2\Gamma\tau_\ell}-(5+4\Gamma\tau_\ell)e^{-4\Gamma\tau_\ell}] \tag{7.93}$$

$$b_\ell + f_\ell = 2\,(1+e^{-2\Gamma\tau_\ell})^2 \tag{7.94}$$

and

$$g_\ell^{(2)} = 1 + e^{-2\Gamma\tau_\ell} \quad .$$

IV) Thermal Light with Lorentzian Spectrum, Arbitrary T/τ_c, Small Detector. In this case, direct substitution gives (7.92) again, where a_ℓ and b_ℓ are given by (7.77,78) and

$$d_\ell = 4g_\ell^{(2)}\left\{g_\ell^{(2)}-2\,\mathrm{sh}^2\gamma[(\ell-1)\gamma+1-e^{-\gamma}/\mathrm{sh}\gamma]|g_\ell|^2\right\} \tag{7.95}$$

and

$$f_\ell = 4[g_\ell^{(2)}]^2 \quad , \qquad g_\ell^{(2)} = 1 + \mathrm{sh}^2\gamma|g_\ell|^2 \quad . \tag{7.96}$$

This shows that the contribution of normalization does not affect the shot-noise term (the third term in (7.92)). Moreover, by examining (7.95) and (7.96), we see that f_ℓ is positive and for small γ, d_ℓ is also positive, meaning that normalization results in an increase of the relative error. For large γ, d_ℓ may become negative and may end by reducing the relative error, as in the case of coherent light.

V) Mixture of Coherent and Thermal Light with Lorentzian Spectrum $T \ll \tau_c$, $\Delta\omega\tau \ll 1$, $\bar{n}_{th} \ll 1$, $\bar{n}_{th} \ll \bar{n}_c$. As was discussed before, the error in measuring the heterodyned correlation function is mainly due to the coherent component; hence

$$\mathrm{Var}[\hat{g}_\ell^{(2)}] = \frac{1}{N}\,\frac{1}{\bar{n}^2} \quad ,$$

resulting in a relative error $e_\ell = \left\{\mathrm{Var}[\hat{g}_\ell^{(2)}]\right\}^{\frac{1}{2}}/[g_\ell^{(2)}-1]$ given by

$$e_\ell = \frac{1}{\sqrt{N}}\,[2\bar{n}_{th}|g_\ell|\cos(\Delta\omega\tau)]^{-1} \quad . \tag{7.97}$$

Comparing this to the relative error in measuring $\hat{g}_\ell^{(2)}$ for a purely thermal light in the shot-noise limit,

$$e_\ell = \frac{1}{\sqrt{N}}\,\frac{1}{\bar{n}_{th}}\,\frac{(1+|g_\ell|^2)^{\frac{1}{2}}}{|g_\ell|^2} \quad , \tag{7.98}$$

we see that heterodyning improves the relative accuracy by a factor

$$2(1 + |g_\ell|^{-2})^{\frac{1}{2}} \quad ,$$

where, for simplicity, we assume $\Delta\omega = 0$ (homodyne case).

This factor is equal to $2\sqrt{2}$ when $\tau_\ell = 0$ ($|g_\ell| = 1$) and increases considerably as τ_ℓ increases.

Clipped Autocorrelation Function

As in (7.66), the variance of the estimator of the singly clipped autocorrelation function is given by

$$\mathrm{Var}[\hat{G}_k^{(2)}(\tau_\ell)] = \frac{2}{N^2}\sum_{m=1}^{N-1} (N-m)\langle n_k(t_{m+\ell})n(t_m)n_k(t_\ell)n(0)\rangle + \frac{1}{N}\langle n_k^2(t_\ell)n^2(0)\rangle - |G_k^{(2)}(\tau_\ell)|^2 \quad . \tag{7.99}$$

The expectations in (7.99) can be determined from the mgf by using the relation

$$\langle n_k(t_1)n(t_2)n_k(t_3)n(t_4)\rangle = \sum_{n=k+1}^{\infty}\sum_{m=k+1}^{\infty}\frac{(-1)^{n+m}}{n!m!}\frac{\partial^n}{\partial s_1^n}\frac{\partial}{\partial s_2}\frac{\partial^m}{\partial s_3^m}\frac{\partial}{\partial s_4} Q(s_1,s_2,s_3,s_4)\Big|_{\substack{s_1 = s_3 = 1 \\ s_2 = s_4 = 0}}$$

and

$$\langle n_k^2(t_1)n^2(t_2)\rangle = \sum_{n=k+1}^{\infty} \frac{(-1)^n}{n!} \frac{\partial^n}{\partial s_1^n}\left(\frac{\partial}{\partial s_2^2} - \frac{\partial}{\partial s_2}\right) Q(s_1,s_2)\Big|_{s_1 = 1,\ s_2 = 0}$$

$$= \langle n^2\rangle - \sum_{n=0}^{k} \frac{(-1)^n}{n!} \frac{\partial^n}{\partial s_1^n}\left(\frac{\partial^2}{\partial s_2^2} - \frac{\partial}{\partial s_2}\right) Q(s_1,s_2)\Big|_{s_1=1,\ s_2=0} \quad .$$

Also for the normalized clipped autocorrelation function, we can write

$$\mathrm{Var}[g_\ell^{(2)}(\tau_\ell)] = \frac{1}{\bar{n}^2\, \bar{n}_k^2}\, \mathrm{Var}[\hat{G}_k^{(2)}(\tau_\ell)] + \Delta_k \quad ,$$

where

$$\Delta_k = [g_k^{(2)}(\tau_\ell)]^2 \left[\frac{\mathrm{Var}(\hat{n})}{\bar{n}^2} + \frac{\mathrm{Var}(\hat{n}_k)}{\bar{n}_k^2} + 2\, \frac{\mathrm{Cov}(\hat{n},\hat{n}_k)}{\bar{n}\,\bar{n}_k}\right]$$

$$- \frac{2}{\bar{n}\,\bar{n}_k}\, g_k^{(2)}(\tau_\ell)\left[\frac{1}{\bar{n}}\langle\hat{G}_k^{(2)}\cdot(\hat{n}-\bar{n})\rangle + \frac{1}{\bar{n}_k}\langle\hat{G}_k^{(2)}\cdot(\hat{n}_k-\bar{n}_k)\rangle\right] \quad . \tag{7.100}$$

Using the above equations, the variance of $\hat{G}_k$ and $\hat{g}_k$ can, in principle, be calculated for any field of a given moment-generating function Q. The computations are obviously very lengthy even in the simplest case of thermal light with a Lorentzian spectrum. A lengthy analytic expression was derived for this case by JAKEMAN et al. [246] in the special case k = 0 and for small T/τ_c and A/A_c. In *the shot-noise limit* and for an arbitrary k ($T \ll \tau_c$, $A \ll A_c$), the variance of $\hat{G}_k$ simplifies considerably and the normalized square error takes the form [245]

$$e_k^2 = \frac{\mathrm{Var}[\hat{g}_k^{(2)}(\tau)]}{[g_k^{(2)}(\tau)-1]^2} = \frac{1}{N}\, \frac{1}{\bar{n}^2}\, \frac{1}{\bar{n}^k}\, \frac{1}{(1+k)^2}\, [(1+k)|g^{(1)}(\tau)|^{-2} + |g^{(1)}(\tau)|^{-4}] . \tag{7.101}$$

Because $\bar{n}$ is assumed much smaller than 1, this variance increases as k increases. The ratio between the normalized variance with k = 1 and that with k = 0 is proportional to $\bar{n}$. Hence, the maximum accuracy is obtained with k = 0, for which

$$e_0^2 = \frac{1}{N}\frac{1}{\bar{n}^2}\,[|g^{(1)}(\tau)|^{-2} + |g^{(1)}(\tau)|^{-4}] \quad .$$

From (7.90), it is obvious that this is the same normalized error as in the unclipped case.

For *coherent light* (7.99) and (7.100) give

$$\mathrm{Var}[\hat{g}_k^{(2)}] = \frac{1}{N\bar{n}}\left(\frac{1}{\bar{n}_k} - 1\right) \quad . \tag{7.102}$$

This can be used to assess the performance of a singly clipped autocorrelator measuring light that is heterodyned with a strong coherent component. In this case, $g_k^{(2)}$ and $\bar{n}_k$ are given by (7.50-53) and the relative error is

$$e_k = \frac{1}{\sqrt{N}}\,E(k,\bar{n}_c)/[2\bar{n}_{th}|g^{(1)}(\tau)|\cos(\Delta\omega\tau)] \quad , \tag{7.103}$$

where the factor

$$E(k,\bar{n}_c) = \left\{\frac{e^{2\bar{n}_c}}{\bar{n}_c^{2k+1}}\,\Gamma(1+k,\bar{n}_c)[k! - \Gamma(1+k,\bar{n}_c)]\right\}^{\frac{1}{2}} \tag{7.104}$$

(Γ is the imcomplete gamma function) represents the change of the relative error over the unclipped case (cf. (7.97)).

This factor has a minimum at $k \simeq \bar{n}_c$. The minimum increases as $\bar{n}_c = k$ increases until the asymptotic value $\sqrt{\pi/2}$ is approached at large values of $\bar{n}_c$. Thus, at worst, clipping increases the relative error by a factor of 1.23. We have seen before that heterodyning improves the accuracy by a factor of $2\sqrt{2}$. Moreover, we have seen that for purely thermal light, in the shot-noise limit, clipping at zero does not decrease accuracy. This means that clipped heterodyning results in improvement by a factor of 2.26.

The accuracy of randomly clipped correlators was discussed by DAUDPOTA [113].

7.1.9 Accuracy of Estimating Unknown Parameters of a Given Spectral Profile

In the previous sections, we have determined the uncertainty of the measured autocorrelation function $\hat{g}^{(2)}(\tau_\ell)$ at an arbitrary time delay τ_ℓ under different conditions. This estimate of uncertainty is of great value in assessing the degree of confidence in the measured results. In many situations, the profile of the autocorrelation function is a priori known except for one or several parameters. We are then more interested in the accuracy of estimating these parameters rather than in the correlation function itself. In the following, we discuss some examples of practical interest.

I) Thermal light, Lorentzian spectrum, $T \ll \tau_c$, $A \ll A_c$, and unknown spectral width. Estimation based on the normalized full correlation function. Here we have

$$g_\ell^{(2)} = 1 + e^{-2\Gamma|\tau_\ell|} \quad .$$

We use the formalism for parameter estimation that was presented in Sec. 2.7.2 (MMSE Nonlinear Fitting), where the measured function x_ℓ corresponds to the correlation function $g_\ell^{(2)}$, and the parameter to be estimated, θ, is in this case the bandwidth Γ. By using (2.149), we see that the bandwidth Γ can be estimated from a set of measurements $g_\ell^{(2)}$ by solving the nonlinear equation

$$\sum_\ell [\hat{g}_\ell^{(2)} - 1 - e^{-2\hat{\Gamma}|\tau_\ell|}]|\tau_\ell|e^{-\hat{\Gamma}|\tau_\ell|}/\Lambda_{\ell\ell} = 0 \quad ,$$

where $\Lambda_{\ell\ell} = \mathrm{Var}(\hat{g}_\ell)$ is given by (7.92).

The estimation error is given by (cf. (2.152))

$$e_\Gamma^2 = \frac{\mathrm{Var}(\hat{\Gamma})}{\Gamma^2} = \left(\sum_{\ell=1}^{L} 4x_\ell^2 e^{-4x_\ell}/\Lambda_{\ell\ell}\right)^{-1} \quad , \quad x_\ell = \Gamma|\tau_\ell| \quad . \qquad (7.105)$$

This error is plotted in Fig.7.11 for L = 1, 20 and ∞. Note that in deriving (7.105) we have assumed that the errors in the different channels are uncorrelated. If the channel correlation is taken into consideration, the error is greater, as is shown in Fig.7.11.

II) Thermal Light, Lorentzian Spectrum, $T \ll \tau_c$, Arbitrary Unknown A/A_c, Shot-Noise Limit. In this problem, we have

$$g_\ell^{(2)} = 1 + c\, e^{-2\Gamma|\tau_\ell|} \quad , \quad c = f(A) \quad , \qquad (7.106)$$

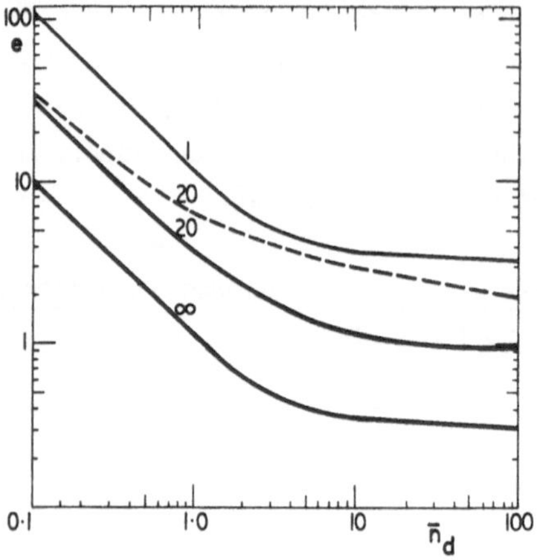

Fig.7.11. Percentage error in estimating the bandwidth Γ by fitting to $g_\ell^{(2)} = 1 + \exp(-2\Gamma|\tau_\ell|)$, as a function of the counting rate per coherence time $\bar{n}_d$, for the indicated number of channels L.——— channels assumed uncorrelated; ----- channel correlation taken into consideration. $\gamma = 0.1$ and $N_c = 10^4$. For $L = 1$, $x_\ell = 0.5$; for $L = 20$, $\{x_\ell\} = \{0.1, 0.2, \ldots, 2\}$. The light is assumed thermal with a Lorentzian spectrum (after SALEH and CARDOSO [423])

with two unknown parameters $\theta_1 = c$ and $\theta_2 = \Gamma$; from (7.89) we have

$$\Lambda_{\ell\ell} = \mathrm{Var}[\hat{g}_\ell^{(2)}] = \frac{1}{N}\frac{1}{\bar{n}^2}\, g_\ell^{(2)} \quad . \tag{7.107}$$

By using a two-dimensional generalization of (2.150), we get

$$e_\Gamma^2 = \frac{1}{4c^2}\,\frac{\sum_{\ell=1}^{L} \xi_\ell^2\, \bar{d}_\ell^2\, \Lambda_{\ell\ell}}{\left(\sum_\ell \xi_\ell\, d_\ell \bar{d}_\ell\right)^2} \quad , \tag{7.108}$$

where

$$\begin{aligned} \xi_\ell &= 1/\Lambda_{\ell\ell} \\ d_\ell &= x_\ell\, e^{-2x_\ell} \\ \bar{d}_\ell &= (x_\ell - \bar{x})e^{-2x_\ell} \\ \bar{x} &= \sum_\ell \xi_\ell x_\ell e^{-4x_\ell} \Big/ \sum_\ell \xi_\ell e^{-4x_\ell} \end{aligned} \tag{7.109}$$

This enables us to compute the error as a function of $c = f(A)$ and study the effect of increasing the detector area for a constant flux of photoelectrons per unit area. As a function of A/A_c, the error should have a behavior similar to that shown in Fig.7.10. In the limit $A \ll A_c$, $c \simeq 1$ and for $L \to \infty$, (7.108) takes the simple form [246]

$$e_\Gamma^2 = \frac{1}{N_c} \frac{21.22}{\bar{n}_d} \quad . \tag{7.110}$$

III). Thermal Light, Lorentzian Spectrum, Unknown Linewidth, Arbitrary T/τ_c, Small A/A_c. In this case, we have

$$g_\ell^{(2)} = 1 + c\, e^{-2\Gamma|\tau_\ell|} \quad , \tag{7.111}$$

where

$$c = \mathrm{sh}^2\gamma \quad , \quad \gamma = \Gamma T \quad . \tag{7.112}$$

The variance of $g_\ell^{(2)}$ is given by (7.92,77,78,95,96) as a function of γ and $\bar{n}$.

This is an estimation problem with one unknown parameter Γ. Hence (2.152) can be used directly and gives

$$e_\Gamma^2 = \left[\sum_\ell 4\, \mathrm{sh}^2\gamma(\cosh\gamma - \mathrm{sh}\gamma - x_\ell \mathrm{sh}\gamma)^2 e^{-4x_\ell}/\Lambda_{\ell\ell}\right]^{-1}$$

$$\Lambda_{\ell\ell} = \mathrm{Var}[\hat{g}_\ell^{(2)}] \quad , \quad x_\ell = \Gamma\tau_\ell \quad . \tag{7.113}$$

A plot of this as a function of γ and $\bar{n}$ should give us an idea about the dependence of the linewidth error on $\gamma = \Gamma T = T/\tau_c$.

Another less-accurate approach is to regard the constant c in (7.111) as an independent unknown parameter and to use the 2-parameter estimation equations (Sec. 2.7.2, page 56). This gives an expression for the error identical to that in the previous case (7.108,109) except that the variance $\Lambda_{\ell\ell}$ should now be given by (7.92,77,78,95,96). This approach is more practical because other factors may affect the parameter c in an unknown way (e.g., the finite area of the detector). Although we are unable to take into consideration the effect of such factors on the variance $\Lambda_{\ell\ell}$, it seems that the only reasonable alternative is to eliminate c. The error can then be computed and the effects of parameters such as $\bar{n}_d$, γ, L, and N_c can be assessed. The results of computations by JAKEMAN et al. [246] are shown in Figs.7.12 and 7.13 and can be summarized in the following:

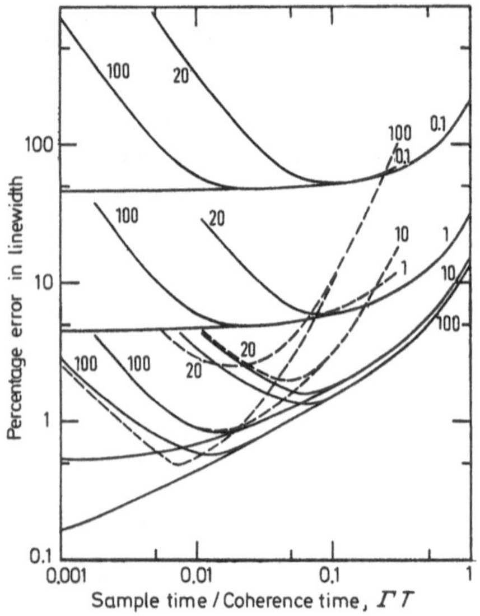

Fig.7.12. Percentage error in linewidth as a function of sampling time for the number of counts per coherence time and number of channels shown on the right-hand and left-hand side of the graph, respectively: —— unclipped; -----clipped at zero. When $\bar{n}_d$ = 0.1 and 1, the clipped and unclipped results coincide for $\Gamma T < 0.1$. $N_c = 10^4$. The light is assumed thermal and the spectrum is Lorentzian (after JAKEMAN et al. [246])

a) Effect of count rate $\bar{n}_d$. For small values of $\bar{n}_d$, the accuracy improves considerably with an increase of $\bar{n}_d$. For a larger $\bar{n}_d$, the improvement becomes less pronounced and saturation takes place for $\bar{n}_d \gtrsim 10$.

b) Effect of sampling time T. For a fixed number of channels L and a fixed total observation time T_0, an increase of the sampling time T results in two opposed effects-time integration of the signal, which tends to reduce the accuracy, and increase of the range of the observed time delay τ_{max} = LT, which tends to improve the accuracy. The optimum sampling-time ratio $\gamma = T/\tau_c$ corresponding to maximum accuracy is τ_{max}/τ_c = 2-3 for $\bar{n}_d$ = 0.1-1 and τ_{max}/τ_c = 1-2 for $\bar{n}_d$ = 10-100. For L = 20 channels, this corresponds to γ = 0.05-0.1 for small $\bar{n}_d$ and γ = 0.1-0.15 for large $\bar{n}_d$.

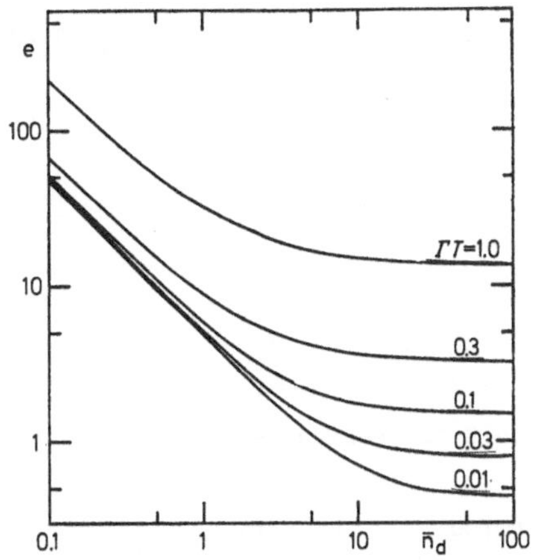

Fig.7.13. Dependence of the percentage error in linewidth obtained by an infinite-channel correlator on the number of photoelectrons per coherence time and the sampling time. $N_c = 10^4$. The light is assumed thermal with a Lorentzian spectrum (after JAKEMAN et al. [246])

c) Effect of detector area. If the mean number of counts per coherence time per coherence area r is fixed and if the detector area is increased, $\bar{n}_d = rA/A_c$ increases proportionally, resulting in a reduction of the error, as is described in a). If the detector area exceeds a coherence area, the integration effects tend to wash out the signal and cause serious increase of the error. Because the effect of $\bar{n}_d$ saturates at $\bar{n}_d = 10$, the optimum strategy becomes: If $r > 10$, choose A such that $\bar{n}_d = 10$ ($A = 10\,A_c/r$). If $r < 10$, take $A = A_c$.

d) Effect of number of channels L. If the total experiment time T_0 and the maximum delay time are fixed, then an increase of the number of channels L is expected to improve the accuracy of fitting. Moreover, it results in a smaller sampling time ($T = \tau_{max}/L$) and hence smaller γ. If L is so large that γ is much smaller than its optimum value, then the improvement of fitting that results from the large number of points is compensated by the increase of error at each point. As a result no net gain of accuracy is achieved by using a number of channels that exceeds a certain limit. Fig.7.12 illustrates this quantitatively.

IV) Thermal Light, Lorentzian Spectrum, $A \ll A_c$, $T \ll \tau_c$. Estimation Based on the Singly Clipped Normalized Autocorrelation Function. In this case, the clipped correlation function is given by (7.42)

$$g_k^{(2)}(\tau_\ell) = 1 + c\, e^{-2\Gamma|\tau_\ell|} \quad , \qquad c = \frac{1+k}{1+\bar{n}} \quad . \tag{7.114}$$

A one-parameter estimation procedure gives

$$e_\Gamma^2 = \frac{1}{c^2}\sum_{\ell=1}^{L} 4x_\ell^2\, e^{-4x_\ell}/\Lambda_{\ell\ell} \quad . \tag{7.115}$$

A two-parameter estimation procedure in which c is treated as an unknown gives again (7.108,109).

The real problem is actually how to determine $\Lambda_{\ell\ell}$, the variance of the clipped normalized correlation function. As was previously mentioned, this has been done analytically only in the case of $k = 0$. The percentage error in this case is plotted in Fig.7.12. This shows that in the shot-noise limit, the effect of zero clipping is negligible, as is expected from the previous discussion in this subsection; but, it becomes more important as $\bar{n}_d$ is increased. Moreover, the minimum error that corresponds to the optimum γ is shifted more rapidly towards lower values of γ as $\bar{n}_d$ is increased.

HUGHES et al. [223] have computer simulated the photoelectron-counting process and computed the statistical accuracy as a function of the parameters involved. Their results show that the optimum choice of the clipping level lies around the value $k = \bar{n}$.

7.1.10 Summary

The results of theoretical, experimental, and computer-simulation work on the statistical accuracy of digital correlators can be summarised in the following principles of operation [355]:

a) The experimental setup should be designed to maximize the energy contained in a coherence area of the detected field.

b) The detector should subtend about one coherence area.

c) The measured correlation function should span about two optical coherence times.

d) Many short measurements normalized independently are preferable to a single long one.

e) The clipping level should be set equal to the average number of counts per sampling time, in which case the loss of accuracy due to clipping is insignificant.

7.2 Estimation of Spectral Parameters Based on Measurement of Probability Distributions

In this section, we consider other photoelectron statistical methods of determining an optical spectral profile or of estimating unknown spectral parameters. This covers measurements of the probability of photoelectron coincidence, the probability distribution of inter-event times, and single and joint photoelectron-counting probabilities. Although the autocorrelation technique is of much wider use than any of these methods, under certain conditions they offer some advantages of speed or simplicity.

7.2.1 Probability of Coincidence

According to Sec. 3.2.1, the probability that two photoelectrons occur simultaneously at the instants t_1 and t_2 is proportional to the intensity autocorrelation,

$$\lambda(t_1,t_2) \sim G_I(t_1,t_2) \quad . \tag{7.116}$$

By measuring this probability, we have direct access to the intensity correlation function. This can be achieved by using a delayed-coincidence circuit [351, 394, 430]. The time delay (t_2-t_1) may be introduced before or after detection, as illustrated in Fig.7.14a,b. The sampling intervals

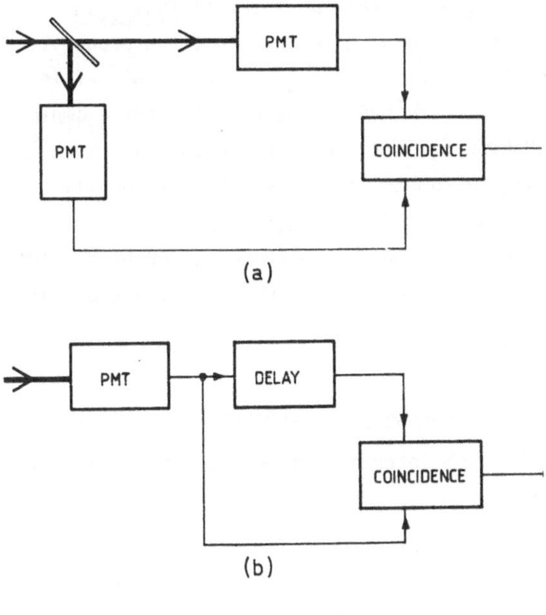

Fig.7.14a and b. Schemes for coincidence experiments

have to be as short as possible in order to ensure that there is a negligible chance of detecting two events in one sampling time. This technique is successful when the coherence times to be measured are short. Nanosecond decay times can be measured with this technique, by use of a variable delay line.

Multichannel Coincidence

Several delay units may be used simultaneously (e.g., by using a series-parallel shift register) and by using several coincidence circuits, multichannel operation becomes possible.

The digital correlator described in Sec. 7.1.2 can be used to obtain multichannel delayed coincidence if very short sampling times are used and if it is operated in the zero-zero double-clipped mode.

Another practical method of measuring the probability of coincidence $\lambda(0,\tau)$ is to detect one pulse and make a histogram of the time delays of all other pulses that succeed it. Let t_j be the time of occurrence of one pulse and $t_j+\tau_{sj}$, $s = 1,2, \ldots S$, be the times of occurrence of the S pulses that follow it. If such an observation is made N times ($j = 1,2, \ldots N$), then because each of the times τ_{sj}, $s = 1,2, \ldots S$, $j = 1,2, \ldots N$ is a time of coincidence, they can be sorted out to obtain one histogram that should approximate $\lambda(t,t+\tau)$ as a function of τ. Note that if S is not large, the result will be quite distorted. For example, if we choose $S = 1$, the resultant histogram gives the inter-event probability, which is quite different from the probability of coincidence (see Secs. 3.5.3 and 7.2.2).

If $\lambda(0,\tau)$ is to be estimated accurately within an interval $[0,\tau_{max}]$, then the number S should be sufficiently large to ensure that the probability of more than S counts falling in the interval $[0,\tau_{max}]$ is negligible. This can always be done by making the count rate sufficiently small. For example, in the instrument of CHOPRA and MANDEL [91], $S = 6$. For a count rate such that $\langle n\rangle\tau_{max} = 1$, the probability of a number of counts larger than 6 is negligible ($<10^{-3}$ for coherent light).

Triple coincidence (and triple correlation) can also be measured [94, 101, 102, 122, 123]. This gives $\lambda(t_1,t_2,t_3) \sim G_I(t_1,t_2,t_3)$ which contains useful information that characterizes non-Gaussian fields.

7.2.2 Time-Interval Probability. Pulse-Separation Technique

The probability distribution of the time interval between successive photoelectrons (cf. Sec. 3.5.3) is given by

$$P_i(\tau) = \left\langle \lambda(0)\lambda(\tau) \exp\left[-\int_0^\tau \lambda(\theta)d\theta\right]\right\rangle / \langle\lambda(0)\rangle \quad . \tag{7.117}$$

We assume that $A \ll A_c$ and therefore $\lambda(t) = AI(t)$.

If the counting rate is such that $\langle\lambda\rangle\tau \ll 1$ (where $\langle\lambda\rangle$ is the mean number of counts per second), then the contribution of the exponential in (7.117) is negligible; hence

$$P_i(\tau) \simeq \langle\lambda(0)\lambda(\tau)\rangle/\langle\lambda(0)\rangle = \langle\lambda\rangle g_I(0,\tau) \quad , \quad \langle\lambda\rangle\tau \ll 1 \quad . \tag{7.118}$$

Consequently, for light beams that have very small degeneracy parameters ($\bar{n}_d = \langle\lambda\rangle\tau_c \ll 1$) the intensity correlation function can be determined from measurement of the inter-event probability distribution, up to maximum delay time τ_{max} such that

$$\langle\lambda\rangle\tau_{max} \ll 1 \quad . \tag{7.119}$$

The time-interval distribution can be measured by using a time-to-amplitude converter (TAC) in which the start and stop pulses are given by two successive photoelectron pulses (Fig.7.15). The start pulse generates a linear ramp waveform that is stopped by the stop pulse. The output of the TAC is therefore a pulse whose height is proportional to the time interval between the photoelectron pulses. These pulses of varying heights are then sorted out and stored in the memory of a multichannel pulse-height analyzer (PHA).

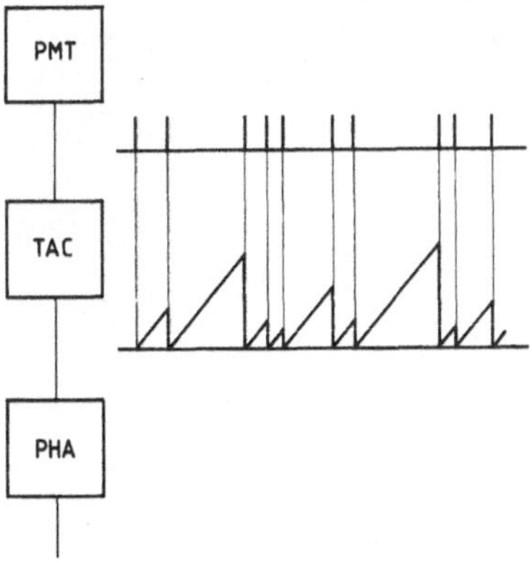

Fig.7.15. Measurement of the PD of inter-event time by using a time-to-amplitude converter (TAC) and a pulse-height analyzer (PHA)

Alternatively, the time interval can be digitally measured by using a fast clock that starts and stops upon the arrival of two successive events (Fig.7.16). The number of counted clock pulses determines the address of the proper channel of a multichannel analyzer (MCA) where one count is stored.

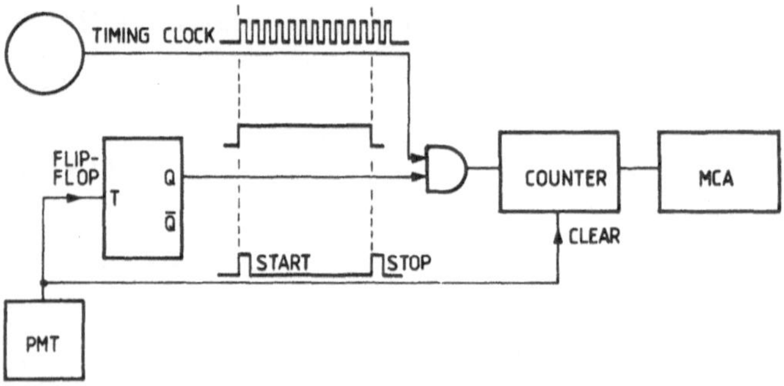

Fig.7.16. Measurement of the PD of inter-event time by using a fast clock and a multichannel analyzer (MCA)

The time-interval probability has been measured in several laboratories using these techniques, e.g. [17, 116, 121, 264, 376, 429, 496].

The Statistical Accuracy

Let the range of values of τ, to be covered, be divided into L intervals of width $\Delta\tau$, where L is the number of available channels. If the measured interval lies between $\ell\Delta T$ and $(\ell+1)\Delta T$, one count is stored in the ℓth channel. After N repetitions a histogram of the probability is obtained from the counts stored in the channels. If N_ℓ is the accumulated number in channel ℓ, then a good estimate for $P(\tau_\ell)$ is given by

$$\hat{P}(\tau_\ell) = N_\ell/N\Delta T \quad . \tag{7.120}$$

This ratio $\hat{P}(\tau_\ell)$ is an unbiased estimate of $P(\tau_\ell)$

$$\langle\hat{P}(\tau_\ell)\rangle = P(\tau_\ell) \quad .$$

The variance of $\hat{P}(\tau_\ell)$ can be determined if we assume that separate measurements are statistically independent. In this case, N_ℓ will have a binomial distribution with variance

$$\mathrm{Var}(N_\ell) = N\, P(\tau_\ell)\Delta T[1 - P(\tau_\ell)\Delta T] \quad ,$$

from which

$$\mathrm{Var}[\hat{P}(\tau_\ell)] = \frac{1}{N\Delta T} P(\tau_\ell)[1-P(\tau_\ell)\Delta T] \simeq \frac{1}{N\Delta T} P(\tau_\ell) \tag{7.121}$$

As N increases, the experimental error decreases in proportion to $1/\sqrt{N}$.

Measurement of Spectral Parameters

For light with an arbitrary degeneracy parameter, (7.118) does not hold and we have to resort to the general equation (7.117) which can be handled in only a limited number of special cases. For example, if the light is thermal and has a Lorentzian spectrum, $P_i(\tau)$ is given by (5.67). The measured distribution $\hat{P}(\tau_\ell)$ can then be used to determine the linewidth Γ by least-square fitting. If the average counting rate is known exactly, then the least-square error in estimating Γ is given by (cf. Sec. 2.7.2)

$$\mathrm{Var}(\hat{\Gamma}) = \left\{ \sum_{\ell=1}^{L} \left[\frac{\partial}{\partial\tau} P(\tau_\ell)\right]^2 / \mathrm{Var}[\hat{P}(\tau_\ell)] \right\}^{-1} \quad . \tag{7.122}$$

By substitution from (5.67) and (7.121), $\mathrm{Var}(\hat{\Gamma})$ can be determined. In the limit $\bar{n}_d \ll 1$,

$$P(\tau_\ell) = \langle\lambda\rangle(1 + e^{-2\Gamma\tau_\ell})$$

and we get

$$\frac{\mathrm{Var}(\hat{\Gamma})}{\Gamma^2} = \frac{1}{4N\langle n\rangle}\left[\sum_\ell x_\ell^2 e^{-4x_\ell}/(1+e^{-2x_\ell})\right]^{-1} \quad , \; x_\ell = \Gamma\tau_\ell \quad , \quad \langle n\rangle = \langle\lambda\rangle\Delta T \quad . \tag{7.123}$$

For an infinite number of channels, this reduces to

$$\frac{\mathrm{Var}(\hat{\Gamma})}{\Gamma^2} \simeq \frac{10}{N\langle n\rangle} \quad . \tag{7.124}$$

By comparing this to the corresponding error in Γ as measured by a digital correlator (7.107), we see that the same order of magnitude of accuracy is achieved. The pulse-separation technique is actually very practical in cases of faster phenomena (short τ_c).

7.2.3 Single-Photoelectron-Counting Probability

As we have already seen, the single-photoelectron-counting distribution P(n) is independent of the spectral profile for short sampling times, $T \ll \tau_c$. For longer sampling times, P(n) does depend on the spectral profile. However, this dependence is generally not mathematically explicit. The distribution P(n) has been found explicitly for only one case: thermal light with Lorentzian spectrum. Comparison of P(n) for several spectra shows that it is not very sensitive to the spectral profile. Therefore, it should not be expected that a reasonable estimate of the spectrum could be reached from measurements of P(n).

Formally, the correlation function can be obtained from the factorial moments of the single-photoelectron-counting distribution by using the relation,

$$G_I(0,T) = \frac{1}{2}\,\frac{\partial^2}{\partial T^2}\,F_n^{(2)} \quad , \tag{7.125}$$

which is obtained from (3.69) and (3.78) [343]. Yet the very-high precision of measurement required to determine the second derivative in (7.125) renders this method ineffective.

The most that can be done with single-photoelectron-counting is to estimate some parameters of an otherwise known spectrum. The simplest example is estimation of the linewidth of thermal light whose spectrum is Lorentzian. According to (5.60), P(n) is a function of Γ, the linewidth, and $\langle n\rangle$, the count rate. If $\langle n\rangle$ is measured, then Γ can be determined by fitting. Actually, the knowledge of one point on the distribution P(n) should be sufficient to determine Γ. The simplest point is P(0). From (5.54),

$$P(0) = e^{\gamma}/[\cosh(z) + (\gamma/2z+z/2\gamma)\sinh(z)] \quad ,$$
$$z^2 = 2\gamma\langle n\rangle + \gamma^2 \quad , \qquad \gamma = \Gamma T \quad . \tag{7.126}$$

If P(0) and $\langle n\rangle$ are measured, γ can be determined easily by solving the nonlinear equation (7.126). The sensitivity of P(0) to variations of γ depends on the mean counts per coherence time $\bar{n}_d$, as is illustrated in Fig.7.17. It is obvious that this sensitivity increases as $\bar{n}_d$ increases. If the sampling time is chosen such that γ is large, the sensitivity also increases. It is illustrative to calculate the statistical error of Γ (or γ) that results from errors in the determination of P(0) and <n>.

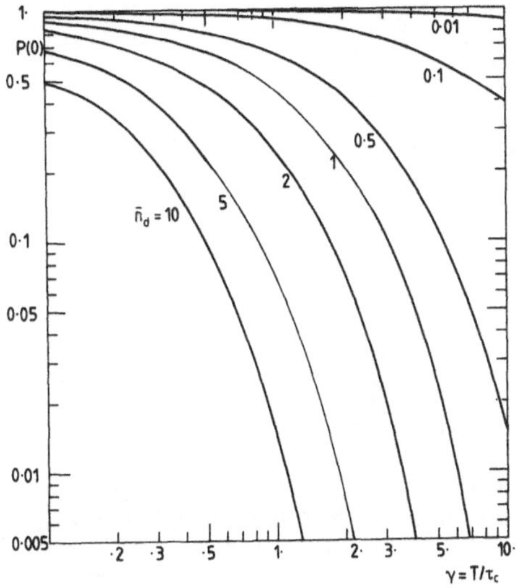

Fig.7.17. Dependence of the probability of zero counts on the sampling time-coherence time ratio and the number of photoelectrons per coherence time. The light is thermal and the spectrum is Lorentzian

Using arguments similar to those which led to (7.121), we get

$$\mathrm{Var}[\hat{P}(0)] = \frac{1}{N} P(0)[1 - P(0)] \quad , \tag{7.127}$$

where $\hat{P}(0)$ is the probability of zero counts as estimated from observing N intervals. Using the results of Sec. 6.2.1, we can write

$$\mathrm{Var}(\hat{n}) = \frac{\langle n\rangle^2}{N}\left(\frac{1}{\langle n\rangle} + \frac{1}{\gamma}\right) \quad , \tag{7.128}$$

where $\hat{n}$ is the experimental estimate for <n>. Assuming that the two errors are statistically uncorrelated, we can write

$$\mathrm{Var}(\gamma) = \left[\frac{\partial\gamma}{\partial P(0)}\right]^2 \mathrm{Var}[\hat{P}(0)] + \left(\frac{\partial\gamma}{\partial\langle n\rangle}\right)^2 \mathrm{Var}(\hat{n}) \quad . \tag{7.129}$$

This enables us to plot curves for the percentage error $e_\gamma = 100 \cdot \mathrm{Var}^{1/2}(\gamma)/\gamma$ as a function of γ for serveral values of $\bar{n}_d$. In Fig.7.18, these errors

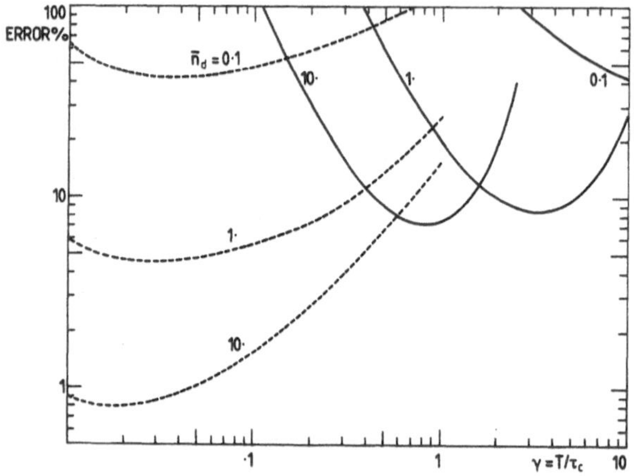

Fig.7.18. The percentage error in linewidth: —— estimated from measurement of the probability of zero counts; - - - - measured with a 100-channel correlator. The radiation is thermal and the spectrum is Lorentzian

are compared to the corresponding errors of the method of autocorrelation (100) channels). We observe that the error reaches a minimum when the sampling time is approximately equal to the coherence time. It is indeed surprising that the accuracy is comparable to the correlation technique. Measurements of P(0) and $\langle n\rangle$ are extremely simple. The rate $\langle n\rangle$ can be measured by using a counter. The probability P(0) can be measured [263] by using a flip-flop (Fig.7.19) which is reset at the arrival of each sampling clock pulse. Unless one or more photoelectrons trigger it, it remains reset and at the arrival of the next sampling clock pulse one count is sent to a counter.

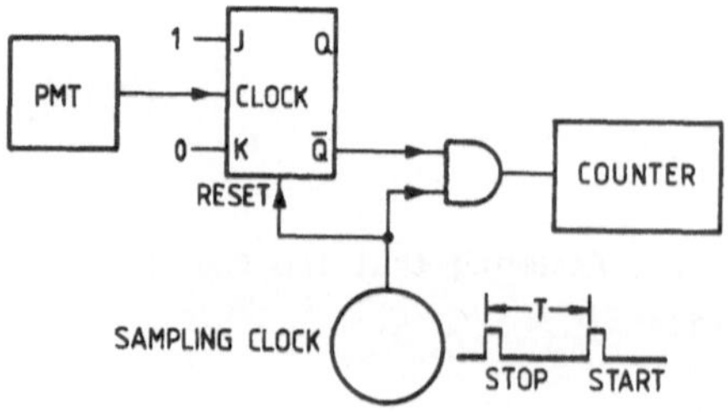

Fig.7.19. Measurement of the probability of zero counts

Because this technique requires sampling times approximately equal to the coherence times, it should be capable of handling much-faster phenomena.

SERRALLACH and ZULAUf [438] demonstrated the experimental feasibility of determining Γ from $P(n)$. JAKEMAN et al. [231] used the second factorial moment $F_n^{(2)}$ for this purpose. Expressions for the statistical accuracy of measuring $F_n^{(m)}$ can be found in [232].

7.2.4 Joint-Photoelectron-Counting Probability

The joint-photoelectron-counting probability $P(n_1,n_2)$ is more sensitive to the spectral profile than the single-photoelectron-counting probability $P(n)$. In the simpler case of short counting time ($T \ll \tau_c$), $P(n_1,n_2)$ for time delay τ is determined by $g^{(1)}(\tau)$. Yet, as is discussed in Sec. 5.3.1, an explicit mathematical relation between $P(n_1,n_2)$ and $g^{(1)}(\tau)$ has been obtained only for thermal light, (5.281). In this case, if $P(n_1,n_2)$ is measured for time delay τ, then $g^{(1)}(\tau)$ can be determined by fitting the experimental results to theoretical curves. ARECCHI et al. [13] measured $P(n_1,n_2)$ for thermal Lorentzian light and confirmed the theoretical expressions (see Fig.5.26).

The distribution $P(n_1,n_2)$ for a certain time delay τ can be measured by using a two-dimensional multichannel analyzer. The number of counts n_1 and n_2 at two intervals separated by the time delay τ defines an address in a two-dimensional memory. Whenever that address is called, one count is added at that location. At the end of the experiment, location (n_1,n_2) contains a number equal to the number of times the joint counts n_1 and n_2 occurred.

Instead of estimating the number $|g^{(1)}(\tau)|$ at a given τ by fitting to a two-dimensional distribution $P(n_1,n_2)$, it appears much easier to measure one point, say $P(0,0)$, from which $|g^{(1)}(\tau)|$ can be obtained by solving a nonlinear equation. From (5.272) and for thermal light,

$$P(0,0) = [(1+\bar{n})^2 - \bar{n}^2|g^{(1)}(\tau)|^2]^{-1} \quad . \tag{7.130}$$

We should then be able to infer $|g^{(1)}(\tau)|$ directly from measurement of $P(0,0)$. This is the basis of the instrument built by FURCINITTI et al. [160]. It is interesting to note that $P(0,0)$ can be measured simply by using a digital correlator in the double-complementary zero-clipped mode. It has the advantages of complementary clipping mentioned in Sec. 7.1.5. Moreover, the measurement of $P(0,0)$ is insensitive to sudden increases of the intensity of the detected light, which may be caused by dust particles intercepting the light beam.

7.2.5 Instants of Occurrence of a Realization of Photoelectron Events

Here we assume that we can record the instants $(t_1, t_2, \ldots t_n)$ at which photoelectrons occur in a time interval [0,T]. These instants contain all the information that could possibly be retrieved from the optical field by an observation in [0,T]. How could information about the spectrum be extracted from $\{t_i\}$? We recall from Sec. 5.3.1 that the JPD of observing $\{t_i\}$ is given by

$$P(\{t_i\}) = Q(1)\bar{n}^n \sum_{\pi_n} \prod_{i=1}^{n} \eta(t_i, t_{\pi_i}) \quad . \tag{7.131}$$

Both Q(1) and $\eta(t,t')$ depend on $\bar{n}$ and on the spectral profile (see (5.327, 28)). In principle, a ML estimator can be used to determine both $\bar{n}$ and the spectral profile. Only in very special cases would the ML method yield a manageable analytic result. Lorentzian light in the limit $T \gg \tau_c$ is one such case. From (5.333),

$$\eta(t,t') = (\alpha T)^{-1} \exp(-\alpha\Gamma|t-t'|) \quad , \tag{7.132}$$

$$\alpha = (1 + 2\bar{n}/\Gamma T)^{1/2} \quad , \tag{7.133}$$

and from (5.54), for large $\gamma = \Gamma T$,

$$Q^{(1)} = 2 \exp\left\{\Gamma T(1-\alpha)\left[1 + \frac{1}{2}\left(\alpha + \frac{1}{\alpha}\right)\right]^{-1}\right\} \quad . \tag{7.134}$$

Therefore, $P(\{t_i\})$ is a function of $\bar{n}$ and Γ which are to be estimated. Maximizing $P(\{t_i\})$ with respect to $\bar{n}$ and Γ, we obtain the ML estimates, $\hat{n}$ and $\hat{\Gamma}$. If we use the fact that $\Gamma T \gg 1$, we obtain

$$\hat{n} \approx n \quad , \tag{7.135}$$

and

$$\beta^2[n+(n^2+\beta^2)^{\frac{1}{2}}]^{-1} + n-\beta = \sum_{\pi_n} \beta\{\ \}_\pi \exp(-\beta\{\ \}_\pi) / \sum_{\pi_n} \exp(-\beta\{\ \}_\pi) \quad , \tag{7.136}$$

where $\beta = \Gamma T\alpha$ and

$$\{\ \}_\pi = (|t_1-t_{\pi_1}| + |t_2-t_{\pi_2}| + \ldots + |t_n-t_{\pi_n}|) \quad . \qquad (7.137)$$

This shows that n is sufficient for estimating the mean intensity. This is the same result we obtained in Sec. 6.5.1 for light with a rectangular spectrum. The width Γ can be obtained by solving the nonlinear equation (7.136) [418].

7.3 Single-Photoelectron-Decay Spectroscopy

The subject of this section is a pulse of light whose average intensity decays rapidly with time. We are interested in estimating the average intensity as a function of time (or at least the lifetime of the pulse) by using photoelectron measurements.

This problem has applications in the study of the lifetime of fluorescent or luminescent radiation from molecules subjected to a pulse of excitation from a spark discharge, an X-ray source, or a strong laser [49, 67, 270]. Fortunately, in such problems, the excitation can be repeated periodically for as many cycles as is convenient. An ensemble of response radiation is thus obtained, from which statistical ensemble averages can be determined.

The most direct method of measuring the profile of the average intensity decay is to count the photoelectrons in many short time intervals following the excitation and record the average for each time interval, over many cycles. If $n^i(\ell)$ is the number of counts in the interval $[(\ell-1)\Delta T, \ell\Delta T]$ in the cycle i, then the average intensity $\langle I(t)\rangle$ can be estimated from $\hat{I}(t) = \hat{\lambda}(t)/A$, where

$$\hat{\lambda}(t_\ell) = \frac{1}{\Delta T}\sum_{i=1}^{N} n^i(\ell) \quad , \qquad \ell = 1,2, \ldots L \quad , \qquad t_\ell = (\ell - \tfrac{1}{2})\Delta T \quad . \qquad (7.138)$$

Here L is the number of channels of the measuring instrument (the number of points on the intensity-time curve), N is the number of cycles observed (size of the ensemble), and A is the area of the detector (assumed small).

For weak radiation, multichannel delayed coincidence is another useful method [348]. A fixed time interval [0,T] following the excitation pulse at t = 0 is divided into L subintervals. Emitted photoelectrons fill some of these subintervals in each cycle of observation. A histogram is constructed for the number of times that each subinterval is filled. This can be done

(Fig.7.20) by using L delay units (a shift register). The ℓth unit delays the incoming photoelectrons by an interval $\ell T/L$. The coincidence between the outcome of these delay units and the excitation pulse at t = 0 is detected by L coincidence units (AND gates). This instrument is actually a slightly modified digital correlator (cf. Fig.7.6).

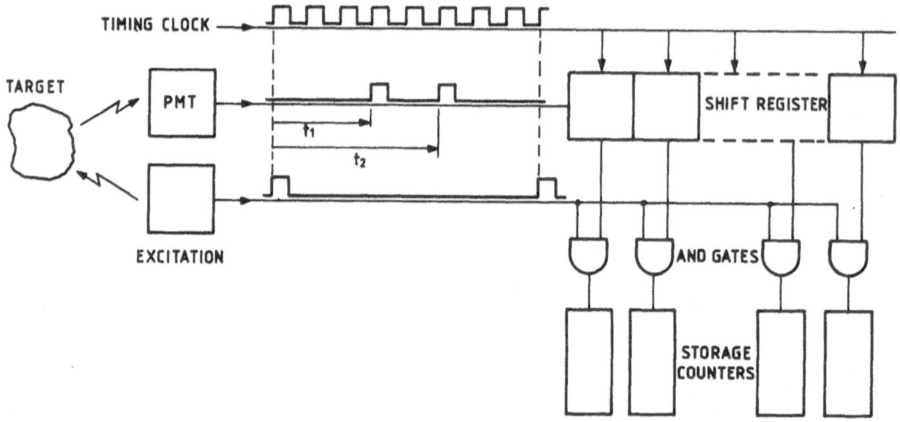

Fig.7.20. Measurement of decay by multichannel delayed coincidence

For weak pulses of very short duration (say, several nanoseconds), the available time is not sufficient for counting in L intervals and processing the data appropriately.

The single-photoelectron-decay technique is based instead on recording the time of arrival of a single photoelectron, the first photoelectron after the excitation. After many cycles of observation, a histogram of these times given the probability distribution of the time of arrival of the first photoelectron, $P_f(t)$. We recall from (3.80) that this probability is related to the light intensity by

$$P_f(t) = \left\langle \lambda(t)\exp\left[-\int_0^t \lambda(t')dt'\right]\right\rangle \quad , \qquad \lambda(t) = AI(t) \quad . \tag{7.139}$$

If $\langle I(t)\rangle$ is to be observed in the time range $[0,t_{max}]$, then, under the condition

$$\int_0^{t_{max}} \langle\lambda(t')\rangle\, dt' \ll 1 \quad , \tag{7.140}$$

the exponential terms in (7.139) can be neglected and we have (cf. also the discussion following (5.68))

$$P_f(t) \approx \langle\lambda(t)\rangle \quad , \qquad t \leq t_{max} \quad . \tag{7.141}$$

This enables us to estimate the desired intensity profile.

Condition (7.140) states that the mean number of counts in the time interval of interest (e.g., the lifetime of the phenomenon) is much less than 1, i.e., the decaying pulse contains an average energy that corresponds to much less than one photoelectron. If the energy in the pulse is such that (7.140) is not satisfied, (7.141) no longer holds and the instrument suffers from the so-called "pile-up distortion."

One method of correcting this distortion [124-126] is to measure $P_c(1,t:[0,T])$, the probability distribution of the time of occurrence of the photoelectron, given that it is the only photoelectron present in the interval [0,T]. According to (3.92),

$$P_c(1,t:[0,T]) = \left\langle \lambda(t)\exp\left[-\int_0^T \lambda(t')dt'\right]\right\rangle \quad . \tag{7.142}$$

For economy of notation, we simply refer to this probability distribution as $P_c(t)$. Under the condition

$$\int_0^T \langle\lambda(t')\rangle\, dt' \ll 1 \quad , \tag{7.143}$$

we obtain again

$$P_c(t) \approx \langle\lambda(t)\rangle \quad , \qquad t \leq T \quad . \tag{7.144}$$

Because the exponential term in (7.142) is not dependent on t, we should expect that the distortion in (7.144) is less than that in (7.141). This actually depends on the statistical properties of the radiation. In Sec. 7.3.1, we examine the performance of the two instruments based on (7.141) and (7.144) (called spectrometer I and II, respectively) in the completely coherent case. In Sec. 7.3.2, we examine the effect of partial coherence of the detected radiation [427].

7.3.1 Coherent Light

Spectrometer I

Assuming that the light is coherent, the expectation value in (7.139) becomes unnecessary and spectrometer I measures

$$P_f(t) = \lambda(t)\exp\left[-\int_0^t \lambda(t')dt'\right] \quad , \tag{7.145}$$

which is a distorted version of $\lambda(t)$. The photoelectrons obey an inhomogeneous Poisson point process (cf. Sec. 3.4).

If $\int_0^{t_{max}}\lambda(t')dt' \ll 1$, then to the first-order approximation

$$P_f(t) \simeq \lambda(t)\left[1 - \int_0^t \lambda(t')dt'\right] \quad , \tag{7.146}$$

i.e., the distortion at time t is proportional to $\int_0^t\lambda(t')dt'$. A theoretical correction [99] can be made by inverting (7.146)

$$\lambda(t) = \frac{P_f(t)}{1 - \int_0^t P_f(t')dt'} \quad . \tag{7.147}$$

Equation (7.145) itself can be inverted by using the iterative relation

$$\lambda^{(i+1)}(t) = P_f(t)\exp\int_0^t \lambda^{(i)}(\theta)\, d\theta \quad , \quad i = 0,1, \ldots \tag{7.148}$$

$$\lambda^{(0)}(\theta) = P_f(t) \quad . \tag{7.149}$$

In many cases, an expression for $\lambda(t)$ is known (from the physics of the problem) and it is desired to estimate one or more unknown parameters. An expression for $P_f(t)$ can then be determined from which the parameters can be estimated by usual fitting procedures.

Note, however, that if many photoelectrons are expected to arrive within a lifetime, most measurements will be crowded in the first part of the decay curve, whereas points on its tail would be very poorly represented.

An analysis of the statistical accuracy involved in such a problem is therefore of interest.

Let N be the number of cycles observed and N_ℓ be the number of cycles in which a photoelectron is recorded in the time interval between t_ℓ and $t_\ell+\Delta t$. A good estimate for $P_f(t_\ell)$ is then given by

$$\Delta t\hat{P}_f(t_\ell) = N_\ell/N \quad . \tag{7.150}$$

Because the cycles are statistically independent, N_ℓ has a binomial distribution with mean $NP_f(t_\ell)\Delta t$ and variance $NP(t_\ell)\Delta t[1-P_f(t_\ell)\Delta t]$. Therefore,

$$\mathrm{Var}[\hat{P}_f(t_\ell)] = \frac{1}{N\Delta t} P_f(t_\ell)\,[1-P_f(t_\ell)\Delta t] \sim \frac{1}{N\Delta t}\; P_f(t_\ell) \quad . \tag{7.151}$$

If $P_f(t)$ depends on one parameter Θ, and if $\hat{P}_f(t_\ell)$, $\ell = 1,2, \ldots L$ is measured by using an L-channel analyser, then the error in estimating Θ, can be calculated from (2.152).

We take the exponentially decaying case, $\lambda(t) = \lambda_0 e^{-t/\tau_0}$ as an example. This is the most common decay (e.g., in molecular luminescence, fluorescence, radioactive decay).

Direct substitution in (7.145) and (2.152) gives

$$P_f(t) = \lambda_0 \exp[-x-\alpha(1-e^{-x})] \quad , \qquad x = t/\tau_0, \qquad \alpha = \lambda_0\tau_0 \tag{7.152}$$

$$\frac{\mathrm{Var}(\tau_0)}{\tau_0^2} = \frac{1}{N}\,\frac{1}{f(\alpha)} \quad , \tag{7.153}$$

$$f(\alpha) = \alpha \sum_\ell [(x_\ell-\alpha)+\alpha(1+x_\ell)e^{-x_\ell}]^2 \exp[-x_\ell-\alpha(1-e^{-x_\ell})]\;\Delta x \quad ,$$

$$x_\ell = t_\ell/\tau_0, \; \Delta x = \Delta t/\tau_0 \quad . \tag{7.154}$$

In the limit of an infinite number of channels, $f(\alpha)$ is approximated by the integral

$$f(\alpha) = \alpha \int_0^\infty [(x-\alpha)+\alpha(1+x)e^{-x}]^2 \exp[-x-\alpha(1-e^{-x})]dx \quad . \tag{7.155}$$

In Fig.7.21, the normalized percentage error $e = 100[\mathrm{Var}(\tau_0)]^{1/2}/\tau_0$ versus $\alpha = \lambda_0\tau_0$ is plotted for $N = 10^4$ cycles. As expected, the error is large when α is very small. This is because, in this case, most cycles contain no photoelectrons. When α is very large, the tail of the curve is missed and the estimation error increases. Fig.7.21 illustrates that the optimum value of α lies in the range 0.4 - 1. One interesting conclusion is that if a very small value of α is chosen (to make the fitting-procedure easier), the statistical error becomes larger. If, for example, we use $\alpha = 0.01$, then we have to increase the total observation time about 50 times to obtain the same accuracy as that in the $\alpha = 0.4$ - 1 case. A compromise has to be made between the duration of the experiment and the computer time needed for the fitting.

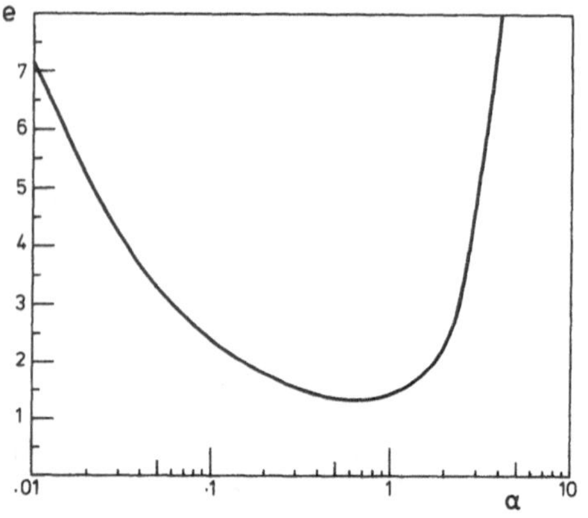

Fig.7.21. Dependence of the percentage error in lifetime, as determined from measurement of the PD of time of arrival of the first photoelectron, on the average number of photoelectrons per lifetime α. The decay is assumed exponential (after SALEH and SELINGER [427])

Spectrometer II

Spectrometer II measures the probability $P_c(t)$, which has the form

$$P_c(t) = \lambda(t)\exp\left[-\int_0^T \lambda(t')dt'\right] \tag{7.156}$$

in the coherent limit. Indeed $P_c(t)$ is proportional to $\lambda(t)$ for all levels of light energy. Yet the instrument is expected to suffer from a lower signal-utilization efficiency because of the number of cycles that are discarded because they do not contain one photoelectron. If N cycles are observed, then, on the average, only

$$N' = N \int_0^T \lambda(t')dt' \cdot \exp\left[-\int_0^T \lambda(t')dt'\right] \tag{7.157}$$

cycles contain one photoelectron and are used in constructing our histogram.

A simple analysis of the statistical accuracy of estimating the lifetime of an exponentially decaying light would help illustrate this numerically.

From (7.156)

$$P_c(t) = \lambda_0 \exp[-\alpha(1-e^{-\beta})]\exp(-x),$$

$$x = t/\tau_0 \ , \quad \alpha = \lambda_0\tau_0, \quad \beta = T/\tau_0 \quad , \tag{7.158}$$

and from (7.157)

$$N' = N\alpha(1-e^{-\beta})\exp[-\alpha(1-e^{-\beta})] \quad . \tag{7.159}$$

Using (7.151) with N replaced by N' from (7.159), (2.152) gives

$$\frac{\mathrm{Var}(\hat{\tau}_0)}{\tau_0^2} = \frac{1}{N}\frac{1}{f(\alpha,\beta)} \tag{7.160}$$

$$f(\alpha,\beta) = \alpha^2(1-e^{-\beta})\exp[-2\alpha(1-e^{-\beta})]\sum_{x_\ell=0}^{\beta}(x_\ell+a)^2 e^{-x_\ell}\Delta x \quad ,$$

$$x_\ell = t_\ell/\tau_0 \ , \quad \Delta x = \Delta t/\tau_0 \tag{7.161}$$

$$a = -\alpha + \alpha(1+\beta)e^{-\beta} \quad . \tag{7.162}$$

In the limit of an infinite number of channels, the summation in (7.161) can be replaced by the integration $\int_0^\beta (x+a)^2 e^{-x}dx$. Using the above equations, we can compute the percentage error in estimating τ_0 as a function of the average number of photoelectrons per lifetime, $\alpha = \lambda_0\tau_0$, for several fixed values of the observation window $\beta = T/\tau_0$. As shown in Fig.7.22 in all cases,

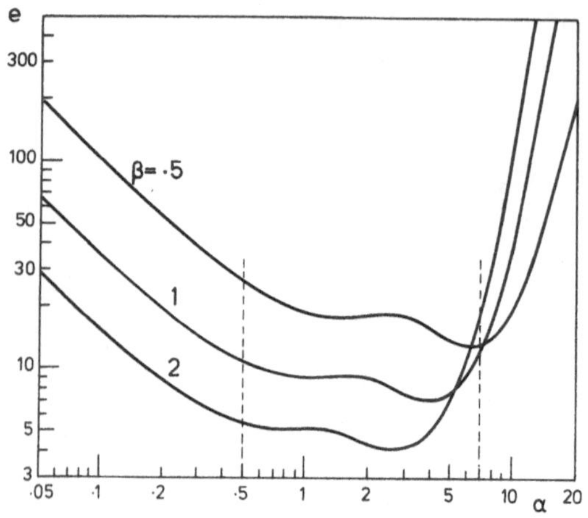

Fig.7.22. Dependence of the percentage error in lifetime τ_0, as determined from the PD of the time of occurrence of a single photoelectron in a time interval T, on the average number of photoelectrons per lifetime α, and the ratio $\beta = T/\tau_0$. The decay is assumed exponential (after SALEH and SELINGER [427])

optimum values of α lie in the range $\alpha = 0.5 - 7$. These values are not much smaller than unity. This is presumably because, for an exponential function, points on the tail are not of much importance in the estimation of the lifetime.

Comparison between Figs.7.21 and 7.22 reveals that spectrometer II is less accurate than spectrometer I. For example, for $N = 10^4$ cycles, spectrometer I has an optimum error of 1.4 % whereas spectrometer II has an optimum error of 4.2 % when $\beta = 2$. This corresponds to an observation time 9 times as long, to achieve the same accuracy.

7.3.2 Effect of Light Coherence

When the radiation is not completely coherent, we have to evaluate the expectation values in (7.139) and (7.142). This is in general extremely difficult. Only in the special case of Gaussian-Markovian light, which decays with a time constant equal to its coherence time, has the problem of photon statistics been studied (cf. Sec. 5.2.7). We content ourselves here with an approximate analysis, which is valid when the light is weak.

In order to avoid ambiguity that may arise when we talk about the coherence time (or the spectral width) of nonstationary light, we consider the following situation: stationary light with intensity $I_0(t)$, average intensity $\langle I_0(t)\rangle = \bar{I}_0$, and normalized intensity correlation function $g^{(2)}(\tau)$ is modulated by a deterministic decaying function $\alpha(t)$, where $\alpha(0)=1$. The resultant light

$$I(t) = I_0(t)\alpha(t) \tag{7.163}$$

is detected by our spectrometers, which aim at estimating $\alpha(t)$. If we assume that

$$\lambda_0 \int_0^{t_{max}} \alpha(t)dt \ll 1 \quad , \qquad \lambda_0 = \bar{I}_0 A \quad ,$$

then we can expand the exponentials in (7.139) and (7.142), take the expectation value, and obtain

$$P_f(t) \simeq \lambda_0\alpha(t) \left[1 - \lambda_0 \int_0^t g_{I_0}(t-\theta)\alpha(\theta)d\theta\right] \tag{7.164}$$

$$P_c(t) \simeq \lambda_0\alpha(t) \left[1 - \lambda_0 \int_0^T g_{I_0}(t-\theta)\alpha(\theta)d\theta\right] \ . \tag{7.165}$$

It is obvious from (7.165) that spectrometer II is no longer distortionless.

In order to pursue (7.164) and (7.165) further, we have to specify the statistics of the stationary field $I_0(t)$. Assuming that the field is thermal (Gaussian), then we can use (4.78).

Let τ_c be the coherence time of the stationary field $I_0(t)$, and consider two limiting cases.

I) $\tau_c \gg \tau_0$. If the coherence time τ_c is much larger than the width of the function $\alpha(t)$, τ_0, then we can safely put $g_{I_0}(t) \simeq 2$ in (7.164) and (7.165) and obtain

$$P_f(t) \simeq \lambda_0\alpha(t) \left[1 - 2 \int_0^t \alpha(t')dt'\right] \ , \tag{7.166}$$

$$P_c(t) \simeq \lambda_0\alpha(t) \left[1 - 2 \int_0^T \alpha(t')dt'\right] \ . \tag{7.167}$$

This shows that spectrometer I has a distortion that is double that in the case of coherent light (deterministic $I_0(t)$). This factor of 2 is due to the photoelectron-bunching effect in the case of thermal light.

Equation (7.167) shows that spectrometer II is distortionless.

II) $\tau_c \ll \tau_0$. In this limit, we can write

$$P_f(t) \simeq \lambda_0\alpha(t)\left[1 - \lambda_0 \int_0^t \alpha(t')dt' - \lambda_0\tau_c\alpha(t)\right] \quad (7.168)$$

$$P_c(t) \simeq \lambda_0\alpha(t)\left[1 - \lambda_0 \int_0^T \alpha(t')dt' - \lambda_0\tau_c\alpha(t)\right] \quad . \quad (7.169)$$

This shows that spectrometer II is slightly distorted. But in the limit $\tau_c = 0$, the third term in (7.168) and (7.169) can be neglected and we obtain the same result as in the coherent case.

We conclude that when the coherence time of the observed light is comparable to the width of the observed pulses, care should be taken to interpret properly the measured profiles. It is fortunate that, in most present applications, the spectral broadening of the field is sufficiently large to make $\tau_c \ll \tau_0$, even in the fastest case ($\tau_0 \sim 1$ ns). The theory presented in this section can be tested, however, by using pseudo-thermal light or laser light scattered from particles in Brownian motion having larger coherence time (e.g., 1 ms) and a controllable pulse width of comparable time.

7.4 Estimation of the Spatial Spectrum (the Spatial Coherence Function)

Throughout Chap. 5 we emphasized the role played by the spatial coherence of light in determining the statistics of the photoelectrons. In previous sections of the present chapter, the effect of spatial coherence on the performance of a photoelectron-counting spectrometer was studied. This section deals with methods of estimating the spatial spectrum itself from photoelectron measurements.

Assume first that only *one photodetector* is available. Is it possible to extract information about the degree of spatial coherence of the illuminating field? We recall from (5.89) that, for thermal light, the second-order factorial moment

$$F_n^{(2)} = \bar{n}^2[1+f(A)] \ , \quad f(A) = \frac{1}{A^2}\iint\limits_{AA} d\underline{r}_1 d\underline{r}_2 |g^{(1)}(\underline{r}_1,\underline{r}_2)|^2 \ , \tag{7.170}$$

depends on $|g^{(1)}(\underline{r}_1,\underline{r}_2)|$ via one parameter, $f(A)$. Higher-order moments depend on $g^{(1)}(\underline{r}_1,\underline{r}_2)$ via more-complicated multiple integrals. The probability distribution $P(n)$ depends on $g^{(1)}(\underline{r}_1,\underline{r}_2)$ in a complicated manner that is, in general, not amenable to mathematical inversion. For practicality, we have to limit ourselves to $F_n^{(2)}$.

If the area of the detector is much smaller than the coherence area of the field ($A \ll A_c$), the parameter $f(A) \simeq 1$ is very insensitive to $g^{(1)}(\underline{r}_1,\underline{r}_2)$. We should not expect to learn about the correlation between fluctuations at two widely spearated points by taking measurements in the neighbourhood of one of them. As A increases, $f(A)$ drops sharply (cf. Fig.5.12). If the profile of the spatial coherence function is known except for one unknown parameter, say A_c, then a mathematical expression for $f(A)$ can be determined (see, e.g., (5.181)) from which A_c can be estimated. In the limit $A \gg A_c$, $f(A) \simeq A_c/A$. Consequently, we can estimate A_c from the measurement of $F_n^{(2)}$. The resolution of the measurement is naturally too coarse to reveal the spatial spectral profile. It should be noted that (7.170) is valid for polarized thermal light in the limit $T \ll \tau_c$. For other situations, other appropriate formulas should be used as described in detail in Chap. 5. For example, if the optical field in question is mixed (heterodyned) with a strong coherent field that has the same central frequency, we have

$$F_n^{(2)} \simeq \bar{n}^2+2\bar{n}_{th}\bar{n}_c f_D(A) \ , \quad f_D(A) = \frac{1}{A^2}\iint\limits_{AA} d\underline{r}_1 \, d\underline{r}_2 \, |g(\underline{r}_1,\underline{r}_2)| \ . \tag{7.171}$$

Unlike (7.170), (7.171) can be used to determine the coherence area of a field of any statistical properties.

More versatility is gained by using *two photodetectors*. The profile of the spatial coherence function itself can then be easily scanned. For example, for thermal light, the cross correlation between photoelectron counts at point detectors 1 and 2 is given by (cf. (5.285))

$$\langle n_1 n_2 \rangle = \bar{n}_1 \bar{n}_2 \, [1 + |g^{(1)}(\underline{r}_1,\underline{r}_2)|^2] \ , \tag{7.172}$$

i.e.,

$$g^{(2)}(\underline{r}_1,\underline{r}_2) = 1 + |g^{(1)}(\underline{r}_1,\underline{r}_2)|^2 \quad , \quad g^{(2)}(\underline{r}_1,\underline{r}_2) = \frac{\langle n_1 n_2 \rangle}{\bar{n}_1 \bar{n}_2} \; . \tag{7.173}$$

By measuring $g^{(2)}$, the absolute value of the spatial coherence function can be directly estimated. The areas of the detectors must be kept smaller than a coherence area of the field.

The phase of $g^{(1)}(\underline{r}_1,\underline{r}_2)$ can be obtained by a heterodyning technique similar to that previously described for the estimation of the phase of $g^{(1)}(\tau)$. The strongly heterodyned intensity correlation function is given by

$$g^{(2)}(\underline{r}_1,\underline{r}_2) \simeq 1 + \frac{2\,\bar{n}_{th}\bar{n}_c}{\bar{n}^2}\,\mathrm{Re}\,\{g^{(1)}(\underline{r}_1,\underline{r}_2)\exp(j\Delta\omega\tau)\} \tag{7.174}$$

independently of the statistics of the field.

An analysis of the performance of a digital cross-correlator is given in Sec. 7.4.1.

Another interesting method of measuring the photoelectron-counting cross correlation $\langle n_1 n_2 \rangle$ by using *one detector* only is to let the light pass to the detector through *two pinholes* [72-76, 129]. If the number of counts collected by the surface of the detector opposite to the pinholes is n_1 and n_2, then the detector would record the sum of counts $n = n_1 + n_2$. The single-fold factorial moment of n is then given by

$$F_n^{(2)} = F_{n_1}^{(2)} + F_{n_2}^{(2)} + 2\,\langle n_1 n_2 \rangle \quad . \tag{7.175}$$

If the pinholes are sufficiently small, then, for thermal light, from (7.170,173)

$$F_n^{(2)} = 2\bar{n}_1^2 + 2\bar{n}_2^2 + 2\bar{n}_1\bar{n}_2\,[1 + |g^{(1)}(\underline{r}_1,\underline{r}_2)|^2] \quad .$$

If the light intensities at the pinholes are equal, then

$$F_n^{(2)} = \frac{3}{2}\,\bar{n}^2\left[1 + \frac{1}{3}\,|g^{(1)}(\underline{r}_1,\underline{r}_2)|^2\right] \quad . \tag{7.176}$$

This is exactly the same result that would be obtained by use of (7.170) and taking the aperture A as two point holes. By changing the positions of the pinholes and measuring the normalized factorial moment $F_n^{(2)}/\bar{n}^2$, we can scan $|g^{(1)}(\underline{r}_1,\underline{r}_2)|^2$. This method is not practical when the coherence area is so large that it is impractical to intercept it with one detector (e.g., in the case of an astronomical intensity interferometer).

7.4.1 The Digital Cross Correlator

A digital photoelectron-counting cross correlator is an instrument that measures $\langle n_1 n_2 \rangle = \bar{n}_1 \bar{n}_2 g^{(2)}(\underline{r}_1,\underline{r}_2,\tau)$. It is based on exactly the same principles as the digital autocorrelator described in Sec. 7.1.2. The digital autocorrelator of Fig.7.3 can easily be modified to compute the cross correlation by connecting the two input lines to the outputs of the two separate photodetectors. If the time dependence of $g^{(2)}(r_1,r_2,\tau)$ is of no interest, a one-channel correlator is sufficient. The correlator measures the number of counts $n_1(m)$ and $n_2(m)$ in detectors 1 and 2, in the time interval m,m = 1,2, ... N. Then it computes the product $n_1(m)n_2(m)$ and stores the sum

$$\hat{G} = \frac{1}{N}\sum_{m=1}^{N} n_1(m)n_2(m) \tag{7.177}$$

A good estimate of the normalized intensity correlation $g^{(2)}(\underline{r}_1,\underline{r}_2)$ can be obtained by using the estimator

$$\hat{g} = \frac{\hat{G}}{\hat{n}_1 \hat{n}_2} \quad , \tag{7.178}$$

where

$$\hat{n}_j = \frac{1}{N}\sum_{m=1}^{N} n_j(m) \quad , \qquad j = 1,2 \tag{7.179}$$

is an estimate of $\bar{n}_j$.

7.4.2 Statistical Accuracy

As we have done in the analysis of the statistical accuracy of a digital correlator (Sec. 7.1.8), we are interested in determining the mean and variance of $\hat{g}$. By expanding $\hat{G}$ and $\hat{n}_j$ around their means, retaining only lower orders or deviations form these means, and assuming that N is very large, we obtain

$$\langle\hat{g}\rangle \approx g^{(2)}(\underline{r}_1,\underline{r}_2) \tag{7.180}$$

$$\mathrm{Var}(\hat{g}) = \frac{\mathrm{Var}(\hat{G})}{\bar{n}_1^2\bar{n}_2^2} + \Delta$$

$$\Delta = 4g^{(2)} + \sum_{j=1,2}\left[\frac{-2g^{(2)}}{\bar{n}_1\bar{n}_2}\,\frac{\langle\hat{G}\hat{n}_j\rangle}{\bar{n}_j} + \frac{[g^{(2)}]^2}{\bar{n}_j^2}\,\mathrm{Var}(\hat{n}_j)\right] + \frac{2}{\bar{n}_1\bar{n}_2}\mathrm{cov}(\bar{n}_1,\bar{n}_2) \quad . \tag{7.181}$$

Using (7.177) and (7.179), and the expressions for multi-fold higher-order moments of thermal light, we obtain (SALEH [414])

$$\mathrm{Var}(\hat{g}) = \frac{1}{N_c}\left[a + \left(\frac{1}{\bar{n}_{d1}} + \frac{1}{\bar{n}_{d2}}\right)b + \frac{1}{\bar{n}_{d1}\bar{n}_2}\,c\right] \quad , \tag{7.182}$$

where a,b, and c are related to the spatial coherence function of the field by

$$\begin{aligned} a &= 2 + z + (4+2z)|g(\underline{r}_1,\underline{r}_2)|^2 - 11z|g(\underline{r}_1,\underline{r}_2)|^4 \\ b &= 1 + 2|g(\underline{r}_1,\underline{r}_2)|^2 - |g(\underline{r}_1,\underline{r}_2)|^4 \\ c &= 1 + |g(\underline{r}_1,\underline{r}_2)|^2 \end{aligned} \tag{7.183}$$

Here, $\bar{n}_{dj}$ is the average number of photoelectrons collected by detector j in a coherence time, N_c is the number of coherence times observed, and z is a parameter that is determined by the temporal coherence of the field,

$$z = \int_{-\infty}^{\infty} |g(t)|^4 dt \Big/ \int_{-\infty}^{\infty} |g(t)|^2 dt \quad , \tag{7.184}$$

where g(t) is the temporal coherence function. For light with a Lorentzian spectrum, z = 1/2.

The ratio $e = 100[\mathrm{Var}(\hat{g})]^{1/2}/[g^{(2)}(\underline{r}_1,\underline{r}_2)-1]$ gives the relative percentage accuracy in measuring $g^{(2)}(\underline{r}_1,\underline{r}_2)$ by a digital cross correlator and is plotted in Fig.7.23 against $|g^{(1)}(\underline{r}_1,\underline{r}_2)|$ for several values of $\bar{n}_d = \bar{n}_{d1} = \bar{n}_{d2}$. The graphs demonstrate that the relative error is much larger for small values of $|g^{(1)}(\underline{r}_1,\underline{r}_2)|$. Attempts to determine the width of the correlation function by locating the point at which it vanishes are expected to be less accurate.

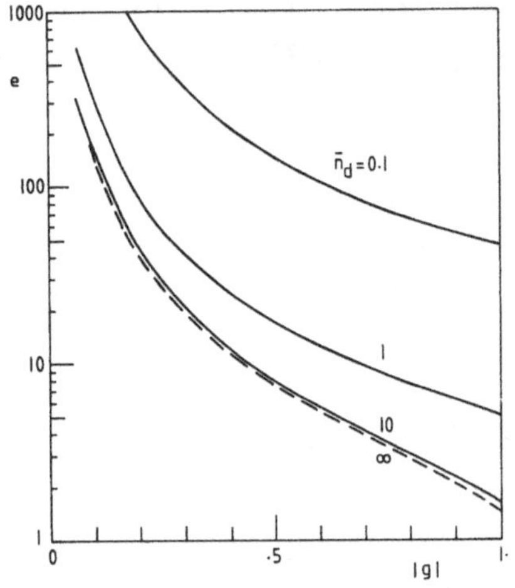

Fig.7.23. Percentage error in estimating the normalized photoelectron-counting correlation function as a function of the spatial degree of coherence $|g|$ for thermal light with Lorentzian spectrum. $N_c = 10^4$

As a function of $\bar{n}_d$, the error decreases rapidly with increase of $\bar{n}_d$ but it eventually saturates at a level beyond which increasing the counting rate has little value in improving the accuracy. It is advantageous to collect as many photoelectrons as possible by increasing the area until the saturation rate is reached.

The areas of the detectors should remain, however, smaller than the area of coherence of the field.

Of more relevance is the error of estimating an unknown parameter θ that determines $|g^{(1)}(\underline{r}_1, \underline{r}_2)|$. This error is given by

$$\mathrm{Var}(\hat{\theta}) \approx \mathrm{Var}(\hat{g}) \left|\frac{\partial g^{(2)}}{\partial\theta}\right|^{-2} \tag{7.185}$$

A good example is the estimation of the radius of an incoherent circular object by measurement of the intensity correlation at an aperture far from the object plane (the Hanbury Brown and Twiss interferometer). From (4.51), the degree of spatial coherence in the observation plane is given by

$$|g(\underline{r}_1, \underline{r}_2)| = J\left(\frac{2\pi a}{\lambda z}|\underline{r}_1 - \underline{r}_2|\right) \quad , \quad J(\theta) = \frac{2J_1(\theta)}{\theta} \quad , \tag{7.186}$$

where z is the distance between the object and aperture planes, a is the radius of the object, and $J_1(.)$ is the Bessel function of order one. By simple substitution in (7.182) and (7.185), we obtain the error graph of Fig.7.24. Minimum error is obtained when the separation between the detectors is approximately $\lambda z/\pi a$. The importance of a detailed study of the statistical accuracy of cross correlators is strongly emphasized.

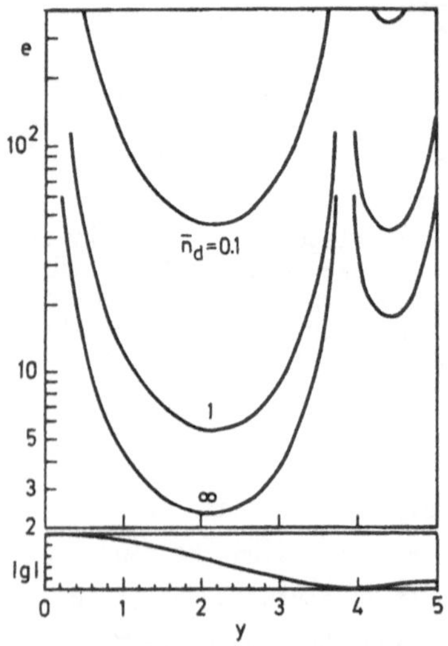

Fig.7.24. Percentage error in estimating the radius of an incoherent object a by measuring photoelectron counting cross correlation at two points separated by $|r_1 - r_2|$ in a plane at a distance z from the source, as a function of $y = 2\pi a\,|r_1 - r_2|/\lambda z$ and for the shown values of mean number of counts per coherence time $\bar{n}_d$. $\gamma = \Gamma T = 0.1$, $N_c = 10^4$ The degree of coherence $|g|$ is also plotted (after SALEH [414])

References

1. J.A. Abate, H.J. Kimble, L. Mandel: Phys. Rev. A 14, 788 (1976)
2. K. Abend: IEEE Trans. IT-12, 64 (1966)
3. J.B. Abbiss, T.W. Chubb, E.R. Pike: Opt. and Laser Techn. 6, 249 (1974)
4. J. Aitchison, J.A.C. Brown: *The Lognormal Distribution* (Cambridge U.P., Cambridge 1957)
5. M.D. Aldrige: J. Appl. Phys. 40, 1720 (1969)
6. O. Andrade, B.J. Rye: J. Phys. D 7, 280 (1974)
7. T. Aoki, Y. Okabe, K. Sakurai: Phys. Rev. A 10, 259 (1974)
8. F.T. Arecchi: Phys. Rev. Lett. 15, 912 (1965)
9. F.T. Arecchi: Nuovo Cimento Suppl. IV, 756 (1966)
10. F.T. Arecchi: In *Quantum Optics Proceedings of the International School of Physics "Enrico Fermi"*, ed. by R.J. Glauber (Academic Press, New York 1969) p. 57
11. F.T. Arecchi, M. Asdente, A.M. Ricca: Phys. Rev. A 14, 383 (1976)
12. F.T. Arecchi, A. Bernê, P. Burlamacchi: Phys. Rev. Lett. 16, 32 (1966)
13. F.T. Arecchi, A. Bernê, A. Sona: Phys. Rev. Lett. 17, 260 (1966)
14. F.T. Arecchi, A. Bernê, A. Sona, B. Burlamacchi: IEEE J. QE-2, 341 (1966)
15. F.T. Arecchi, V. Degiorgio: In *Laser Handbook*, ed. by F.T. Arecchi and E.O. Schulz-Dubois (North Holland, Amsterdam 1972) p. 191
16. F.T. Arecchi, V. Degiorgio, B. Querzola: Phys. Rev. Lett. 19, 1168 (1967)
17. F.T. Arecchi, E. Gatti, A. Sona: Phys. Lett. 20, 27 (1966)
18. F.T. Arecchi, M. Giglio, A. Sona: Phys. Lett. 25A, 341 (1967)
19. F.T. Arecchi, M. Giglio, W. Tartari: Phys. Rev. 163, 186 (1967)
20. F.T. Arecchi, G.S. Rodari, A. Sona: Phys. Lett. 25A, 59 (1967)
21. J.A. Armstrong, A.W. Smith: Phys. Rev. Lett. 14, 68 (1965)
22. J.A. Armstrong, A.W. Smith: In *Progress in Optics*, ed. by E. Wolf, Vol. 6 (North Holland, Amsterdam 1967) p. 211
23. A. Arnedo, F. Rocca: Z. Physik 269, 205 (1974)
24. R. Ash, N.C. Ford: Rev. Sci. Instrum. 44, 506 (1973)
25. R. Barakat: J. Opt. Soc. Am. 53, 317 (1963)
26. R. Barakat: Opt. Commun. 8, 14 (1973)
27. R. Barakat: Opt. Acta 21, 903 (1974)
28. R. Barakat: J. Opt. Soc. Am. 66, 211 (1976)
29. R. Barakat, J. Blake: Phys. Rev. A 13, 1122 (1976)
30. D.C. Barber, J.R. Mallard: Phys. Med. Biol. 16, 635 (1971)
31. I. Bar-David: Proc. IEEE 56, 2167 (1968)
32. I. Bar-David: IEEE Trans. IT-15, 31 (1969)
33. I. Bar-David: Proc. IEEE 59, 1612 (1971)
34. I. Bar-David: J. Opt. Soc. Am. 63, 166 (1973)
35. M.S. Bartlett: *An Introduction to Stochastic Processes* (Cambridge Univ. Press, London, New York 1956)
36. G. Bêdard: Phys. Rev. 151, 1038 (1966)
37. G. Bêdard: Phys. Lett. 21, 32 (1966)
38. G. Bêdard: Phys. Rev. 161, 1304 (1967)
39. G. Bêdard: J. Opt. Soc. Am. 57, 1201 (1967)

40. G. Bédard: Phys. Lett. 24A, 613 (1967)
41. G. Bédard: Proc. Phys. Soc. 90, 131 (1967)
42. G. Bédard, J.C. Chang, L. Mandel: Phys. Rev. 160, 1496 (1967)
43. C. Bendjaballah: J. Phys. A 6, 837 (1973)
44. C. Bendjaballah: Opt. Commun. 17, 55 (1976)
45. C. Bendjaballah, F. Perrot: C.R. Acad. Sci. (Paris) 271 B, 1085 (1970)
46. C. Bendjaballah, F. Perrot: Opt. Commun. 3, 21 (1971)
47. C. Bendjaballah, F. Perrot: Appl. Phys. Lett. 18, 532 (1971)
48. C. Bendjaballah, F. Perrot: J. Appl. Phys. 44, 5130 (1973)
49. W.R. Bennet, Jr., P.J. Kindlmann, G.N. Mercer: Appl. Opt. (Suppl.) 2, 34 (1965)
50. M.J. Beran: *Statistical Continuum Theories* (Wiley-Interscience, New York 1968)
51. M.J. Beran, G.B. Parrent: *Theory of Partial Coherence*, 2nd ed. (Society of Photo-optical Instrumentation Engineers 1974)
52. B.J. Berne, R. Pecora: *Dynamic Light Scattering* (Wiley, New York 1976)
53. M. Bertolotti: In *Photon Correlation and Light Beating Spectroscopy*, ed. by H.Z. Cummins, E.R. Pike (Plenum, New York 1974) pp. 41-74
54. M. Bertolotti, B. Crosignani: J. Phys. A 3, L37 (1970)
55. M. Bertolotti, F. Scudieri, S. Verginelli: Appl. Opt. 15, 1842 (1976)
56. F.J. Beutler, O.A.Z. Leneman: Acta Math. 116, 159 (1966)
57. A. Bijaoui: Astron. Astrophys. 35, 31 (1974)
58. J. Blake, R. Barakat: J. Phys. A 6, 1196 (1973)
59. J. Blake, R. Barakat: Can. J. Phys. 53, 1215 (1975)
60. J. Blake, R. Barakat: Opt. Commun. 16, 303 (1976)
61. V. Bluemel, L.M. Narducci, R. Tuft: J. Opt. Soc. Am. 62, 1309 (1972)
62. E. Boileau, B. Picinbono: J. Opt. Soc. Am. 58, 1238 (1968)
63. V.A. Bogdanovich, V.N. Prokof'yev: Radio Eng. Electron. Phys. 18, 1171 (1973)
64. L.P. Bolgiano, Jr., L.F. Jelsma: Proc. IEEE 52, 218 (1964)
65. M. Born, E. Wolf: *Principles of Optics*, 5th ed. (Pergamon Oxford 1975)
66. G.D. Boyd, J.P. Gordon: Bell Syst. Tech. J. 40, 489 (1961)
67. K.A. Bridgett, T.A. King: Proc. Phys. Soc. 92, 75 (1967)
68. E. Brookner: Proc. IEEE 58, 1767 (1970)
69. E. Brookner: IEEE Trans. COM-22, 265 (1974)
70. M. Brown: In *Stochastic Point Processes: Statistical Analysis, Theory, and Application*, ed. by P.A.W. Lewis (Wiley-Interscience, New York 1972) pp. 62-89
71. E.A. Bucher: Appl. Opt. 11, 884 (1972)
72. J. Bures, C. Delisle: Can. J. Phys. 49, 1255 (1971)
73. J. Bures, C. Delisle, A. Zardecki: Can. J. Phys. 49, 3064 (1971)
74. J. Bures, C. Delisle, A. Zardecki: Can. J. Phys. 50, 760 (1972)
75. J. Bures, C. Delisle, A. Zardecki: Can. J. Phys. 50, 1307 (1972)
76. J. Bures, C. Delisle, A. Zardecki: Phys. Rev. A 6, 2237 (1972)
77. B.I. Cantor, L. Matin, M.C. Teich: Appl. Opt. 14, 2819 (1975)
78. B.I. Cantor, M.C. Teich: J. Opt. Soc. Am. 65, 786 (1975)
79. C.D. Cantrell: Phys. Lett. 29A, 469 (1969)
80. C.D. Cantrell: Phys. Rev. A 1, 672 (1970)
81. C.D. Cantrell: J. Math. Phys. 12, 1005 (1971)
82. C.D. Cantrell: Phys. Rev. A 3, 728 (1971)
83. C.D. Cantrell, J.R. Fields: Phys. Rev. A 7, 2063 (1973)
84. W.H. Carter, E. Wolf: J. Opt. Soc. Am. 63, 1619 (1973)
85. W.H. Carter, E. Wolf: J. Opt. Soc. Am. 65, 1067 (1975)
86. R.F. Chang, V. Korenman, C.O. Alley, R.W. Detenbeck: Phys. Rev. 178, 612 (1969)
87. E.B. Champagne: Appl. Opt. 5, 1843 (1966)
88. S.H. Chen, P. Tartaglia: Opt. Commun. 6, 119 (1972)
89. S.H. Chen, P. Tartaglia, N. Polonsky-Ostrowsky: J. Phys. A 5, 1619 (1972)
90. S.H. Chen, W.B. Veldkamp, C.C. Lai: Rev. Sci. Instr. 46, 1356 (1975)

91. S. Chopra, L. Mandel: Rev. Sci. Instr. 43, 1489 (1972)
92. S. Chopra, L. Mandel: IEEE J. QE-8, 324 (1972)
93. S. Chopra, L. Mandel: In *Coherence and Quantum Optics*, ed. by L. Mandel, E. Wolf (Plenum, New York 1973) p. 805
94. S. Chopra, L. Mandel: Phys. Rev. Lett. 30, 60 (1973)
95. B. Chu: *Laser Light Scattering* (Academic Press, New York 1974)
96. J.R. Clark, S. Karp: Proc. IEEE 58, 1964 (1970)
97. W.G. Clark, E.L. O'Neill: J. Opt. Soc. Am. 61, 934 (1971)
98. D. Clarke, J.F. Grainger: *Polarized Light and Optical Measurements* (Pergamon, Oxford 1971)
99. P.B. Coates: J. Phys. E 1, 878 (1968)
100. M. Corti, A. De Agostini, V. Degiorgio: Rev. Sci. Instr. 45, 888 (1974)
101. M. Corti, V. Degiorgio: Opt. Commun. 11, 1 (1974)
102. M. Corti, V. Degiorgio: Phys. Rev. A 14, 1475 (1976)
103. D.R. Cox: *Renewal Theory* (Methuen, London 1962)
104. D.R. Cox, P.A.W. Lewis: *The Statistical Analysis of Series of Events* (Methuen, London 1966)
105. D.R. Cox, H.D. Miller: *The Theory of Stochastic Processes*, 2nd ed. (Methuen, London 1965)
106. H. Cramér, M.R. Leadbetter: *Stationary and Related Stochastic Processes* (Wiley, New York 1967)
107. B. Crosignani, P. DiPorto, M. Bertolotti: *Statistical Properties of Scattered Light* (Academic Press, New York 1975)
108. T.F. Curran, M. Ross: Proc. IEEE 53, 1770 (1965)
109. H.Z. Cummins, E.R. Pike: *Photon Correlation and Light Beating Spectroscopy* (Plenum Press, New York 1974)
110. J.C. Dainty: J. Opt. Soc. Am. 62, 595 (1972)
111. J.C. Dainty (ed.): *Laser Speckle and Related Phenomena*. Topics in Appl. Phys., Vol. 9 (Springer, Berlin, Heidelberg, New York 1975)
112. D.J. Daley, D. Vere-Jones: In *Stochastic Point Processes: Statistical Analysis, Theory, and Applications*, ed. by P.A.W. Lewis (Wiley, New York 1972)
113. Q.I. Daudpota: Opt. Commun. 17, 143 (1976)
114. L. D'Auria, J.P. Delort, C. Puech, E. Spitz: Opt. Commun. 6, 30 (1972)
115. W.B. Davenport, W.L. Root: *Random Signals and Noise* (McGraw-Hill, New York 1958)
116. F. Davidson: Phys. Rev. 185, 446 (1969)
117. F. Davidson, J. Amoss: J. Opt. Soc. Am. 63, 30 (1973)
118. F. Davidson, C.M. Chao, F.K. Tittel, J.P. Hohimer: IEEE J. QE-10, 409 (1974)
119. F. Davidson, A. Gonzalez-del-Valle: J. Opt. Soc. Am. 65, 655 (1975)
120. F. Davidson, R.S. Iyer: Appl. Opt. 13, 2171 (1974)
121. F. Davidson, L. Mandel: Phys. Lett. 25A, 700 (1967)
122. F. Davidson, L. Mandel: Phys. Lett. 27A, 579 (1968)
123. F. Davidson, L. Mandel: J. Appl. Phys. 39, 62 (1968)
124. C.C. Davis, T.A. King: Rev. Sci. Instr. 41, 407 (1970)
125. C.C. Davis, T.A. King: J. Phys. A 3, 101 (1970)
126. C.C. Davis, T.A. King: J. Phys. E 5, 1072 (1972)
127. V. Degiorgio, J.B. Lastovka: Phys. Rev. A 4, 2033 (1971)
128. O.E. DeLange: Proc. IEEE 58, 1683 (1970)
129. C. Delisle, J. Bures: Can. J. Phys. 49, 1940 (1971)
130. I. De Lotto, P.F. Manfredi, P. Principi: Energia Nucl. (Milan) 11, 557 (1964)
131. D. Dialetis: J. Phys. A 2, 229 (1969)
132. P. Diament, M.C. Teich: J. Opt. Soc. Am. 60, 682 (1970)
133. P. Diament, M.C. Teich: J. Opt. Soc. Am. 60,1489 (1970)
134. P. Diament, M.C. Teich: Appl. Opt. 10, 1664 (1971)
135. J.L. Doob: *Stochastic Processes* (Wiley, New York 1953)
136. L.E. Estes, J.D. Kuppenheimer, L.M., Narducci: Phys. Rev. A 1, 710 (1970)

137. W. Feller: In *Anniversary Volume for Courant* (Wiley Interscience, New York 1948) pp. 105
138. W. Feller: *An Introduction to Probability Theory and its Applications*, 3rd ed.,Vol. 1 (Wiley, New York 1968)
139. W. Feller: *An Introduction to Probability Theory and its Applications*, 2nd ed.,Vol. 2 (Wiley, New York 1971)
140. G.L. Fillmore: Phys. Rev. 182, 1384 (1969)
141. G.L. Fillmore, G. Lachs: IEEE Trans. IT-15, 465 (1969)
142. D. Fink: Appl. Opt. 14, 689 (1975)
143. C. Flammer: *Spheroidal Wave Functions* (Stanford University Press, Stanford 1957)
144. A.T. Forrester, R.A. Gudmudsen, P.O. Johnson: Phys. Rev. 99, 1691 (1955)
145. R. Forrester, Jr., D. Snyder: IEEE Trans. Commun. Systems 21, 1037 (1973)
146. M. Françon, S. Mallick: In *Progress in Optics*, ed. by E. Wolf, Vol. 6 (North-Holland, Amsterdam 1967) p. 71
147. M. Françon, S. Slansky: *Cohérence en Optique* (Ed. Centre Nat. Rech. Sci., Paris 1965)
148. A. Fraser: Rev. Sci. Instr. 42, 1539 (1971)
149. S. Fray, F.A. Johnson, R. Jones, T.D. McLean, E.R. Pike: Phys. Rev. 153, 357 (1967)
150. C. Freed, H.A. Haus: Phys. Rev. Lett. 15, 943 (1965)
151. C. Freed, H.A. Haus: Phys. Rev. A 141, 287 (1966)
152. D.L. Fried: J. Opt. Soc. Am. 57, 169 (1967)
153. D.L. Fried: Appl. Opt. 11, 1268 (1972)
154. D.L. Fried: Appl. Opt. 13, 1282 (1974)
155. D.L. Fried: Appl. Opt. 13, 2463 (1974)
156. D.L. Fried, R.A. Schmeltzer: Appl. Opt. 6, 1729 (1967)
157. D.L. Fried, J.B. Seidman: Appl. Opt. 6, 245 (1967)
158. B.R. Frieden: In *Progress in Optics*, ed. by E. Wolf, Vol. 9 (North-Holland, Amsterdam 1971) p. 311
159. J. Funke: Czech. J. Phys. B 24, 245 (1974)
160. P. Furcinitti, J.D. Kuppenheimer, L.M. Narducci, R.A. Tuft: J. Opt. Soc. Am. 62, 792 (1972)
161. D. Gabor: J. Phys. E 8, 73 (1975)
162. D. Gabor: J. Phys. E 8, 161 (1975)
163. R.M. Gagliardi: IEEE Trans. IT-18, 208 (1972)
164. R.M. Gagliardi, S. Karp: IEEE Trans. COM-17, 208 (1969)
165. R.M. Gagliardi, S. Karp: *Optical Communications* (Wiley Interscience, New York 1976)
166. N.C. Gallagher, B. Liu: Optik 42, 65 (1975)
167. H. Gamo: J. Appl. Phys. 34, 875 (1963)
168. R.J. Glauber: Phys. Rev. 130, 2529 (1963)
169. R.J. Glauber: Phys. Rev. 131, 2766 (1963)
170. R.J. Glauber: In *Quantum Electronics*, ed. by N. Bloemberger, P. Grivet (Columbia University Press, New York 1964) p. 111
171. R.J. Glauber: In *Quantum Optics and Electronics*, ed. by C. de Witt, A. Blandin, C. Cohen-Tannoudji (Gordon and Breach, New York 1965) p. 65
172. R.J. Glauber: In *Physics of Quantum Electronics*, ed. by P.L. Kelley, B. Lax, P. Tannenwald (McGraw-Hill, New York 1966) pp. 788
173. R.J. Glauber: In *Proc. Symp. Modern Optics* (Polytechnic Press, New York 1967) p. 1
174. R.J. Glauber: In *Fundamental Problems in Statistical Mechanics II*, ed. by E.G.D. Cohen (North-Holland, Amsterdam 1968) p. 140
175. R.J. Glauber: In *Quantum Optics*, ed. by S.M. Kay, A. Maitland (Academic Press, New York) p. 53
176. R.J. Glauber: In *Laser Handbook*, ed. by F.T. Arecchi, E.O. Schulz-Dubois (North-Holland, Amsterdam 1972) p. 1

177. Y.A. Gol'dshteyn, V.I. Emin: Radio Eng. Electron. Phys. 18, 1424 (1973)
178. J.W. Goodman: IEEE J. QE-1, 180 (1965)
179. J.W. Goodman: Proc. IEEE 53, 1688 (1965)
180. J.W. Goodman: *Introduction to Fourier Optics* (McGraw-Hill, New York 1968)
181. F.E. Goodwin: IEEE J. QE-3, 524 (1967)
182. F.E. Goodwin, L.P. Bolgiano, Jr.: Proc. IEEE 53, 1745 (1965)
183. J.P. Gordon: Proc. IRE 50, 1898 (1962)
184. I.S. Gradshteyn, I.M. Ryzhik: *Table of Integrals Series and Products* (Academic Press, New York 1965)
185. J. Grandell: In *Stochastic Point Processes: Statistical Analysis, Theory, and Applications*, ed. by P.A.W. Lewis (Wiley Interscience, New York 1972) pp. 90-121
186. H.S. Green, E. Wolf: Proc. Phys. Soc. (London) A 66, 1129 (1953)
187. H. Haken: In *Handbuch der Physik*, ed. by L. Genzel, Vol. 25/2c (Springer, Berlin, Heidelberg, New York 1970)
188. H. Haken: In *Laser Handbook*, ed. by F.T. Arecchi, E.O. Schulz-Dubois, Vol. 1 (North-Holland, Amsterdam 1972) p. 115
189. H. Haken: Rev. Mod. Phys. 47, 67 (1975)
190. F.R. Hallet, A.L. Gray, A. Rybakowski, J.L. Hunt, J.R. Stevens: Can. J. Phys. 50, 2368 (1972)
191. R. Hanbury Brown: Ann. Rev. Astronomy Astrophys. 6, 13 (1968)
192. R. Hanbury Brown: *The Intensity Interferometer: Its Applications to Astronomy* (Taylor and Francis Ltd., London, Halsted Press, Wiley, New York and Toronto 1974)
193. R. Hanbury Brown, R.Q. Twiss: Nature 177, 27 (1956)
194. R. Hanbury Brown, R.Q. Twiss: Nature 178, 1046 (1956)
195. R. Hanbury Brown, R.Q. Twiss: Proc. R. Soc. A 248, 199 (1958)
196. R. Hanbury Brown, R.Q. Twiss: Proc. R. Soc. A 248, 222 (1958)
197. H.A. Haus: In *Quantum Optics: Proceedings of the International School of Physics "Enrico Fermi"*, ed. by R.J. Glauber (Academic Press, New York 1969) p. 111
198. C.W. Helstrom: Proc. Phys. Soc. 83, 777 (1964)
199. C.W. Helstrom: J. Opt. Soc. Am. 57, 353 (1967)
200. C.W. Helstrom: Inform. Contr. 10, 254 (1967)
201. C.W. Helstrom: Inform. Contr. 13, 156 (1968)
202. C.W. Helstrom: *Statistical Theory of Signal Detection*, 2nd ed. (Pergamon, London 1968)
203. C.W. Helstrom: IEEE Trans. AES-5, 562 (1969)
204. C.W. Helstrom: J. Statist. Phys. 1, 231 (1969)
205. C.W. Helstrom: J. Opt. Soc. Am. 60, 521 (1970)
206. C.W. Helstrom: IEEE Trans. AES-7, 210 (1971)
207. C.W. Helstrom: In *Progress in Optics*, ed. by E. Wolf, Vol. 10 (North-Holland, Amsterdam 1972) p. 289
208. C.W. Helstrom: *Quantum Detection and Estimation Theory* (Academic Press, New York 1976)
209. C.W. Helstrom, J.W.S. Liu, J.P. Gordon: Proc. IEEE 58, 1578 (1970)
210. H.J. Hindin: Proc. IEEE 56, 2194 (1968)
211. H.J. Hindin: Proc. IEEE 61, 1501 (1973)
212. H.J. Hindin: Proc. IEEE 62, 638 (1974)
213. H.J. Hindin: Proc. IEEE 63, 536 (1975)
214. E. Hisdal: J. Opt. Soc. Am. 55, 1446 (1965)
215. E. Hisdal: J. Opt. Soc. Am. 57, 35 (1967)
216. E. Hisdal: J. Opt. Soc. Am. 59, 921 (1969)
217. H. Hodara: Proc. IEEE 53, 696 (1965)
218. R. Horák, L. Mišta, J. Peřina: J. Phys. A 4, 231 (1971)
219. E.V. Hoversten: In *Laser Handbook*, ed. by F.T. Arecchi, E.O. Schulz-Dubois, Vol. 2 (North-Holland, Amsterdam 1972) p. 1805

220. E.V. Hoversten, R.O. Harger, S.J. Halme: Proc. IEEE 58, 1626 (1970)
221. E.V. Hoversten, D.L. Snyder, R.O. Harger, K. Kurimoto: IEEE Trans. COM-22, 17 (1974)
222. W. Hubbard: IEEE Trans. COM-19, 221 (1971)
223. A.J. Hughes, E. Jakeman, C.J. Oliver, E.R. Pike: J. Phys. A 6, 1327 (1973)
224. S. Jacobs: Electronics 36, 29 (1963)
225. A.K. Jaiswal, C.L. Mehta: Phys. Rev. 186, 1355 (1969)
226. A.K. Jaiswal, C.L. Mehta: Phys. Rev. A 2, 168 (1970)
227. E. Jakeman: J. Phys. A 3, 201 (1970)
228. E. Jakeman: J. Phys. A 5, 49 (1972)
229. E. Jakeman: In *Photon Correlation and Light Beating Spectroscopy*, ed. by by H.Z. Cummins, E.R. Pike (Plenum Press, New York 1974) pp. 75-149
230. E. Jakeman, J.G. McWhirter: J. Phys. A 9, 785 (1976)
231. E. Jakeman, C.J. Oliver, E.R. Pike: J. Phys. A 1, L406 (1968)
232. E. Jakeman, C.J. Oliver, E.R. Pike: J. Phys. A 1, 497 (1968)
233. E. Jakeman, C.J. Oliver, E.R. Pike: J. Phys. A 3, L45 (1970)
234. E. Jakeman, C.J. Oliver, E.R. Pike: J. Phys. A 4, 827 (1971)
235. E. Jakeman, C.J. Oliver, E.R. Pike: Phys. Lett. 34A, 101 (1971)
236. E. Jakeman, C.J. Oliver, E.R. Pike: Adv. Phys. 42, 349 (1975)
237. E. Jakeman, C.J. Oliver, E.R. Pike, M. Lax, M. Zwanziger: J. Phys. A 3, L55 (1970)
238. E. Jakeman, C.J. Oliver, E.R. Pike, P.N. Pusey: J. Phys. A 5, L93 (1972)
239. E. Jakeman, C.J. Oliver, E.R. Pike, P.N. Pusey: J. Phys. A 6, L35 (1973)
240. E. Jakeman, E.R. Pike: J. Phys. A 1, 128 (1968)
241. E. Jakeman, E.R. Pike: J. Phys. A 1, 690 (1968)
242. E. Jakeman, E.R. Pike: J. Phys. A 2, L411 (1969)
243. E. Jakeman, E.R. Pike: J. Phys. A 2, 115 (1969)
244. E. Jakeman, E.R. Pike, G. Parry, B. Saleh: Opt. Commun. 19, 359 (1976)
245. E. Jakeman, E.R. Pike, S. Swain: J. Phys. A 3, L55 (1970)
246. E. Jakeman, E.R. Pike, S. Swain: J. Phys. A 4, 517 (1971)
247. E. Jakeman, P.N. Pusey: J. Phys. A 6, 88 (1973)
248. E. Jakeman, P.N. Pusey: Phys. Lett. 44, 456 (1973)
249. E. Jakeman, P.N. Pusey: J. Phys. A 8, 369 (1975)
250. A.H. Jazwinski: *Stochastic Processes and Filtering Theory* (Academic Press, New York 1970)
251. R. Jodoin, L. Mandel: J. Opt. Soc. Am. 61, 191 (1971)
252. F.A. Johnson, R. Jones, T.P. McLean, E.R. Pike: Phys. Rev. Lett. 16, 589 (1966)
253. F.A. Johnson, T.P. McLean, E.R. Pike: In *Physics of Quantum Electronics*, ed. by P.L. Kelley, B. Lax, and P. Tannenwald (McGraw-Hill, New York 1966) p. 706
254. R.C. Jones: Proc. IRE 47, 1498 (1959)
255. M. Kac, A.J.F. Siegert: J. Appl. Phys. 18, 383 (1947)
256. Y. Kano: J. Phys. Soc. Japan 40, 1122 (1976)
257. S. Karp, J.R. Clark: IEEE Trans. IT-16, 672 (1970)
258. S. Karp, R.M. Gagliardi: IEEE Trans. IT-16, 142 (1970)
259. S. Karp, E.L. O'Neill, R.M. Gagliardi: Proc. IEEE 58, 1611 (1970)
260. P.L. Kelley, W.H. Kleiner: Phys. Rev. A 136, 316 (1964)
261. H.C. Kelly: IEEE J. QE-7, 541 (1971)
262. H.C. Kelly: J. Phys. A 5, 104 (1972)
263. H.C. Kelly, J.C. Blake: J. Phys. A 4, 103 (1971)
264. H.C. Kelly, J.C. Blake: J. Phys. A 5, L7 (1972)
265. J.R. Kerr, P.J. Titterton, A.R. Kraemer, C.R. Cooke: Proc. IEEE 58, 1691 (1970)
266. T.S. Kinsel: Proc. IEEE 58, 1666 (1970)
267. I. Kitazima: Opt. Commun. 10, 137 (1974)
268. J. Klauder, E.C.G. Sudarshan: *Fundamentals of Quantum Optics* (W.A. Benjamin, New York 1968)
269. J.C. Kluyver: Proc. Sect. K. ned. Akad. Wet. 8, 341 (1905)

270. A.E.W. Knight, B.K. Selinger: Australian J. Chem. 26, 1 (1973)
271. A. Kollin: Ann. Physik, Lpz. 21, 813 (1934)
272. D.E. Koppel: J. Appl. Phys. 42, 3216 (1971)
273. D.E. Koppel: Phys. Rev. A 10, 1938 (1975)
274. D.E. Koppel, D.W. Schaefer: Appl. Phys. Lett. 22, 36 (1973)
275. V. Korenman: Phys. Rev. 138 B, 1012 (1967)
276. P.I. Kuznetsov, R.L. Stratonovich: Izv. Akad. Nauk. SSSR, Ser. Math. 20, 167 (1956)
277. G. Lachs: Phys. Rev. 138, B 1012 (1965)
278. G. Lachs: J. Appl. Phys. 38, 3429 (1967)
279. G. Lachs: J. Appl. Phys. 42, 602 (1971)
280. G. Lachs: IEEE J. QE-10, 590 (1974)
281. G. Lachs, E. Jankowich: Proc. IEEE 56, 744 (1968)
282. G. Lachs, S.R. Laxpati: J. Appl. Phys. 44, 3332 (1973)
283. G. Lachs, M.C. Miner: IEEE Trans. AE-11, 234 (1975)
284. G. Lachs, J.A. Quarato: IEEE Trans. AE-11, 266 (1975)
285. G. Lachs, N.F. Ruggieri: J. Appl. Phys. 43, 1297 (1972)
286. G. Lachs, N.F. Ruggieri: IEEE Trans. AES-9, 860 (1973)
287. G. Lachs, D.R. Voltmer: J. Appl. Phys. 47, 346 (1976)
288. J.N. Lahti, C.M. Nagel: Appl. Opt. 9, 115 (1970)
289. W.E. Lamb, Jr., M.O. Scully: In *Polarization, Matière et Rayonnement*, ed. by Société Française de Physique (Presses Univ. de France, Paris 1969) p. 363
290. L.D. Landau, E.M. Lifshitz: *Statistical Physics*, 2nd ed. (Pergamon Press, Oxford 1959)
291. M.E. Lasser: IEEE Spectrum 3, 73 (1966)
292. B.P. Lathi: *An Introduction to Random Signals and Communication Theory* (Intertext, London 1968)
293. R.S. Lawrence, J.W. Strohbehn: Proc. IEEE 58, 1523 (1970)
294. M. Lax: In *Statistical Physics*, ed. by M. Chrêtien, E.P. Gross, S.D. Desser, Vol. 2 (Gordon and Breach, New York 1968)
295. S.R. Laxpati, G. Lachs: J. Appl. Phys. 43, 4773 (1972)
296. E. Ledinegg: Acta Phys. Austriaca 20, 253 (1965)
297. G.M. Lee, R.B. Fluchel, E.A. Paddon, D.C. Torretta: IEEE Trans. COM-22, 55 (1974)
298. R.H. Lehmberg: Phys. Rev. 167, 1152 (1968)
299. J.W. Liu: IEEE Trans. IT-16, 319 (1970)
300. R.F. Lucy, K. Lang, C.J. Peters, K. Duval: Appl. Opt. 6, 1333 (1967)
301. J. Lyons, G.T. Troup: Phys. Lett 31 A, 182 (1970)
302. O. Macchi: IEEE Trans. IT-17, 2 (1971)
303. O. Macchi: C.R. Acad. Sc. (Paris) 272A, 437 (1971)
304. O. Macchi, B.C. Picinbono: IEEE Trans. IT-18, 562 (1972)
305. P.J. Magill, R.P. Soni: Phys. Rev. Lett. 16, 911 (1966)
306. L. Mandel: Proc. Phys. Soc. 72, 1037 (1958)
307. L. Mandel: Proc. Phys. Soc. 74, 233 (1959)
308. L. Mandel: J. Opt. Soc. Am. 51, 1342 (1961)
309. L. Mandel: In *Progress in Optics*, ed. by E. Wolf, Vol. 2 (North-Holland, Amsterdam 1963) p. 181
310. L. Mandel: Proc. Phys. Soc. 81, 1104 (1963)
311. L. Mandel: J. Opt. Soc. Am. 56, 1200 (1966)
312. L. Mandel: J. Opt. Soc. Am. 57, 613 (1967)
313. L. Mandel: Phys. Rev. 181, 75 (1969)
314. L. Mandel: J. Appl. Phys. 43, 258 (1972)
315. L. Mandel: J. Opt. Soc. Am. 66, 968 (1976)
316. L. Mandel, C.L. Mehta: Nuovo Cimento B 61, 149 (1969)
317. L. Mandel, E.C.G. Sudarshan, E. Wolf: Proc. Phys. Soc. 84, 435 (1964)
318. L. Mandel, E. Wolf: Rev. Mod. Phys. 37, 231 (1965)
319. L. Mandel, E. Wolf: *Selected Papers on Coherence and Fluctuations of Light*, Vol. I and II (Dover Publ., New York 1970)
320. L. Mandel, E. Wolf: J. Opt. Soc. Am. 65, 413 (1975)

321. L. Mandel, E. Wolf: J. Opt. Soc. Am. 66, 529 (1976)
322. A.S. Marathay, G.B. Parrent: J. Opt. Soc. Am. 60, 243 (1970)
323. E.W. Marchand, E. Wolf: J. Opt. Soc. Am. 62, 379 (1972)
324. W. Martienssen, E. Spiller: Am. J. Phys. 32, 919 (1964)
325. W. Martienssen, E. Spiller: Phys. Rev. Lett. 16, 531 (1966)
326. W. Martienssen, E. Spiller: Phys. Rev. 145, 285 (1966)
327. J. McFadden: J. Roy. Statist. Soc. B 24, 364 (1962)
328. J. McFadden, W. Weissblum: J. Roy. Statist. Soc. B 25, 413 (1963)
329. T.S. McKechnie: Opt. Quant. Elect. 8, 61 (1976)
330. T.P. McLean, E.R. Pike: Phys. Lett. 15, 318 (1965)
331. R.W. McMillan, N.P. Barnes: Appl. Opt. 15, 2501 (1976)
332. C.L. Mehta: In *Lectures on Theoretical Physics*, ed. by W.E. Brittin, Vol. 7c (University of Colorado Press, Boulder 1965) p. 345
333. C.L. Mehta: Nuovo Cimento 45, 280 (1966)
334. C.L. Mehta: J. Math. Phys. 8, 1798 (1967)
335. C.L. Mehta: J. Opt. Soc. Am. 58, 1233 (1968)
336. C.L. Mehta: In *Progress in Optics*, ed. by E. Wolf, Vol. 8 (North-Holland, Amsterdam 1970) p. 373
337. C.L. Mehta, S. Gupta: Phys. Rev. A 11, 1634 (1975)
338. C.L. Mehta, A.K. Jaiswal: Phys. Rev. A 2, 2570 (1970)
339. C.L. Mehta, E. Wolf, A.P. Balachandran: J. Math. Phys. 7, 133 (1966)
340. M.L. Mehta, C.L. Mehta: J. Opt. Soc. Am. 63, 826 (1973)
341. H. Melchior, M.B. Fisher, F.R. Arams: Proc. IEEE 58, 1466 (1970)
342. D. Meltzer, W. Davis, L. Mandel: Appl. Phys. Lett. 17, 242 (1970)
343. D. Meltzer, L. Mandel: IEEE J. QE-6, 661 (1970)
344. D. Meltzer, L. Mandel: Phys. Rev. A 3, 1763 (1971)
345. M. Menzel, B. Stoffregen: Optik 46, 203 (1976)
346. D. Middleton: IRE Trans. PGIT 3, 86 (1957)
347. K.S. Miller: *Complex Stochastic Processes* (Addison-Wesely, Reading 1974)
348. S. Minami, T. Akaki, T. Uchida, K. Kimota: Jap. J. Appl. Phys. 14, Suppl. 14-1, 39 (1975)
349. R.L. Mitchell: J. Opt. Soc. Am. 58, 1267 (1968)
350. H. Morawitz: Phys. Rev. 139, A 1072 (1965)
351. B.L. Morgan, L. Mandel: Phys. Rev. Lett. 16, 1012 (1966)
352. G.A. Morton: Appl. Opt. 7, 1 (1968)
353. H.M. Nussenzveig: *Introduction to Quantum Optics* (Gordon and Breach, London, New York, Paris 1973)
354. M. Ohta, T. Koizumi: Proc. IEEE 57, 1231 (1969)
355. C.J. Oliver: In *Photon Correlation and Light Beating Spectroscopy*, ed. by H.Z. Cummins, E.R. Pike (Plenum Press, New York, London 1974) pp. 151-223
356. E.L. O'Neill: *Introduction to Statistical Optics* (Addison-Wesley, Reading 1963)
357. S. Pancharatnam: Proc. Indian Acad. Sci. 57, 231 (1963)
358. A. Papoulis: *Probability, Random Variables and Stochastic Processes* (McGraw-Hill, New York 1965)
359. A. Papoulis: *Systems and Transforms with Applications in Optics* (McGraw-Hill, New York 1968)
360. A. Papoulis: J. Franklin Inst. 296, 433 (1973)
361. A. Paoulis: IEEE Trans. COM-22, 162 (1974)
362. G. Parrent, P. Roman: Nuovo Cimento 15, 370 (1960)
363. G. Parry: In *Laser Speckle and Related Phenomena*, ed. by J.C. Dainty. Topics in Applied Physics, Vol. 9 (Springer, Berlin, Heidelberg, New York 1975)
364. E. Parzen: *Stochastic Processes* (Holden-Day, San Francisco 1962)
365. P.R. Pearl, G.J. Troup: Phys. Lett. 27A, 560 (1968)
366. P.R. Pearl, G.J. Troup: Opto. Elect. 1, 152 (1969)

367. K. Pearson: Draper's Company Research Memoirs, Biometric Series 3 (Dulau and Co., London 1906)
368. J. Peřina: Phys. Lett. 24A, 333 (1967)
369. J. Peřina: *Coherence of Light* (Van Nostrand Reinhold Co., London 1971)
370. J. Peřina, R. Horák: J. Phys. A 2, 702 (1969)
371. J. Peřina, L. Mišta: Opt. Acta 21, 329 (1974)
372. J. Peřina, V. Peřinová: Opt. Acta 12, 333 (1965)
373. J. Peřina, V. Peřinová, G. Lachs, Z. Braunerová: Czech. J. Phys. B 23, 1008 (1973)
374. J. Peřina, V. Peřinova, L. Mista: Opt. Commun. 3, 89 (1971)
375. W.N. Peters, R.J. Arguello: IEEE J. QE-3, 532 (1967)
376. D.T. Phillips, H. Kleinman, S.P. Davis: Phys. Rev. 153, 113 (1967)
377. B. Picinbono: Phys. Lett. 29A, 614 (1969)
378. B. Picinbono, C. Bendjaballah, J. Pouget: J. Math. Phys. 11, 2166 (1970)
379. B. Picinbono, M. Rousseau: Phys. Rev. A 1, 635 (1970)
380. E.R. Pike: In *Quantum Optics: Proceedings of the International School of Physics "Enrico Fermi"*, ed. by R.J. Glauber (Academic Press, New York 1969) p. 160
381. E.R. Pike: Nuovo Cimento (Special issue) 1, 277 (1969)
382. E.R. Pike: In *Quantum Optics*, ed. by S.M. Kay, A. Maitland (Academic Press, London, New York 1970)
383. E.R. Pike: J. Phys. Paris 33 C1, 177 (1972)
384. E.R. Pike, E. Jakeman: In *Advances in Quantum Electronics*, ed. by D.W. Goodwin, Vol. 2 (Academic Press, London, New York 1974) p. 1
385. M.J. Piovoso, L.P. Bolgiano, Jr.: Proc. IEEE 1519 (1967)
386. W.K. Pratt: IEEE Trans. COM-14, 664 (1966)
387. W.K. Pratt: *Laser Communication Systems* (Wiley, New York 1969)
388. G. Present, D.B. Scarl: Appl. Opt. 11, 120 (1972)
389. H.D. Pruett: Appl. Opt. 11, 2529 (1972)
390. P.N. Pusey, W.I. Goldberg: Phys. Rev. A 3, 766 (1971)
391. P.N. Pusey, E. Jakeman: J. Phys. A 8, 392 (1975)
392. P.N. Pusey, D.W. Schaefer, D.E. Koppel: J. Phys. A 7, 530 (1974)
393. Lord Rayleigh: Phil. Mag. 37, 321 (1919)
394. G.A. Rebka, R.V. Pound: Nature 180, 1035 (1957)
395. I.S. Reed: IRE Trans. IT-8, 194 (1962)
396. B. Reiffen, H. Sherman: Proc. IEEE 51, 1316 (1963)
397. S.O. Rice: Bell Syst. Tech. J. 23, 1 (1944)
398. S.O. Rice: Bell Syst. Tech. J. 23, 165 (1944)
399. S.O. Rice: Bell Syst. Tech. J. 23, 282 (1944)
400. S.O. Rice: Bell Syst. Tech. J. 24, 46 (1945)
401. H. Risken: Z. Physik 186, 85 (1965)
402. H. Risken: Z. Physik 191, 302 (1966)
403. H. Risken: In *Progress in Optics*, ed. by E. Wolf, Vol. 8 (North-Holland, Amsterdam 1970) p. 239
404. F. Rocca: Phys. Rev. D 8, 4403 (1973)
405. S. Rosenberg, M.C. Teich: J. Appl. Phys. 43, 1256 (1972)
406. S. Rosenberg, M.C. Teich: Appl. Opt. 12, 2625 (1973)
407. S. Rosenberg, M.C. Teich: IEEE IT-19, 807 (1973)
408. M. Ross: *Laser Receivers, Devices, Techniques and Systems* (Wiley, New York 1966)
409. M. Rousseau: C.R. Acad. Sc. Paris 268, 1477 (1969)
410. M. Rousseau: J. de Phys. 30, 675 (1969)
411. M. Rousseau: J. Statist. Phys. 8, 341 (1973)
412. M. Rousseau: J. Phys. A 8, 1265 (1975)
413. A.A.M. Saleh: IEEE J. QE-3, 540 (1967)
414. B.A. Saleh: J. Phys. A 6, 980 (1973)
415. B.E.A. Saleh: J. Phys. A 6, L161 (1973)
416. B.E.A. Saleh: Proc. IEEE 62, 530 (1974)
417. B.E.A. Saleh: IEEE Trans. IT-20, 262 (1974)

418. B.E.A. Saleh: J. Phys. A 7, 1360 (1974)
419. B.E.A. Saleh: J. Appl. Phys. 46, 943 (1975)
420. B.E.A. Saleh: Opt. Commun. 13, 120 (1975)
421. B.E.A. Saleh: Appl. Phys. 8, 269 (1975)
422. B.E.A. Saleh: IEEE Trans. AES-14 (1978)
423. B.E.A. Saleh, M.F. Cardoso: J. Phys. A 6, 1897 (1973)
424. B.E.A. Saleh, J. Hendrix: J. Phys. A 8, 1134 (1975)
425. B.E.A. Saleh, J. Minkowski: J. Phys. A 6, L165 (1973)
426. B.E.A. Saleh, J. Minkowski: J. Phys. A 8, 120 (1975)
427. B.E.A. Saleh, B.K. Selinger: Appl. Opt. 16, 1408 (1977)
428. M. Sargent III, O. Scully, W.E. Lamb: *Laser Physics* (Addison-Wesley, Reading 1974)
429. D.B. Scarl: Phys. Rev. Lett. 17, 663 (1966)
430. D.B. Scarl: Phys. Rev. 175, 1661 (1968)
431. D.W. Schaefer: In *Laser Applications to Optics and Spectroscopy: Physics of Quantum Electronics*, ed. by S.F. Jacobs, M. Sargent III, J.F. Scott, M.O. Scully, Vol. 2 (Addison-Wesley, Reading 1975) p. 245
432. D.W. Schaefer, B.J. Berne: Phys. Rev. Lett. 28, 475 (1972)
433. D.W. Schaefer, P.N. Pusey: Phys. Rev. Lett. 29, 843 (1972)
434. D.W. Schaefer, P.N. Pusey: In *Coherence and Quantum Optics*, ed. by L. Mandel, E. Wolf (Plenum Press, New York 1973) p. 839
435. J.A. Schell, G. Lachs: IEEE Trans. AES-7, 1207 (1971)
436. F. Scudieri, M. Bertolotti, R. Bartolino: Appl. Opt. 13, 181 (1974)
437. M.O. Scully, M. Sargent III: Phys. Today 25, 38 (1972)
438. E.E. Serrallach, M. Zulauf: J. Appl. Math. Phys. 26, 669 (1975)
439. J.H. Shapiro: Appl. Opt. 13, 2462 (1974)
440. G.C. Sherman, A.J. Devaney, L. Mandel: Opt. Commun. 6, 115 (1972)
441. W.A. Shurcliff: *Polarized Light* (Harvard University Press, Cambridge 1962)
442. A.J.F. Siegert: IRE IT-3, 4 (1954)
443. A.J.F. Siegert: IRE IT-6, 38 (1957)
444. A.E. Siegman: Proc. IEEE 54, 1350 (1966)
445. A.E. Siegman: Appl. Opt. 5, 1588 (1966)
446. A.E. Siegman, S.E. Harris, B.J. McMurtry: In *Optical Masers*, ed. by J. Fox (Wiley, New York 1963) pp. 511-527
447. D. Slepian: Bell Syst. Tech. J. 37, 165 (1958)
448. D. Slepian, H.O. Pollack: Bell Syst. Tech. J. 40, 43 (1961)
449. D. Slepian, E. Sonnenblick: Bell Syst. Tech. J. 44, 1745 (1965)
450. A.W. Smith, J.A. Armstrong: Phys. Rev. Lett. 16, 1169 (1966)
451. D.L. Snyder: IEEE Trans. IT-18, 91 (1972)
452. D.L. Snyder: IEEE Trans. IT-18, 558 (1972)
453. D.L. Snyder: *Random Point Processes* (Wiley Interscience, New York 1975)
454. D.L. Snyder, I.B. Rhodes: IEEE Trans. COM-20, 1139 (1972)
455. S. Solimeno, E. Corti, B. Nicoletti: J. Opt. Soc. Am. 60, 1245 (1970)
456. S.K. Srinivasan: Phys. Lett. 47A, 151 (1974)
457. S.K. Srinivasan: *Stochastic Point Processes and Their Applications* (Charles Griffin and Co. Ltd., London 1974)
458. S.K. Srinivasan, S. Sukavanam: Phys. Lett. 35A, 81 (1971)
459. S.K. Srinivasan, S. Sukavanam: J. Phys. A 5, 682 (1972)
460. S.K. Srinivasan, S. Sukavanam, E.C.G. Sudarshan: J. Phys. A 6, 1910 (1973)
461. R.L. Stratonovich: *Topics in the Theory of Random Noise*, Vol. 1 (Gordon and Breach, New York 1963)
462. J.A. Stratton, P.M. Morse, L.J. Chu, J.D.C. Little, F.J. Corbató: *Spheroidal Wave Functions* (Technology Press of MIT, Cambridge, and Wiley, New York 1956)
463. J.W. Strohbehn: Proc. IEEE 56, 1301 (1968)
464. J.W. Strohbehn: In *Progress in Optics*, ed. by E. Wolf, Vol. 9 (North-Holland, Amsterdam 1971) p. 75

465. W. Swindell: *Polarized Light* (Dowden, Hutchinson and Ross, Inc., Stroudsberg, Pennsylvania 1975)
466. N. Takai: Opt. Commun. 14, 24 (1975)
467. P. Tartaglia, S.H. Chen: Opt. Commun. 7, 379 (1973)
468. P. Tartaglia, T.A. Postol, S.H. Chen: J. Phys. A 6, L35 (1973)
469. V.T. Tatarski: *Wave Propagation in a Turbulent Medium* (McGraw-Hill, New York 1961)
470. M.C. Teich: Proc. IEEE 56, 37 (1968)
471. M.C. Teich: Appl. Phys. Lett. 14, 201 (1969)
472. M.C. Teich: Proc. IEEE 57, 786 (1969)
473. M.C. Teich, P. Diament: Phys. Lett. 30A, 93 (1969)
474. M.C. Teich, P. Diament: J. Appl. Phys. 41, 415 (1970)
475. M.C. Teich, W.J. McGill: Phys. Rev. Lett. 36, 754 (1976)
476. M.C. Teich, S. Rosenberg: Opto-Electron. 3, 63 (1971)
477. M.C. Teich, S. Rosenberg: Appl. Opt. 12, 2616 (1973)
478. M.C. Teich, R.Y. Yen: J. Appl. Phys. 43, 2480 (1972)
479. J.B. Thomas: *An Introduction to Statistical Communication Theory* (Wiley, New York 1969)
480. P.J. Titterton, J.P. Speck: Appl. Opt. 12, 425 (1973)
481. U.M. Titulaer, R.J. Glauber: Phys. Rev. 140, B 676 (1965)
482. G. Toraldo di Francia: J. Opt. Soc. Am. 59, 799 (1969)
483. N. Tornau, B. Echtermeyer: Ann. Phys. Leipz. 29, 289 (1973)
484. G.J. Troup: Proc. Phys. Soc. 86, 39 (1965)
485. G.J. Troup: Phys. Lett. 17, 264 (1965)
486. G.J. Troup: Phys. Lett. 87, 361 (1966)
487. G.J. Troup: *Optical Coherence Theroy, Recent Developments* (Methuen, London 1967)
488. G.J. Troup: *Progress in Quantum Electronics*, ed. by J.H. Sanders, S. Stenholm (Pergamon Press, New York 1972)
489. G.J. Troup, J. Lyons: Phys. Lett. 29A, 705 (1969)
490. G.J. Troup, J. Lyons: Laser and Unconventional Optics Journal No. 33, 3 (1971)
491. G.J. Troup, R.G. Turner: Rep. Progr. Phys. 37, 771 (1974)
492. R.G. Tull: Appl. Opt. 7, 2023 (1968)
493. R.Q. Twiss: Opt. Acta 16, 423 (1969)
494. G.A. Vanasse, H. Sakai: In *Progress in Optics*, ed. by E. Wolf, Vol. 6 (North-Holland, Amsterdam 1967) p. 259
495. R.W.W. Van Resandt: Rev. Sci. Instr. 45, 1507 (1974)
496. F.C. Van Rijswijk, C. Smit: Physica 49, 549 (1970)
497. F.C. Van Rijswijk, U.L. Smith: Physica 83A, 121 (1976)
498. H.L. Van Trees: *Detection, Estimation and Modulation Theroy*, Part 1 (Wiley, New York 1968)
499. H.L. Van Trees: *Detection, Estimation and Modulation Theory*, Part 3, *Radar/Sonar Signal Processing – Gaussian Signals in Noise* (Wiley, (New York 1971)
500. R. Vasudevan, S.K. Srinivasan: Nuovo Cimento 8, 278 (1972)
501. J.F. Vinson: *Optische Kohärenz* (Akademie Verlag, Berlin 1971)
502. T. Wang, J.W. Strohbehn: J. Opt. Soc. Am. 64, 994 (1974)
503. W.E. Webb, J.T. Marino: Appl. Opt. 14, 1413 (1975)
504. H.A. Whale: J. Atmosph. Terrest. Phys. 36, 1045 (1974)
505. N. Wiener: Acta Math. 55, 118 (1930)
506. J.T. Winthrop: J. Opt. Soc. Am. 61, 15 (1971)
507. E. Wolf: Nuovo Cimento 12, 884 (1954)
508. E. Wolf: Proc. Roy. Soc. A 230, 246 (1955)
509. E. Wolf: Nuovo Cimento 13, 1165 (1959)
510. E. Wolf: In *Optical Masers*, ed. by J. Fox (Wiley, New York 1963) p. 29
511. E. Wolf: In *Quantum Electronics III*, ed. by P. Grivet, N. Bloembergen (Dunod, Paris 1964 and Columbia Univ. Press, New York 1964) p. 13
512. E. Wolf: Opt. Acta 13, 281 (1966)
513. E. Wolf, W.H. Carter: Opt. Commun. 13, 205 (1975)

514. E. Wolf, W.H. Carter: Opt. Commun. 16, 297 (1976)
515. E. Wolf, C.L. Mehta: Phys. Rev. Lett. 13, 705 (1964)
516. W. Wonneberger: Physica 78, 22 (1974)
517. R. Yen, P. Diament, M.C. Teich: IEEE Trans. IT-18, 302 (1972)
518. A. Zardecki: J. Math. Phys. 11, 244 (1970)
519. A. Zardecki: Can. J. Phys. 49, 1724 (1971)
520. A. Zardecki, J. Bures, C. Delisle: Phys. Rev. A 6, 1209 (1972)
521. A. Zardecki, C. Delisle: Can. J. Phys. 51, 1017 (1973)

Subject Index

Airy function 106, 306
Alignment of fields for heterodyning 218, 304
Amplitude modulation 286
Analytic signal 41
Angular correlation function 105
- diameter of stars, measurement 107, 119, 419
- , frequency domain 90, 95
- spectrum 90,95
Aperture integration effect 331
A posteriori conditional mean 55
- probability distribution 55
Area of coherence 102
- detector 167, 189, 193, 218, 247, 259, 304, 321
Array detectors 261, 333
Arrival time, Point process 60
- - , photoelectrons 170, 186, 199, 206, 211, 224, 227, 230, 236, 237
- - ; Poisson process 66
- - , Poisson process, doubly stochastic 78
Atmospheric scintillation 242, 327
- turbulence 242, 327
Autocorrelation *see* correlation

Background radiation 284
Bandpass Lorentzian spectrum 29
- signal 41
- stochastic process 44
Bandwidth 102
Bayes estimator 54
- strategy 53
Beating frequency 126, 203
Bessel probability distribution 12, 19, 81
- function of imaginary argument 237
Bias, estimator 54
Binary coding 287
Bit-error probability, BPLM 318, 325
- - , PDBM 317, 325, 331
- - , PGBM 315, 324, 331
- - , PPM 321
Bose-Einstein probability 13, 81, 168
Bunching, photons 171, 206, 224, 230, 238

Central-limit theorem 20, 145, 176
Central moments 7, 15
Chaotic light *see* thermal light
Characteristic function 10, 21
Chi-square distribution 12, 19
Circular polarization 287
Circularly symmetric complex random variable 21
- - - Gaussian random variable 22
Classical electrodynamic theory 86
Clipping gate 369
- , ramp 372
- , random 371

Coherence area 102
- function 94
- - , propagation of 104
- , longitudinal 112
- , spatial 101
- , temporal 101
- time 101
- volume 112
Coherency matrix 121
- - , intensity 123
Coherent light 97, 167
Coincidence 59, 395, 406
- , multichannel 396, 406
Complementary clipping 369
Complete coherence 97
- - , higher order 100
- incoherence 98
Complex analytic signal 41
- degree of coherence 96
- degree of spectral coherence 97
- envelope 42
- random variables 20
- representation, bandpass signal 41
- - , bandpass stochastic process 46
- stochastic process 38
Conditional mean 55
- probability density 14
Confluent hypergeometric function 146
Convolution 19, 82, 175, 178, 199, 210, 222
Correlation 15, 25, 38, 44, 94, 350
- , clipped 359
- , complementary clipped 369
- , doubly clipped 360
- , normalized 358
- , randomly clipped 371
- , scaled 373
Correlator 354
- , clipped 368
- , hardware 356
- , randomly clipped 371
- , scaled 373
- , sequential processing 357
- , software 355
Counters 356, 402, 403
Counting statistics 63
- - , photoelectrons 168, 177, 185, 194, 199, 201, 205, 210, 213, 215, 224, 227, 229, 235, 237, 240, 244, 251, 269
- - , Poisson process 68
- - , Poisson process, doubly stochastic 72, 76
Counting time 166
Covariance 15
Cross correlator 417
Cross-power spectrum 95
Cross-spectral density 28, 95
- purity 100
- - , effect of propagation 110
Cumulants 8, 12, 16
- , Gaussian RV 12, 18
- , - complex RV 22
- , Poisson process 71
- , Poisson process, doubly stochastic 75
Cumulant generating function 10, 16
- - - , Gaussian RV 18
- - - , Poisson process 69
- - - , Poisson process, doubly stochastic 72

Dark current 272
Dead-time effect 272
Degenerecy parameter 167
Degree of coherence 96
- higher-order coherence 99

Degree polarization 122
- spectral coherence 97
Degrees of freedom 193, 200, 201, 212
Density of events 58
Detection, direct 288, 312
- , heterodyne 302, 336
- , homodyne 302, 308, 336
- , photoelectric 161
- , quantum-limited 289, 297, 339, 345
- , radar signal 314, 322, 328, 333, 336, 337
Detectivity 296
Detector area 167, 189, 193, 218, 247, 259, 304, 321
Diffraction 91
- , Fraunhofer 91
- , Fresnel 91
- grating 349
- of spatially incoherent light 106
Diffused light 152
Digital communication systems 286, 312, 336
Direct detection 288, 312
- - receiver 288
Diversity 333
Doubly stochastic Poisson point process 72
Duty cycle, binary modulation 316

Electromagnetic field 89
- theory 86, 89
Error function 142, 229, 237
- probability 52
- - , BPLM 318, 325
- - , PDBM 317, 325, 331
- - , PGBM 315. 324. 331
- - , PPM 321

Estimation, intensity 289
- , linewidth 388
- , optical signals 288
- , parameters of an intensity distribution 300
- , spatial spectrum 414
- , spectrum 350
Estimator bias 54
- variance 54
Excess photon noise 168
Expected value 7, 14
Exponential-integral function 235, 236
Exponental probability distribution 12, 19, 81

Fabry-Perot interferometer 349
Factorial moments 9, 13, 16
- - , Poisson process 71
- - , Poisson process, doubly stochastic 74
- moment-generating function 10, 17
- - - , Poisson process 70
- - - , Poisson process, doubly stochastic 72
Factorization 100, 121
Far field radiation 91
Flat photoelectron-counting distribution 242
Fokker-Planck equation 35, 139
Forward Kolmogrov operator 35, 344
- recurrence times 58, 60
- - - , Poisson process 66
- - - , Poisson process, doubly stochastic 78
Fourier transform 41
Fraunhofer diffraction 91
Frequency modulation 286
Fresnel diffraction 91

Gain, photomultiplier 279
Gamma probability distribution 12, 19
Gauss-Markov process 37
Gaussian bandpass stochastic process 49
- beam 90, 113
- -Gaussian scattering 234
- hypergeometric function 252
- light 114
- light, quasi-stationary 132
- random variable 12, 22
- random variables 17, 24
- spectrum 29, 352
- stochastic process 29, 35
Generating functional, point process 62
- - , Poisson point process 65
Geometric probability distribution 13, 81, 168

Hamiltonian 161
Hanbury Brown and Twiss interferometer 1, 119, 417
Helmholtz equation 89
Heterodyne detection 302, 336
Higher-order correlation function 94, 99
- , complete coherence 100
Homodyne detection 302, 308, 336
Homogeneous optical field 95
- Poisson point process 65

Imaging 90, 105
Impulse response 91
Incoherent light 98
Incomplete gamma function 213, 366, 388
Information in optical channels 347
- rate 320
Instantaneous intensity 103
Instants of occurrence, random events 57, 62, 67, 78
- - , probability density 62, 67, 78
- - , photoelectrons 263, 339, 404
Integrated intensity 166
Intensity-coherency matrix 123
Intensity correlation function 103
- , estimation 289
- fluctuations, laser light 141, 144
- - , partially polarized light 122
- - , quasi-stationary Gaussian light 133
- - , superposition of coherent and thermal light 126
- - , thermal light 116
- interferometer 119, 417
- , light 102
- modulation 231, 286, 297
Inter-event times 58
- - , probability distribution 61, 66, 78
Interference fringes 108, 149
Interferometer, Young 108
- , Michelson 113
- , Hanbury Brown and Twiss 119, 417
Inverse Poisson transform 83

Joint Gaussian random variables 17
- probability distribution 11
- - - , light intensity 103
- - - , optical fields 93
- - - , photoelectron counts 252, 269

Karhunen-Loève expansion 30, 40, 50, 171, 189, 208, 255

Kolmogrov equation 35
- forward operator 35, 344

Laguerre probability distribution 13, 81
Laplace transform 80
Laser, Van der Pol model 137
- light 137, 229
- linewidth 144
- , multimode 145
- , pumping parameter 139
Legendre polynomial 240
Lifetime, fluorescence 405
- , luminescence 405
- , radioactive-decay 405
Likelihood function 52
- ratio 53
Linear optical system, coherent 90
- - - , incoherent 107
Linewidth 102
Log-normal distribution 12, 19, 243
- - , permanence 149, 249
Log-normally modulated light 242
Longitudinal coherence 112
Lorentzian spectrum 29, 32, 37, 202, 215, 258, 352

M-ary coding 287
Markov process 27
Markov-Gauss process 37
Maximum *a posteriori* estimator 55, 300
- likelihood estimation 54
- - strategy 52
Maxwell's equations 86, 89
Mean value 7
Measurement, angular diameter of stars 107, 119, 419
- , autocorrelation function 354
- , cross correlation function 93, 417
- , decaying intensity 406
- , dead-time effect 277
- , joint photoelectron-counting probability 403
- , normalized correlation function 358
- , probability of coincidence 395
- , probability of zero photoelectron counts 402
- , time-interval probability 396
Mercer's theorem 31
Michelson interferometer 113
Minimum detectable signal 295
- mean-square error 55
Mixing efficiency 218, 314
Mixture, coherent and thermal light 125
- , - partially polarized thermal light 131
- , - phase-fluctuating light 154
Modes, circular aperture 192
- , spatial 191, 193, 196
- , temporal 171, 188, 191, 195
Modified Bessel function of first kind 214
- spherical Bessel function of first kind 185
Modulated light 231
Modulation systems 286
Moments 7, 12, 16
- , Gaussian RV 12, 18
- , complex RV 20, 21, 24
- , Poisson process 70
- , Poisson process, doubly stochastic 73
Moment-generating function 9, 12, 13, 16

Moment-generating function, complex RV 20, 21, 22, 23
- - , Gaussian RV 12, 18
- - , Poisson process 69
- - , Poisson process, doubly stochastic 72
- functional 26
Multichannel analyzer 398, 403
Multicoincidence, measurement 396
- , photoelectrons 263
- , probability 59
Multifold photoelectron statistics 249, 263, 339, 404
Mutual coherence function 94

Negative-binomial probability distribution 13, 81
Noncentral chi-square distribution 12, 19, 81, 214
Nonideal effects in photodetectors 271
Nonlinear fitting 55
Nonstationary process 38
Normal distribution 12, 19
Normalized coherence function 96, 99
- cross-power spectrum 96
- error 289
- higher-order coherence function 99
- intensity correlation function 104

Occurrence times, random events 57, 62, 67, 78
- - , photoelectrons 263, 339, 404
On-off keying 287
Optical communication systems 284
- system 90
- - , coherent 90
- - , incoherent 107
Ornstein-Uhlenbeck process 37
Orthogonal functions 30
Oscillator, Van der Pol 137

Parabolic cylinder function 237
Parameter estimation 54
Partially coherent light 93
- polarized light 120, 198
Perturbation theory 162
Phase, complex random variable 21
- , fluctuating light 152
- , lock loop 301
- modulation 286
- screen 152
- , shift keying 287, 336
Photodetection theory 161
Photodetector 161
Photoelectron counting statistics 168, 177, 185, 194, 199, 201, 205, 210, 213, 215, 224, 227, 229, 235, 237, 240, 244, 251, 269
- - - , multifold 252, 258, 271
Photoelectrons 165
Photomultiplier gain 279
- response function 279
Photon bunching 171, 206, 224, 230, 238
Photons 165
Pile-up distortion 407
Plane wave 89
Point process, definition 57
- spread function 91
Poisson distribution 13, 68, 81
- process 65
- transform, 79
- - , inversion 83
Polarization 92
- modulation 286
Polarized thermal light 114, 167, 250, 290, 293, 351, 363, 377, 384, 389

Polarizer 92
Power spectrum 28
Poynting vector 102
Probability distribution 6, 11
Prolate spheroidal wave functions 33
Propagation of light 89, 104
- , coherence functions 104
- , cross-spectral purity 110
- through atmospheric turbulence 327
Pulse-amplitude modulation 287
Pulse-delay binary modulation 287, 316, 324, 332
Pulse-gated binary modulation 287, 314, 322, 328, 333
Pulse-height analyzer 398
Pulse-polarization modulation 287, 318, 324
Pulse-position modulation 288, 319, 324, 332
Pulse-separation technique 396
Pumping parameter, laser 139

Quadrature components 43, 49
Quantum detection and estimation theory 347
- efficiency 164
- electrodynamic theory 87
- limited detection 289, 297, 339, 345
- noise 168
Quasi-monochromatic light 86
Quasi-stationary Gaussian light 132, 223
- stochastic process 49

Radar signal detection 314, 322, 328, 333, 336, 337
Ramp clipping 372
Random variables 6
Randomly clipped autocorrelation 371
Rayleigh probability distribution 12, 19, 148
Rectangular spectrum 29, 33, 188
Resolution, spectrometers 349
Retarder 92
Rician distribution 12, 19
Rough surface, scattering from 114, 152, 156, 197

Sampling time 166, 354
Scaled autocorrelation 373
Scattering of laser light 114, 145, 151, 197, 350
Scintillation, atmospheric 242, 327
Semiclassical theory 87
- - , laser 137
- - , photodetection 161
Shot noise 168
Single-photoelectron decay spectroscopy 405
Sinusoidal-wave intensity modulation 239
Skewness coefficient 9, 149, 171
Spatial coherence function 100
- integration factors 220
- modes of an optical field 191, 193, 196
- spectrum 414
Speckle 156
Spectrum 27
- , Brillouin 352
- , Gaussian 29, 352
- , Lorentzian 29, 32, 37, 188, 202, 215, 258, 352
- , rectangular 29, 33
- , Sum of Lorentzian 197, 352
- , truncated 197
- , white 29, 35

Square-wave intensity modulation 239
Stationary stochastic process 26
- optical field 94
Statistical accuracy, estimating correlation function 375
- - , - clipped correlation function 386
- - , - normalized correlation function 381
- - , - parameters of a spectrum 388, 399, 402
- - , - probability of zero counts 401
- - , - radius of an incoherent object 419
- - , - spatial spectrum 417
- - , - time-interval probability 398
- independence 14
Steepest-descent method 244
Stochastic process 25
Superposition, coherent and thermal light 125, 203, 268, 291, 302, 321, 352, 365, 380, 385
- , - partially polarized thermal light 131, 221

Temporal coherence 101
- modes of an optical field 171, 188, 191, 195
Test of hypotheses 52
Thermal light, partially polarized 120, 198
- - , polarized 114, 167, 250, 290, 293, 351, 363, 377, 384, 389
- - , transient 135, 227
Time arrival, Point process 60
- - , photoelectrons 170, 186, 199, 206, 211, 227, 230, 236, 237
- - , Poisson process 66
- - , Poisson process, doubly stochastic 78
Time-bandwidth product 32, 33, 166
Time-shift invariance 27
Time statistics, point process 59
- - , photoelectrons 170, 186, 199, 206, 210, 224, 227, 230, 236, 237, 263, 273
- - , Poisson process 66
- - , Poisson process, doubly stochastic 77
Time-to-amplitude convertor 397
Transformation of random variables 11, 18
Transient thermal light 135, 227
Triangular-wave intensity modulation 239
Triggered counting 64
Truncated Gaussian probability distribution 141
Turbulence, atmospheric 242, 327

Unbiased estimator 54
Uniform probability distribution 19, 21
Unpolarized light 122, 124, 200

Van Cittert-Zernike theorem 107
Van der Pol oscillator 137
Variance 8
Vector field 89
Visibility 109
Volume of coherence 112

Wave equation 89
White Gaussian stochastic process 35

Whittaker's function 235, 237
Wide-sense stationarity 27
Wiener-Lêvy process 37

Young's interferometer 108

Topics in Applied Physics

Founded by Helmut K. V. Lotsch

Volume 2

Laser Spectroscopy

of Atoms and Molecules
Editor: *H. Walther*
137 figures, 22 tables. XVI, 383 pages. 1976
ISBN 3-540-07324-8

Contents:
H. Walther: Atomic and Molecular Spectroscopy with Lasers
E. D. Hinkley, K. W. Nill, F. A. Blum: Infrared Spectroscopy with Tunable Lasers
K. Shimoda: Double-Resonance Spectroscopy of Molecules by Means of Lasers
J. M. Cherlow, S. P. S. Porto: Laser Raman Spectroscopy of Gases
B. Decomps, M. Dumont, M. Ducloy: Linear and Nonlinear Phenomena in Laser Optical Pumping
K. M. Evenson, F. R. Petersen: Laser Frequency Measurements, the Speed of Light, and the Meter

Volume 9

Laser Speckle and Related Phenomena

Editor: *J. C. Dainty*
133 figures. XIII, 286 pages. 1975
ISBN 3-540-07498-8

Contents:
J. C. Dainty: Introduction
J. W. Goodman: Statistical Properties of Laser Speckle Patterns
G. Parry: Speckle Patterns in Partially Coherent Light
T. S. McKechnie: Speckle Reduction
M. Françon: Information Processing Using Speckle Patterns
A. E. Ennos: Speckle Interferometry
J. C. Dainty: Stellar Speckle Interferometry

Volume 13

High-Resolution Laser Spectroscopy

Editor: *K. Shimoda*
132 figures. XIII, 378 pages. 1976
ISBN 3-540-07719-7

Contents:
K. Shimoda: Introduction
K. Shimoda: Line Broadening and Narrowing Effects
P. Jacquinot: Atomic Beam Spectroscopy
V. S. Letokhov: Saturation Spectroscopy
J. L. Hall, J. A. Magyar: High Resolution Saturated Absorption Studies of Methane and Some Methyl-Halides
V. P. Chebotayev: Three-Level Laser Spectroscopy
S. Haroche: Quantum Beats and Time-Resolved Fluorescence Spectroscopy
N. Bloembergen, M. D. Levenson: Doppler-Free Two-Photon Absorption Spectroscopy

Volume 18

Ultrashort Light Pulses

Picosecond Techniques and Applications
Editor: *S. L. Shapiro*
173 figures. XI, 389 pages. 1977
ISBN 3-540-08103-8

Contents:
S. L. Shapiro: Introduction – A Historical Overview
D. J. Bradley: Methods of Generation
E. P. Ippen, C. V. Shank: Techniques for Measurement
D. H. Auston: Picosecond Nonlinear Optics
D. von der Linde: Picosecond Interactions in Liquids and Solids
K. B. Eisenthal: Picosecond Relaxation Processes in Chemistry
A. J. Campillo, S. L. Shapiro: Picosecond Relaxation Measurements in Biology

Springer-Verlag
Berlin
Heidelberg
New York